Applied Machine Learning Using mlr3 in R

mlr3 is an award-winning ecosystem of R packages that have been developed to enable state-of-the-art machine learning capabilities in R. **Applied Machine Learning Using mlr3 in R** gives an overview of flexible and robust machine learning methods, with an emphasis on how to implement them using mlr3 in R. It covers various key topics, including basic machine learning tasks, such as building and evaluating a predictive model; hyperparameter tuning of machine learning approaches to obtain peak performance; building machine learning pipelines that perform complex operations such as pre-processing followed by modelling followed by aggregation of predictions; and extending the mlr3 ecosystem with custom learners, measures, or pipeline components.

Features:

- In-depth coverage of the mlr3 ecosystem for users and developers
- Explanation and illustration of basic and advanced machine learning concepts
- Ready to use code samples that can be adapted by the user for their application
- Convenient and expressive machine learning pipelining enabling advanced modelling
- Coverage of topics that are often ignored in other machine learning books

The book is primarily aimed at researchers, practitioners, and graduate students who use machine learning or who are interested in using it. It can be used as a textbook for an introductory or advanced machine learning class that uses R, as a reference for people who work with machine learning methods, and in industry for exploratory experiments in machine learning.

Applied Machine Learning Using mlr3 in R

Edited by
Bernd Bischl, Raphael Sonabend,
Lars Kotthoff, and Michel Lang

CRC Press
Taylor & Francis Group
Boca Raton London New York

CRC Press is an imprint of the
Taylor & Francis Group, an **informa** business

First edition published 2024
by CRC Press
2385 Executive Center Drive, Suite 320, Boca Raton, FL 33431, U.S.A.

and by CRC Press
4 Park Square, Milton Park, Abingdon, Oxon, OX14 4RN

CRC Press is an imprint of Taylor & Francis Group, LLC

ISBN: 978-1-032-51567-0 (hbk)
ISBN: 978-1-032-50754-5 (pbk)
ISBN: 978-1-003-40284-8 (ebk)

DOI: 10.1201/9781003402848

Typeset in Latin Modern font
by KnowledgeWorks Global Ltd.

Publisher's note: This book has been prepared from camera-ready copy provided by the authors.

Table of contents

Preface   xi

Editors   xiii

Contributors   xv

1 Introduction and Overview   **1**
Lars Kotthoff, Raphael Sonabend, Natalie Foss, and Bernd Bischl
- 1.1 Installation Guidelines 1
- 1.2 How to Use This Book 2
- 1.3 mlr3book Code Style 4
- 1.4 mlr3 by Example 5
- 1.5 The `mlr3` Ecosystem 6
  - 1.5.1 R6 for Beginners 7
  - 1.5.2 data.table for Beginners 8
- 1.6 Essential mlr3 Utilities 9
- 1.7 Design Principles 11

I   Fundamentals   **13**

2 Data and Basic Modeling   **15**
Natalie Foss and Lars Kotthoff
- 2.1 Tasks 16
  - 2.1.1 Constructing Tasks 16
  - 2.1.2 Retrieving Data 18
  - 2.1.3 Task Mutators 20
- 2.2 Learners 22
  - 2.2.1 Training 22
  - 2.2.2 Predicting 24
  - 2.2.3 Hyperparameters 26
  - 2.2.4 Baseline Learners 31
- 2.3 Evaluation 32
  - 2.3.1 Measures 32
  - 2.3.2 Scoring Predictions 33
  - 2.3.3 Technical Measures 34
- 2.4 Our First Regression Experiment 36
- 2.5 Classification 37
  - 2.5.1 Our First Classification Experiment 37
  - 2.5.2 TaskClassif 38
  - 2.5.3 LearnerClassif and MeasureClassif 41
  - 2.5.4 `PredictionClassif`, Confusion Matrix, and Thresholding . . 43

2.6   Task Column Roles 48
2.7   Supported Learning Algorithms 50
2.8   Conclusion 52
2.9   Exercises 52

3   Evaluation and Benchmarking   53
Giuseppe Casalicchio and Lukas Burk

3.1   Holdout and Scoring 54
3.2   Resampling 56
       3.2.1   Constructing a Resampling Strategy 58
       3.2.2   Resampling Experiments 59
       3.2.3   ResampleResult Objects 61
       3.2.4   Custom Resampling 64
       3.2.5   Stratification and Grouping 66
3.3   Benchmarking 69
       3.3.1   benchmark() 70
       3.3.2   BenchmarkResult Objects 72
3.4   Evaluation of Binary Classifiers 73
       3.4.1   Confusion Matrix 73
       3.4.2   ROC Analysis 75
3.5   Conclusion 79
3.6   Exercises 80

II   Tuning and Feature Selection   83

4   Hyperparameter Optimization   85
Marc Becker, Lennart Schneider, and Sebastian Fischer

4.1   Model Tuning 86
       4.1.1   Learner and Search Space 86
       4.1.2   Terminator 88
       4.1.3   Tuning Instance with `ti` 89
       4.1.4   Tuner 90
       4.1.5   Logarithmic Transformations 94
       4.1.6   Analyzing and Using the Result 95
4.2   Convenient Tuning with `tune` and `auto_tuner` 98
4.3   Nested Resampling 100
       4.3.1   Nested Resampling with an `AutoTuner` 102
       4.3.2   The Right (and Wrong) Way to Estimate Performance 103
4.4   More Advanced Search Spaces 105
       4.4.1   Scalar Parameter Tuning 105
       4.4.2   Defining Search Spaces with `ps` 106
       4.4.3   Transformations and Tuning Over Vectors 107
       4.4.4   Hyperparameter Dependencies 110
       4.4.5   Recommended Search Spaces with `mlr3tuningspaces` 111
4.5   Conclusion 114
4.6   Exercises 114

5   Advanced Tuning Methods and Black Box Optimization   116
Lennart Schneider and Marc Becker

5.1   Error Handling and Memory Management 116

	5.1.1	Encapsulation and Fallback Learner 116
	5.1.2	Memory Management 118
5.2	Multi-Objective Tuning 119
5.3	Multi-Fidelity Tuning via Hyperband 121
	5.3.1	Hyperband and Successive Halving 122
	5.3.2	mlr3hyperband 123
5.4	Bayesian Optimization 125
	5.4.1	Black Box Optimization 126
	5.4.2	Building Blocks of Bayesian Optimization 129
	5.4.3	Automating BO with OptimizerMbo 137
	5.4.4	Bayesian Optimization for HPO 138
	5.4.5	Noisy Bayesian Optimization 140
	5.4.6	Practical Considerations in Bayesian Optimization 142
5.5	Conclusion 144
5.6	Exercises 144

6 Feature Selection   **146**
Marvin N. Wright

6.1	Filters 146
	6.1.1	Calculating Filter Values 148
	6.1.2	Feature Importance Filters 149
	6.1.3	Embedded Methods 149
	6.1.4	Filter-Based Feature Selection 150
6.2	Wrapper Methods 152
	6.2.1	Simple Forward Selection Example 153
	6.2.2	The FSelectInstance Classes 155
	6.2.3	The FSelector Class 155
	6.2.4	Starting the Feature Selection 156
	6.2.5	Optimizing Multiple Performance Measures 157
	6.2.6	Nested Resampling 158
6.3	Conclusion 159
6.4	Exercises 159

III   Pipelines and Preprocessing   **161**

7 Sequential Pipelines   **163**
Martin Binder and Florian Pfisterer

7.1	PipeOp: Pipeline Operators 164
7.2	Graph: Networks of PipeOps 166
7.3	Sequential Learner-Pipelines 168
	7.3.1	Learners as PipeOps and Graphs as Learners 169
	7.3.2	Inspecting Graphs 170
	7.3.3	Configuring Pipeline Hyperparameters 171
7.4	Conclusion 172
7.5	Exercises 173

8 Non-sequential Pipelines and Tuning   **174**
Martin Binder, Florian Pfisterer, Marc Becker, and Marvin N. Wright

| 8.1 | Selectors and Parallel Pipelines . . . . 175 |
| 8.2 | Common Patterns and ppl() . . . . 178 |

8.3   Practical Pipelines by Example 178
    8.3.1   Bagging with "greplicate" and "subsample" 179
    8.3.2   Stacking with po("learner_cv") 181
8.4   Tuning Graphs 185
    8.4.1   Tuning Graph Hyperparameters 185
    8.4.2   Tuning Alternative Paths with po("branch") 186
    8.4.3   Tuning with po("proxy") 189
    8.4.4   Hyperband with Subsampling 191
    8.4.5   Feature Selection with Filter Pipelines 192
8.5   Conclusion 194
8.6   Exercises 195

9 Preprocessing   **196**
Janek Thomas
9.1   Data Cleaning 197
9.2   Factor Encoding 198
9.3   Missing Values 200
9.4   Pipeline Robustify 203
9.5   Transforming Features and Targets 204
9.6   Functional Feature Extraction 206
9.7   Conclusion 209
9.8   Exercises 210

IV   Advanced Topics   **211**

10 Advanced Technical Aspects of mlr3   **213**
Michel Lang, Sebastian Fischer, and Raphael Sonabend
10.1 Parallelization 213
    10.1.1 Parallelization of Learners 216
    10.1.2 Parallelization of Resamplings and Benchmarks 218
    10.1.3 Parallelization of Tuning 220
    10.1.4 Nested Resampling Parallelization 221
    10.1.5 Parallelization of Predictions 224
10.2 Error Handling 225
    10.2.1 Encapsulation 226
    10.2.2 Fallback Learners 228
10.3 Logging 230
10.4 Data Backends 232
    10.4.1 Databases with DataBackendDplyr 232
    10.4.2 Parquet Files with DataBackendDuckDB 235
10.5 Extending mlr3 and Defining a New `Measure` 236
10.6 Conclusion 238
10.7 Exercises 239

11 Large-Scale Benchmarking   **240**
Sebastian Fischer, Michel Lang, and Marc Becker
11.1 Getting Data with OpenML 241
    11.1.1 Datasets 242
    11.1.2 Task 244
    11.1.3 Task Collection 246

11.2 Benchmarking on HPC Clusters 247
    11.2.1 Experiment Registry Setup 248
    11.2.2 Job Submission 250
    11.2.3 Job Monitoring, Error Handling, and Result Collection 252
    11.2.4 Custom Experiments with batchtools 254
11.3 Statistical Analysis 256
11.4 Conclusion 258
11.5 Exercises 258

12 Model Interpretation   **259**
Susanne Dandl, Przemysław Biecek, Giuseppe Casalicchio,
and Marvin N. Wright
12.1 The iml Package 260
    12.1.1 Feature Importance 261
    12.1.2 Feature Effects 262
    12.1.3 Surrogate Models 263
    12.1.4 Shapley Values 268
12.2 The counterfactuals Package 269
    12.2.1 What-If Method 270
    12.2.2 MOC Method 271
12.3 The DALEX Package 273
    12.3.1 Global EMA 275
    12.3.2 Local EMA 278
12.4 Conclusions 281
12.5 Exercises 282

13 Beyond Regression and Classification   **283**
Raphael Sonabend, Patrick Schratz, and Damir Pulatov
13.1 Cost-Sensitive Classification 284
    13.1.1 Cost-Sensitive Measure 284
    13.1.2 Thresholding 285
13.2 Survival Analysis 286
    13.2.1 TaskSurv 288
    13.2.2 LearnerSurv, PredictionSurv, and Predict Types 289
    13.2.3 MeasureSurv 292
    13.2.4 Composition 293
    13.2.5 Putting It All Together 295
13.3 Density Estimation 296
    13.3.1 TaskDens 296
    13.3.2 LearnerDens and PredictionDens 297
    13.3.3 MeasureDens and Putting It All Together 298
13.4 Cluster Analysis 299
    13.4.1 TaskClust 300
    13.4.2 LearnerClust and PredictionClust 301
    13.4.3 MeasureClust 303
    13.4.4 Visualization 304
    13.4.5 Putting It All Together 307
13.5 Spatial Analysis 308
    13.5.1 TaskClassifST and TaskRegrST 309
    13.5.2 Spatiotemporal Cross-Validation 310
    13.5.3 Spatial Prediction 312

13.6 Conclusion 314
13.7 Exercises 314

14 Algorithmic Fairness   **316**
Florian Pfisterer
14.1 Bias and Fairness 317
14.2 Group Fairness Notions 317
14.3 Auditing a Model For Bias 319
14.4 Fair Machine Learning 320
14.5 Conclusion 322
14.6 Exercises 323

References   **325**

Index   **335**

Preface

This book is the culmination of many years worth of software design, coding, writing, and editing. It is very important to us that all our contributors are credited appropriately.

Citation details of packages in the `mlr3` ecosystem can be found in their respective GitHub repositories.

When you are citing this book, cite chapters directly; citations can be found at the end of each chapter. If you need to reference the full book, use:

```
Bischl, B., Sonabend, R., Kotthoff, L., & Lang, M. (Eds.). (2024).
"Applied Machine Learning Using mlr3 in R". CRC Press.
https://mlr3book.mlr-org.com
```

```
@book{Bischl2024
    title = {Applied Machine Learning Using {m}lr3 in {R}}
    editor = {Bernd Bischl and Raphael Sonabend and Lars Kotthoff
     and Michel Lang},
    url = {https://mlr3book.mlr-org.com},
    year = {2024},
    isbn = {9781032507545},
    publisher = {CRC Press}
}
```

See the front page of the book website (https://mlr3book.mlr-org.com) for full licensing details.

We hope you enjoy reading this book.

Bernd, Raphael, Lars, Michel

Editors

Bernd Bischl is a professor of Statistical Learning and Data Science at LMU Munich and co-director of the Munich Center for Machine Learning. He studied Computer Science, Artificial Intelligence and Data Science and holds a PhD in Statistics. His research interests include AutoML, model selection, interpretable ML and the development of statistical software. He wrote the initial version of `mlr` in 2012 and 2013 and still leads the team of developers of `mlr3`, now largely focusing on design, code review and strategic development.

Raphael Sonabend is the CEO and co-founder of OSPO Now and a visiting researcher at Imperial College London. They hold a PhD in statistics, specializing in machine learning applications for survival analysis. They wrote the `mlr3` packages `mlr3proba` and `mlr3benchmark`.

Lars Kotthoff is an associate professor of Computer Science at the University of Wyoming, US. He has studied and held academic appointments in Germany, UK, Ireland, and Canada. Lars has been contributing to `mlr` for about a decade. His research aims to automate machine learning and other areas of AI.

Michel Lang is the scientific coordinator of the Research Center Trustworthy Data Science and Security. He has a PhD in Statistics and has been developing statistical software for over a decade. He joined the `mlr` team in 2014 and wrote the initial version of `mlr3`.

Contributors

Marc Becker
Ludwig-Maximilians-Universität
  München, München, Germany

Przemysław Biecek
MI2.AI, Warsaw University of
  Technology
University of Warsaw, Warszawa, Poland

Martin Binder
Ludwig-Maximilians-Universität
  München
Munich Center for Machine Learning
  (MCML), München, Germany

Bernd Bischl
Ludwig-Maximilians-Universität
  München
Munich Center for Machine Learning
  (MCML), München, Germany

Lukas Burk
Ludwig-Maximilians-Universität
  München
Leibniz Institute for Prevention
  Research and Epidemiology – BIPS
Munich Center for Machine Learning
  (MCML), München, Germany

Giuseppe Casalicchio
Ludwig-Maximilians-Universität
  München
Munich Center for Machine Learning
  (MCML), München, Germany
Essential Data Science Training GmbH,
  München, Germany

Susanne Dandl
Ludwig-Maximilians-Universität
  München
Munich Center for Machine Learning
  (MCML), München, Germany

Sebastian Fischer
Ludwig-Maximilians-Universität
  München, München, Germany

Natalie Foss
University of Wyoming, Laramie,
  United States

Lars Kotthoff
University of Wyoming, Laramie,
  United States

Michel Lang
Research Center Trustworthy Data
  Science and Security
TU Dortmund University, Dortmund,
  Germany

Florian Pfisterer
Ludwig-Maximilians-Universität
  München, München, Germany

Damir Pulatov
University of Wyoming, Laramie,
  United States

Lennart Schneider
Ludwig-Maximilians-Universität
  München
Munich Center for Machine Learning
  (MCML), München, Germany

Patrick Schratz
Friedrich Schiller University Jena,
  Thuringia, Germany

Raphael Sonabend
OSPO Now, London,
  United Kingdom

Janek Thomas
Ludwig-Maximilians-Universität
    München
Munich Center for Machine Learning
    (MCML), München, Germany
Essential Data Science Training GmbH,
    München, Germany

Marvin N. Wright
Leibniz Institute for Prevention
    Research and Epidemiology – BIPS,
    München, Germany
University of Bremen, Bremen, Germany
University of Copenhagen, Copenhagen,
    Denmark

1

Introduction and Overview

Lars Kotthoff
University of Wyoming

Raphael Sonabend
Imperial College London

Natalie Foss
University of Wyoming

Bernd Bischl
Ludwig-Maximilians-Universität München, and Munich Center for Machine Learning (MCML)

Welcome to the **Machine Learning in R** universe. In this book, we will guide you through the functionality offered by `mlr3` step by step. If you want to contribute to our universe, ask any questions, read documentation, or just chat with the team, head to https://github.com/mlr-org/mlr3 which has several useful links in the README.

The `mlr3` (Lang et al. 2019) package and the wider `mlr3` ecosystem provide a generic, object-oriented, and extensible framework for regression (Section 2.1), classification (Section 2.5), and other machine learning tasks (Chapter 13) for the R language (R Core Team 2019). On the most basic level, the unified interface provides functionality to train, test, and evaluate many machine learning algorithms. You can also take this a step further with hyperparameter optimization, computational pipelines, model interpretation, and much more. `mlr3` has similar overall aims to `caret` and `tidymodels` for R, `scikit-learn` for Python, and MLJ for Julia. In general, `mlr3` is designed to provide more flexibility than other ML frameworks while still offering easy ways to use advanced functionality. While `tidymodels` in particular makes it very easy to perform simple ML tasks, `mlr3` is more geared towards advanced ML.

Before we can show you the full power of `mlr3`, we recommend installing the `mlr3verse` package, which will install several, important packages in the `mlr3` ecosystem.

```
install.packages("mlr3verse")
```

1.1 Installation Guidelines

There are many packages in the `mlr3` ecosystem that you may want to use as you work through this book. All our packages can be installed from GitHub and

R-universe[1]; the majority (but not all) packages can also be installed from CRAN. We recommend adding the mlr-org R-universe to your R options so you can install all packages with `install.packages()`, without having to worry about which package repository it comes from. To do this, install `usethis` and run the following:

```
usethis::edit_r_profile()
```

In the file that opens add or change the `repos` argument in `options` so it looks something like the code below (you might need to add the full code block below or just edit the existing `options` function).

```
options(repos = c(
  mlrorg = "https://mlr-org.r-universe.dev",
  CRAN = "https://cloud.r-project.org/"
))
```

Save the file, restart your R session, and you are ready to go!

If you want the latest development version of any of our packages, run

```
remotes::install_github("mlr-org/{pkg}")
```

with {pkg} replaced with the name of the package you want to install. You can see an up-to-date list of all our extension packages at https://github.com/mlr-org/mlr3/wiki/Extension-Packages.

1.2   How to Use This Book

The `mlr3` ecosystem is the result of many years of methodological and applied research. This book describes the resulting features and discusses best practices for ML, technical implementation details, and in-depth considerations for model optimization. This book may be helpful for both practitioners who want to quickly apply machine learning (ML) algorithms and researchers who want to implement, benchmark, and compare their new methods in a structured environment. While we hope this book is accessible to a wide range of readers and levels of ML expertise, we do assume that readers have taken at least an introductory ML course or have the equivalent expertise and some basic experience with R. A background in computer science or statistics is beneficial for understanding the advanced functionality described in the later chapters of this book, but not required. A comprehensive ML introduction for those new to the field can be found in James et al. (2014). Wickham and Grolemund (2017) provides a comprehensive introduction to data science in R.

The book is split into the following four parts:

Part I: Fundamentals In this part of the book we will teach you the fundamentals of `mlr3`. This will give you a flavor of the building blocks of the `mlr3` universe and the

---

[1]R-universe is an alternative package repository to CRAN. The bit of code below tells R to look at both R-universe and CRAN when trying to install packages. R will always install the latest version of a package.

basic tools you will need to tackle most machine-learning problems. We recommend that all readers study these chapters to become familiar with `mlr3` terminology, syntax, and style. In Chapter 2 we will cover the basic classes in `mlr3`, including `Learner` (machine learning implementations), `Measure` (performance metrics), and `Task` (machine learning task definitions). Chapter 3 will take evaluation a step further to include discussions about resampling – robust strategies for measuring model performance – and benchmarking – experiments for comparing multiple models.

Part II: Tuning and Feature Selection In this part of the book, we will look at more advanced methodology that is essential to developing powerful ML models with good predictive ability. Chapter 4 introduces hyperparameter optimization, which is the process of tuning model hyperparameters to obtain better model performance. Tuning is implemented via the `mlr3tuning` package, which also includes methods for automating complex tuning processes, including nested resampling. The performance of ML models can be improved by tuning hyperparameters but also by carefully selecting features. Chapter 6 introduces feature selection with filters and wrappers implemented in `mlr3filters` and `mlr3fselect`. For readers interested in taking a deep dive into tuning, Chapter 5 discusses advanced tuning methods including error handling, multi-objective tuning, and tuning with Hyperband and Bayesian optimization methods.

Part III: Pipelines and Preprocessing In Part III we introduce `mlr3pipelines`, which allows users to implement complex ML workflows easily. In Chapter 7 we will show you how to build a pipeline out of discrete configurable operations and how to treat complex pipelines as if they were any other machine learning model. In Chapter 8 we will build on the previous chapter by introducing non-sequential pipelines, which can have multiple branches that carry out operations concurrently. We will also demonstrate how to tune pipelines, including how to tune which operations should be included in the pipeline. Finally, in Chapter 9 we will put pipelines into practice by demonstrating how to solve common problems that occur when fitting ML models to messy data.

Part IV: Advanced Topics In the final part of the book, we will look at advanced methodology and technical details. This part of the book is more theory-heavy in some sections to help ground the design and implementation decisions. We will begin by looking at advanced technical details in Chapter 10 that are essential reading for advanced users who require parallelization, custom error handling, or large databases. Chapter 11 will build on all preceding chapters to introduce large-scale benchmarking experiments that compare many models, tasks, and measures; including how to make use of `mlr3` extension packages for loading data, using high-performance computing clusters, and formal statistical analysis of benchmark experiments. Chapter 12 will discuss different packages that are compatible with `mlr3` to provide model-agnostic interpretability for feature importance and local explainability of individual predictions. Chapter 13 will then delve into detail on domain-specific methods that are implemented in our extension packages including survival analysis, density estimation, spatio-temporal analysis, and more. Readers may choose to selectively read sections in this chapter depending on your use case (i.e., if you have domain-specific problems to tackle), or to use these as introductions to new domains to explore. Finally, Chapter 14 will introduce algorithmic fairness, which includes specialized measures and methods to identify and reduce algorithmic biases.

We have marked sections that are particularly complex with respect to either technical or methodological detail and could be skipped on a first read with the following information box:

> **i   This section covers advanced ML or technical details.**

Each chapter includes examples, API references, and explanations of methodologies. At the end of each part of the book, we have included exercises for you to test yourself on what you have learned; you can find the solutions to these exercises at https://mlr3book.mlr-org.com/solutions.html. We have marked more challenging (and possibly time-consuming) exercises with an asterisk, "*".

If you want more detail about any of the tasks used in this book or links to all the `mlr3` dictionaries, see the appendices in the online version of the book at https://mlr3book.mlr-org.com/.

Reproducibility

At the start of each chapter we run `set.seed(123)` and use `renv1.0.0` to manage package versions, you can find our lockfile at https://github.com/mlr-org/mlr3book /blob/submission/book/renv.lock.

1.3   mlr3book Code Style

Throughout this book we will use the following code style:

1. We always use `=` instead of `<-` for assignment.

2. Class names are in `UpperCamelCase`

3. Function and method names are in `lower_snake_case`

4. When referencing functions, we will only include the package prefix (e.g., `pkg::function`) for functions outside the `mlr3` universe or when there may be ambiguity about in which package the function lives. Note you can use `environment(function)` to see which namespace a function is loaded from.

5. We denote packages, fields, methods, and functions as follows:
   - `package` (highlighted in the first instance)
   - `package::function()` or `function()` (see point 4)
   - `$field` for fields (data encapsulated in an R6 class)
   - `$method()` for methods (functions encapsulated in an R6 class)
   - `Class` (for R6 classes primarily, these can be distinguished from packages by context)

Now let us see this in practice with our first example.

1.4   mlr3 by Example

The `mlr3` universe includes a wide range of tools taking you from basic ML to complex experiments. To get started, here is an example of the simplest functionality – training a model and making predictions.

```
library(mlr3)
task = tsk("penguins")
split = partition(task)
learner = lrn("classif.rpart")

learner$train(task, row_ids = split$train)
learner$model
```

```
n= 231

node), split, n, loss, yval, (yprob)
      * denotes terminal node

1) root 231 129 Adelie (0.441558 0.199134 0.359307)
  2) flipper_length< 206.5 144  44 Adelie (0.694444 0.298611 0.006944)
    4) bill_length< 43.05 98   3 Adelie (0.969388 0.030612 0.000000) *
    5) bill_length>=43.05 46   6 Chinstrap (0.108696 0.869565 0.021739) *
  3) flipper_length>=206.5 87   5 Gentoo (0.022989 0.034483 0.942529) *
```

```
prediction = learner$predict(task, row_ids = split$test)
prediction
```

```
<PredictionClassif> for 113 observations:
    row_ids      truth  response
          1     Adelie    Adelie
          2     Adelie    Adelie
          3     Adelie    Adelie
---
        328  Chinstrap Chinstrap
        331  Chinstrap    Adelie
        339  Chinstrap Chinstrap
```

```
prediction$score(msr("classif.acc"))
```

```
classif.acc
      0.9558
```

In this example, we trained a decision tree on a subset of the penguins dataset, made predictions on the rest of the data and then evaluated these with the accuracy measure. In Chapter 2 we will break this down in more detail.

The `mlr3` interface also lets you run more complicated experiments in just a few lines of code:

```
library(mlr3verse)

tasks = tsks(c("breast_cancer", "sonar"))

glrn_rf_tuned = as_learner(ppl("robustify") %>>% auto_tuner(
    tnr("grid_search", resolution = 5),
    lrn("classif.ranger", num.trees = to_tune(200, 500)),
    rsmp("holdout")
))
glrn_rf_tuned$id = "RF"

glrn_stack = as_learner(ppl("robustify") %>>% ppl("stacking",
    lrns(c("classif.rpart", "classif.kknn")),
    lrn("classif.log_reg")
))
glrn_stack$id = "Stack"

learners = c(glrn_rf_tuned, glrn_stack)
bmr = benchmark(benchmark_grid(tasks, learners,
        rsmp("cv", folds = 3)))

bmr$aggregate(msr("classif.acc"))
```

```
        task_id learner_id classif.acc
1: breast_cancer         RF      0.9737
2: breast_cancer      Stack      0.9386
3:         sonar         RF      0.8406
4:         sonar      Stack      0.7246
```

In this (much more complex!) example we chose two tasks and two learners and used automated tuning to optimize the number of trees in the random forest learner (Chapter 4), and a machine learning pipeline that imputes missing data, collapses factor levels, and stacks models (Chapter 7 and Chapter 8). We also showed basic features like loading learners (Chapter 2) and choosing resampling strategies for benchmarking (Chapter 3). Finally, we compared the performance of the models using the mean accuracy with three-fold cross-validation.

You will learn how to do all this and more in this book.

1.5 The `mlr3` Ecosystem

Throughout this book, we often refer to `mlr3`, which may refer to the single `mlr3` base package but usually refers to all packages in our ecosystem; this should be clear from context. The `mlr3` *package* provides the base functionality that the rest of the ecosystem depends on for building more advanced machine learning tools. Figure 1.1 shows the packages in our ecosystem that extend `mlr3` with capabilities for preprocessing, pipelining, visualizations, additional learners, additional task types, and much more.

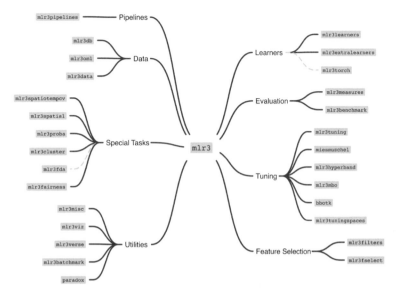

Figure 1.1: Overview of the mlr3 ecosystem, the packages with gray dashed lines are still in development, all others have a stable interface.

A complete and up-to-date list of extension packages can be found at https://mlr-org.com/ecosystem.html.

As well as packages within the mlr3 ecosystem, software in the mlr3verse also depends on the following popular and well-established packages:

- R6: The class system predominantly used in mlr3.
- data.table: High-performance extension of R's data.frame.
- digest: Cryptographic hash functions.
- uuid: Generation of universally unique identifiers.
- lgr: Configurable logging library.
- mlbench and palmerpenguins: Machine learning datasets.
- future / future.apply / parallelly: For parallelization (Section 10.1).
- evaluate: For capturing output, warnings, and exceptions (Section 10.2).

We build on R6 for object orientation and data.table to store and operate on tabular data. As both are core to mlr3, we *briefly* introduce both packages for beginners; in-depth expertise with these packages is not necessary to work with mlr3.

1.5.1 R6 for Beginners

R6 is one of R's more recent paradigms for object-oriented programming. If you have experience with any (class) object-oriented programming, then R6 should feel familiar. We focus on the parts of R6 that you need to know to use mlr3.

Objects are created by constructing an instance of an R6Class variable using the $new() initialization method. For example, say we have implemented a class called Foo, then foo = Foo$new(bar = 1) would create a new object of class Foo and set the `bar` argument of the constructor to the value 1. In practice, we implement a lot of sugar functionality (Section 1.6) in `mlr3` that makes construction and access a bit more convenient.

Some `R6` objects may have mutable states that are encapsulated in their *fields*, which can be accessed through the dollar, `$`, operator. Continuing the previous example, we can access the `bar` value in the `foo` object by using `foo$bar` or we could give it a new value, e.g. `foo$bar = 2`. These fields can also be "active bindings", which perform additional computations when referenced or modified.

In addition to fields, *methods* allow users to inspect the object's state, retrieve information, or perform an action that changes the internal state of the object. For example, in `mlr3`, the `$train()` method of a learner changes the internal state of the learner by building and storing a model. Methods that modify the internal state of an object often return the object itself. Other methods may return a new R6 object. In both cases, it is possible to "chain" methods by calling one immediately after the other using the `$`-operator; this is similar to the `%>%`-operator used in `tidyverse` packages. For example, `Foo$bar()$hello_world()` would run the `$bar()` method of the object `Foo` and then the `$hello_world()` method of the object returned by `$bar()` (which may be `Foo` itself).

Fields and methods can be public or private. The public fields and methods define the API to interact with the object. In `mlr3`, you can safely ignore private methods unless you are looking to extend our universe by adding a new class (Chapter 10).

Finally, `R6` objects are `environments`, and as such have reference semantics. This means that, for example, `foo2 = foo` does not create a new variable called `foo2` that is a copy of `foo`. Instead, it creates a variable called `foo2` that references `foo`, and so setting `foo$bar = 3` will also change `foo2$bar` to 3 and vice versa. To copy an object, use the `$clone(deep = TRUE)` method, so to copy `foo`: `foo2 = foo$clone(deep = TRUE)`.   `$clone()`

For a longer introduction, we recommend the `R6` vignettes found at https://r6.r-lib.org/; more details can be found at https://adv-r.hadley.nz/r6.html.

1.5.2 data.table for Beginners

The package `data.table` implements `data.table()`, which is a popular alternative to R's `data.frame()`. We use `data.table` because it is blazingly fast and scales well to bigger data.

As with `data.frame`, `data.tables` can be constructed with `data.table()` or `as.data.table()`:

```
library(data.table)
# converting a matrix with as.data.table
as.data.table(matrix(runif(4), 2, 2))
```

```
       V1     V2
1: 0.8586 0.4891
2: 0.8874 0.7181
```

```
# using data.table
dt = data.table(x = 1:6, y = rep(letters[1:3], each = 2))
dt
```

```
   x y
1: 1 a
2: 2 a
3: 3 b
4: 4 b
5: 5 c
6: 6 c
```

data.tables can be used much like data.frames, but they provide additional functionality that makes complex operations easier. For example, data can be summarized by groups with a by argument in the [operator and they can be modified in-place with the := operator.

```
# mean of x column in groups given by y
dt[, mean(x), by = "y"]
```

```
   y  V1
1: a 1.5
2: b 3.5
3: c 5.5
```

```
# adding a new column with :=
dt[, z := x * 3]
dt
```

```
   x y  z
1: 1 a  3
2: 2 a  6
3: 3 b  9
4: 4 b 12
5: 5 c 15
6: 6 c 18
```

Finally data.table also uses reference semantics so you will need to use copy() to clone a data.table. For an in-depth introduction, we recommend the vignette "Introduction to Data.table" (2023).

1.6 Essential mlr3 Utilities

mlr3 includes a few important utilities that are essential to simplifying code in our ecosystem.

Sugar Functions

Most objects in `mlr3` can be created through convenience functions called helper functions or sugar functions. They provide shortcuts for common code idioms, reducing the amount of code a user has to write. For example `lrn("regr.rpart")` returns the learner without having to explicitly create a new R6 object. We heavily use sugar functions throughout this book and provide the equivalent "full form" for complete detail at the end of each chapter. The sugar functions are designed to cover the majority of use cases for most users, knowledge about the full `R6` backend is only required if you want to build custom objects or extensions.

Many object names in `mlr3` are standardized according to the convention: `mlr_<type>_<key>`, where `<type>` will be `tasks`, `learners`, `measures`, and other classes that will be covered in the book, and `<key>` refers to the ID of the object. To simplify the process of constructing objects, you only need to know the object key and the sugar function for constructing the type. For example: `mlr_tasks_mtcars` becomes `tsk("mtcars")`; `mlr_learners_regr.rpart` becomes `lrn("regr.rpart")`; and `mlr_measures_regr.mse` becomes `msr("regr.mse")`. Throughout this book, we will refer to all objects using this abbreviated form.

Dictionaries

`mlr3` uses dictionaries to store R6 classes, which associate keys (unique identifiers) with objects (R6 objects). Values in dictionaries are often accessed through sugar functions that retrieve objects from the relevant dictionary, for example `lrn("regr.rpart")` is a wrapper around `mlr_learners$get("regr.rpart")` and is thus a simpler way to load a decision tree learner from `mlr_learners`. We use dictionaries to group large collections of relevant objects so they can be listed and retrieved easily. For example, you can see an overview of available learners (that are in loaded packages) and their properties with `as.data.table(mlr_learners)` or by calling the sugar function without any arguments, e.g. `lrn()`.

mlr3viz

`mlr3viz` includes all plotting functionality in `mlr3` and uses `ggplot2` under the hood. We use `theme_minimal()` in all our plots to unify our aesthetic, but as with all `ggplot` outputs, users can fully customize this. `mlr3viz` extends `fortify` and `autoplot` for use with common `mlr3` outputs including `Prediction`, `Learner`, and `BenchmarkResult` objects (which we will introduce and cover in the next chapters). We will cover major plot types throughout the book. The best way to learn about `mlr3viz` is through experimentation; load the package and see what happens when you run `autoplot` on an `mlr3` object. Plot types are documented in the respective manual page that can be accessed through `?autoplot.<class>`; for example, you can find different types of plots for regression tasks by running `?autoplot.TaskRegr`.

1.7   Design Principles

> **i   This section covers advanced ML or technical details.**

The `mlr` package (Bischl et al. 2016) was first released on CRAN in 2013, with the core design and architecture dating back further. Over time, the addition of many features led to a complex design that made it too difficult for us to extend further. In hindsight, we saw that some design and architecture choices in `mlr` made it difficult to support new features, in particular with respect to ML pipelines. So in 2018, we set about working on a reimplementation, which resulted in the first release of `mlr3` on CRAN in July 2019.

Learning from our history, we now follow these design principles in the `mlr3` ecosystem:

- **Object-oriented programming**. We embrace R6 for a clean, object-oriented design, object state changes, and reference semantics. This means that the state of common objects (e.g. tasks (Section 2.1) and learners (Section 2.2)) is encapsulated within the object, for example, to keep track of whether a model has been trained, without the user having to worry about this. We also use inheritance to specialize objects, e.g. all learners are derived from a common base class that provides basic functionality.
- **Tabular data**. Embrace `data.table` for its top-notch computational performance as well as tabular data as a structure that can be easily processed further.
- **Unified tabular input and output data formats.** This considerably simplifies the API and allows easy selection and "split-apply-combine" (aggregation) operations. We combine `data.table` and R6 to place references to non-atomic and compound objects in tables and make heavy use of list columns.
- **Defensive programming and type safety**. All user input is checked with `checkmate` (Lang 2017). We use `data.table`, which has behavior that is more consistent than several base R methods (e.g., indexing `data.frame`s simplifies the result when the `drop` argument is omitted). And we have extensive unit tests!
- **Light on dependencies**. One of the main maintenance burdens for `mlr` was to keep up with changing learner interfaces and behavior of the many packages it depended on. We require far fewer packages in `mlr3`, which makes installation and maintenance easier. We still provide the same functionality, but it is split into more packages that have fewer dependencies individually.
- **Separation of computation and presentation**. Most packages of the `mlr3` ecosystem focus on processing and transforming data, applying ML algorithms, and computing results. Our core packages do not provide visualizations because their dependencies would make installation unnecessarily complex, especially on headless servers (i.e., computers without a monitor where graphical libraries are not installed). Hence, visualizations of data and results are provided in `mlr3viz`.

Part I

Fundamentals

2

Data and Basic Modeling

Natalie Foss
University of Wyoming

Lars Kotthoff
University of Wyoming

In this chapter, we will introduce the `mlr3` objects and corresponding R6 classes that implement the essential building blocks of machine learning. These building blocks include the data (and the methods for creating training and test sets), the machine learning algorithm (and its training and prediction process), the configuration of a machine learning algorithm through its hyperparameters, and evaluation measures to assess the quality of predictions.

In the simplest definition, machine learning (ML) is the process of learning models of relationships from data. Supervised learning is a subfield of ML in which datasets consist of labeled observations, which means that each data point consists of features, which are variables to make predictions from, and a target, which is the quantity that we are trying to predict. For example, predicting a car's miles per gallon (target) based on the car's properties (features) such as horsepower and the number of gears is a supervised learning problem, which we will return to several times in this book. In `mlr3`, we refer to datasets, and their associated metadata as tasks (Section 2.1). The term "task" is used to refer to the prediction problem that we are trying to solve. Tasks are defined by the features used for prediction and the targets to predict, so there can be multiple tasks associated with any given dataset. For example, predicting miles per gallon (mpg) from horsepower is one task, predicting horsepower from mpg is another task, and predicting the number of gears from the car's model is yet another task.

Supervised learning can be further divided into regression – which is the prediction of numeric target values, e.g. predicting a car's mpg – and classification – which is the prediction of categorical values/labels, e.g., predicting a car's model. Chapter 13 also discusses other tasks, including cost-sensitive classification and unsupervised learning. For any supervised learning task, the goal is to build a model that captures the relationship between the features and target, often with the goal of training the model to learn relationships about the data so it can make predictions for new and previously unseen data. A model is formally a mapping from a feature vector to a prediction. A prediction can take many forms depending on the task; for example, in classification this can be a predicted label, or a set of predicted probabilities or scores. Models are induced by passing training data to machine learning algorithms, such as decision trees, support vector machines, neural networks, and many more. Machine learning algorithms are called learners in `mlr3` (Section 2.2) as, given data, they learn models. Each learner has a parameterized space that potential models are drawn from and during the training process, these parameters are fitted to best match the data. For example, the parameters could be the coefficients used for

Margin notes: Machine Learning/ Supervised Learning · Regression Classification · Model · Learners

individual features when training a linear regression model. During training, most machine learning algorithms are "fitted"/"trained" by optimizing a loss function that quantifies the mismatch between ground truth target values in the training data and the predictions of the model.

For a model to be most useful, it should generalize beyond the training data to make "good" predictions (Section 2.2.2) on new and previously "unseen" (by the model) data. The simplest way to test this is to split data into training data and test data – where the model is trained on the training data and then the separate   Train/Test
test data is used to evaluate models in an unbiased way by assessing to what extent   Data
the model has learned the true relationships that underlie the data (Chapter 3).
This evaluation procedure estimates a model's generalization error, i.e., how well we   Generaliza-
expect the model to perform in general. There are many ways to evaluate models   tion Error
and split data for estimating generalization error (Section 3.2).

This brief overview of ML provides the basic knowledge required to use `mlr3` and is summarized in Figure 2.1. In the rest of this book, we will provide introductions to methodology when relevant. For texts about ML, including detailed methodology and underpinnings of different algorithms, we recommend Hastie, Friedman, and Tibshirani (2001), James et al. (2014), and Bishop (2006).

In the next few sections, we will look at the building blocks of `mlr3` using regression as an example, we will then consider how to extend this to classification in Section 2.5.

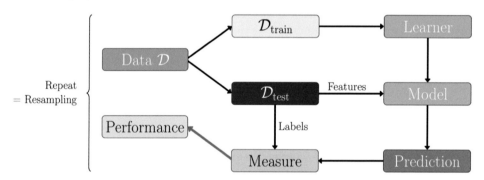

Figure 2.1: General overview of the machine learning process.

2.1   Tasks

Tasks are objects that contain the (usually tabular) data and additional metadata that define a machine learning problem. The metadata contains, for example, the name of the target feature for supervised machine learning problems. This information is extracted automatically when required, so the user does not have to specify the prediction target every time a model is trained.

2.1.1   Constructing Tasks

`mlr3` includes a few predefined machine learning tasks in the `mlr_tasks` Dictionary.   `mlr_tasks`

```
mlr_tasks
```

```
<DictionaryTask> with 21 stored values
Keys: ames_housing, bike_sharing, boston_housing, breast_cancer,
  german_credit, ilpd, iris, kc_housing, moneyball, mtcars,
  optdigits, penguins, penguins_simple, pima, ruspini, sonar,
  spam, titanic, usarrests, wine, zoo
```

tsk() To get a task from the dictionary, use the tsk() function and assign the return value to a new variable. Below we retrieve tsk("mtcars"), which uses the mtcars dataset:

```
tsk_mtcars = tsk("mtcars")
tsk_mtcars
```

```
<TaskRegr:mtcars> (32 x 11): Motor Trends
* Target: mpg
* Properties: -
* Features (10):
  - dbl (10): am, carb, cyl, disp, drat, gear, hp, qsec, vs, wt
```

Running tsk() without any arguments will list all the tasks in the dictionary, this also works for all other sugar constructors that you will encounter throughout the book.

> 💡 Help Pages
>
> Usually in R, the help pages of functions can be queried with ?. The same is true of R6 classes, so if you want to find the help page of the mtcars task you could use ?mlr_tasks_mtcars. We have also added a $help() method to many of our classes, which allows you to access the help page from any instance of that class, for example: tsk("mtcars")$help().

TaskRegr To create your own regression task, you will need to construct a new instance of as_task_regr() TaskRegr. The simplest way to do this is with the function as_task_regr() to convert a data.frame type object to a regression task, specifying the target feature by passing this to the target argument. By example, we will ignore that mtcars is already available as a predefined task in mlr3. In the code below, we load the datasets::mtcars dataset, subset the data to only include columns "mpg", "cyl", "disp", print the modified data's properties, and then set up a regression task called "cars" (id = "cars") in which we will try to predict miles per gallon (target = "mpg") from the number of cylinders ("cyl") and displacement ("disp"):

```
data("mtcars", package = "datasets")
mtcars_subset = subset(mtcars, select = c("mpg", "cyl", "disp"))
str(mtcars_subset)
```

```
'data.frame':   32 obs. of  3 variables:
 $ mpg : num  21 21 22.8 21.4 18.7 18.1 14.3 24.4 22.8 19.2 ...
 $ cyl : num  6 6 4 6 8 6 8 4 4 6 ...
 $ disp: num  160 160 108 258 360 ...
```

```
tsk_mtcars = as_task_regr(mtcars_subset, target = "mpg",
        id = "cars")
```

The data can be in any tabular format, e.g. a `data.frame()`, `data.table()`, or `tibble()`. The `target` argument specifies the prediction target column. The `id` argument is optional and specifies an identifier for the task that is used in plots and summaries; if omitted the variable name of the data will be used as the `id`.

💡 UTF8 Column Names

As many machine learning models do not work properly with arbitrary UTF8, names, `mlr3` defaults to throwing an error if any of the column names passed to `as_task_regr()` (and other task constructors) contain a non-ASCII character or do not comply with R's variable naming scheme. Therefore, we recommend converting names with `make.names()` if possible. You can bypass this check by setting `options(mlr3.allow_utf8_names = TRUE)` (but do not be surprised if errors occur later, especially when passing objects to other packages).

Printing a task provides a summary, and in this case, we can see the task has 32 observations and 3 columns (32 x 3), of which `mpg` is the target, there are no special properties (`Properties: -`), and there are 2 features stored in double-precision floating point format.

```
tsk_mtcars
```

```
<TaskRegr:cars> (32 x 3)
* Target: mpg
* Properties: -
* Features (2):
  - dbl (2): cyl, disp
```

We can plot the task using the `mlr3viz` package, which gives a graphical summary of the distribution of the target and feature values:

```
library(mlr3viz)
autoplot(tsk_mtcars, type = "pairs")
```

2.1.2 Retrieving Data

We have looked at how to create tasks to store data and metadata, now we will look at how to retrieve the stored data.

Various fields can be used to retrieve metadata about a task. The dimensions, for example, can be retrieved using `$nrow` and `$ncol`:

```
c(tsk_mtcars$nrow, tsk_mtcars$ncol)
```

```
[1] 32  3
```

The names of the feature and target columns are stored in the `$feature_names` and `$target_names` slots, respectively.

```
c(Features = tsk_mtcars$feature_names,
  Target = tsk_mtcars$target_names)
```

```
Features1 Features2    Target
   "cyl"     "disp"     "mpg"
```

The columns of a task have unique `character`-valued names and the rows are identified by unique natural numbers, called row IDs. They can be accessed through the `$row_ids` field:

```
head(tsk_mtcars$row_ids)
```

```
[1] 1 2 3 4 5 6
```

Row IDs are not used as features when training or predicting but are metadata that allow access to individual observations. Note that row IDs are not the same as row numbers. This is best demonstrated by example (which is shown below), we create a regression task from random data, print the original row IDs, which correspond to row numbers 1-5, then filter three rows (we will return to this method just below) and print the new row IDs, which no longer correspond to the row numbers.

```
task = as_task_regr(data.frame(x = runif(5), y = runif(5)),
  target = "y")
task$row_ids
```

```
[1] 1 2 3 4 5
```

```
task$filter(c(4, 1, 3))
task$row_ids
```

```
[1] 1 3 4
```

This design decision allows tasks and learners to transparently operate on real database management systems, where primary keys are required to be unique, but not necessarily consecutive. See Section 10.4 for more information on using databases as data backends for tasks.

The data contained in a task can be accessed through `$data()`, which returns a `data.table` object. This method has optional `rows` and `cols` arguments to specify subsets of the data to retrieve.

```
# retrieve all data
tsk_mtcars$data()
```

```
     mpg cyl  disp
 1: 21.0   6 160.0
 2: 21.0   6 160.0
 3: 22.8   4 108.0
 4: 21.4   6 258.0
 5: 18.7   8 360.0
---
28: 30.4   4  95.1
```

```
29: 15.8    8 351.0
30: 19.7    6 145.0
31: 15.0    8 301.0
32: 21.4    4 121.0
```

```
# retrieve data for rows with IDs 1, 5, and 10 and all feature
# columns
tsk_mtcars$data(rows = c(1, 5, 10),
cols = tsk_mtcars$feature_names)
```

```
   cyl  disp
1:   6 160.0
2:   8 360.0
3:   6 167.6
```

> 💡 Accessing Rows by Number
>
> You can work with row numbers instead of row IDs by using the `$row_ids`
> field to extract the row ID corresponding to a given row number:
>
> ```
> # select the 2nd row of the task by extracting the second
> # row_id:
> tsk_mtcars$data(rows = task$row_ids[2])
> ```

You can always use "standard" R methods to extract summary data from a task,
for example, to summarize the underlying data:

```
summary(as.data.table(tsk_mtcars))
```

```
     mpg             cyl             disp
Min.   :10.4    Min.   :4.00    Min.   : 71.1
1st Qu.:15.4    1st Qu.:4.00    1st Qu.:120.8
Median :19.2    Median :6.00    Median :196.3
Mean   :20.1    Mean   :6.19    Mean   :230.7
3rd Qu.:22.8    3rd Qu.:8.00    3rd Qu.:326.0
Max.   :33.9    Max.   :8.00    Max.   :472.0
```

2.1.3   Task Mutators

After a task has been created, you may want to perform operations on the task such
as filtering down to subsets of rows and columns, which is often useful for manually
creating train and test splits or to fit models on a subset of given features. Above
we saw how to access subsets of the underlying dataset using `$data()`, however,
this will not change the underlying task. Therefore, we provide mutators, which   Mutators
modify the given `Task` in place, which can be seen in the examples below.

Subsetting by features (columns) is possible with `$select()` with the desired
feature names passed as a character vector and subsetting by observations (rows) is
performed with `$filter()` by passing the row IDs as a numeric vector.

```
tsk_mtcars_small = tsk("mtcars") # initialize with the full task
tsk_mtcars_small$select("cyl") # keep only one feature
tsk_mtcars_small$filter(2:3) # keep only these rows
tsk_mtcars_small$data()
```

```
    mpg cyl
1: 21.0   6
2: 22.8   4
```

As R6 uses reference semantics (Section 1.5.1), you need to use $clone() if you want to modify a task while keeping the original object intact.

```
# the wrong way
tsk_mtcars = tsk("mtcars")
tsk_mtcars_wrong = tsk_mtcars
tsk_mtcars_wrong$filter(1:2)
# original data affected
tsk_mtcars$head()
```

```
   mpg am carb cyl disp drat gear  hp  qsec vs    wt
1:  21  1    4   6  160  3.9    4 110 16.46  0 2.620
2:  21  1    4   6  160  3.9    4 110 17.02  0 2.875
```

```
# the right way
tsk_mtcars = tsk("mtcars")
tsk_mtcars_right = tsk_mtcars$clone()
tsk_mtcars_right$filter(1:2)
# original data unaffected
tsk_mtcars$head()
```

```
    mpg am carb cyl disp drat gear  hp  qsec vs    wt
1: 21.0  1    4   6  160 3.90    4 110 16.46  0 2.620
2: 21.0  1    4   6  160 3.90    4 110 17.02  0 2.875
3: 22.8  1    1   4  108 3.85    4  93 18.61  1 2.320
4: 21.4  0    1   6  258 3.08    3 110 19.44  1 3.215
5: 18.7  0    2   8  360 3.15    3 175 17.02  0 3.440
6: 18.1  0    1   6  225 2.76    3 105 20.22  1 3.460
```

To add extra rows and columns to a task, you can use $rbind() and $cbind() respectively:

```
tsk_mtcars_small$cbind( # add another column
  data.frame(disp = c(150, 160))
)
tsk_mtcars_small$rbind( # add another row
  data.frame(mpg = 23, cyl = 5, disp = 170)
)
tsk_mtcars_small$data()
```

```
    mpg cyl disp
1: 21.0   6   150
2: 22.8   4   160
3: 23.0   5   170
```

2.2 Learners

Objects of class `Learner` provide a unified interface to many popular machine learning algorithms in R. The `mlr_learners` dictionary contains all the learners available in `mlr3`. We will discuss the available learners in Section 2.7; for now, we will just use a regression tree learner as an example to discuss the `Learner` interface. As with tasks, you can access learners from the dictionary with a single sugar function, in this case, `lrn()`.

Learner

mlr_learners

lrn()

```
lrn("regr.rpart")
```

```
<LearnerRegrRpart:regr.rpart>: Regression Tree
* Model: -
* Parameters: xval=0
* Packages: mlr3, rpart
* Predict Types:  [response]
* Feature Types: logical, integer, numeric, factor, ordered
* Properties: importance, missings, selected_features, weights
```

All `Learner` objects include the following metadata, which can be seen in the output above:

* `$feature_types`: the type of features the learner can handle.
* `$packages`: the packages required to be installed to use the learner.
* `$properties`: the properties of the learner. For example, the "missings" properties means a model can handle missing data, and "importance" means it can compute the relative importance of each feature.
* `$predict_types`: the types of prediction that the model can make (Section 2.2.2).
* `$param_set`: the set of available hyperparameters (Section 2.2.3).

To run a machine learning experiment, learners pass through two stages (Figure 2.2):

* Training: A training `Task` is passed to the learner's `$train()` function which trains and stores a model, i.e., the learned relationship of the features to the target.
* Predicting: New data, potentially a different partition of the original dataset, is passed to the `$predict()` method of the trained learner to predict the target values.

Training

Predicting

2.2.1 Training

In the simplest use case, models are trained by passing a task to a learner with the `$train()` method:

$train()

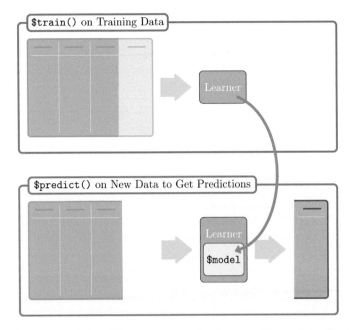

Figure 2.2: Overview of the different stages of a learner. Top – data (features and a target) are passed to an (untrained) learner. Bottom – new data are passed to the trained model which makes predictions for the "missing" target column.

```
# load mtcars task
tsk_mtcars = tsk("mtcars")
# load a regression tree
lrn_rpart = lrn("regr.rpart")
# pass the task to the learner via $train()
lrn_rpart$train(tsk_mtcars)
```

$model   After training, the fitted model is stored in the $model field for future inspection and prediction:

```
# inspect the trained model
lrn_rpart$model
```

```
n= 32

node), split, n, deviance, yval
      * denotes terminal node

1) root 32 1126.00 20.09
  2) cyl>=5 21   198.50 16.65
    4) hp>=192.5 7    28.83 13.41 *
    5) hp< 192.5 14    59.87 18.26 *
  3) cyl< 5 11   203.40 26.66 *
```

We see that the regression tree has identified features in the task that are predictive of
the target (mpg) and used them to partition observations. The textual representation
of the model depends on the type of learner. For more information on any model
see the learner help page, which can be accessed in the same way as tasks with the
help() field, e.g., lrn_rpart$help().

2.2.1.1   Partitioning Data

When assessing the quality of a model's predictions, you will likely want to partition
your dataset to get a fair and unbiased estimate of a model's generalization error.
In Chapter 3, we will look at resampling and benchmark experiments, which will go
into more detail about performance estimation but for now, we will just discuss the
simplest method of splitting data using the partition() function. This function   `partition()`
creates index sets that randomly split the given task into two disjoint sets: a training
set (67% of the total data by default) and a test set (the remaining 33% of the total
data not in the training set).

```
splits = partition(tsk_mtcars)
splits
```

```
$train
 [1]  1  3  4  5  8 10 21 25 32  6  7 11 14 15 16 17 22 31 19 20 26

$test
 [1]  2  9 27 30 12 13 23 24 29 18 28
```

When training we will tell the model to only use the training data by passing the
row IDs from partition to the row_ids argument of $train():

```
lrn_rpart$train(tsk_mtcars, row_ids = splits$train)
```

Now we can use our trained learner to make predictions on new data.

2.2.2   Predicting

Predicting from trained models is as simple as passing your data as a Task to the
$predict() method of the trained Learner.   `$predict()`

Carrying straight on from our last example, we will call the $predict() method of
our trained learner and again will use the row_ids argument, but this time to pass
the IDs of our test set:

```
prediction = lrn_rpart$predict(tsk_mtcars, row_ids = splits$test)
```

The $predict() method returns an object inheriting from Prediction, in this case   `Prediction`
PredictionRegr as this is a regression task.

                                                                                 `PredictionRe`

```
prediction
```

```
<PredictionRegr> for 11 observations:
    row_ids truth response
          2  21.0     24.9
          9  22.8     24.9
         27  26.0     24.9
---
         29  15.8     24.9
         18  32.4     24.9
         28  30.4     24.9
```

The `row_ids` column corresponds to the row IDs of the predicted observations. The `truth` column contains the ground truth data if available, which the object extracts from the task, in this case: `tsk_mtcars$truth(splits$test)`. Finally, the `response` column contains the values predicted by the model. The Prediction object can easily be converted into a `data.table` or `data.frame` using `as.data.table()`/`as.data.frame()` respectively.

All data in the above columns can be accessed directly, for example, to get the first two predicted responses:

```
prediction$response[1:2]
```

```
[1] 24.9 24.9
```

Similarly to plotting `Tasks`, `mlr3viz` provides an `autoplot()` method for Prediction objects.

```
library(mlr3viz)
prediction = lrn_rpart$predict(tsk_mtcars, splits$test)
autoplot(prediction)
```

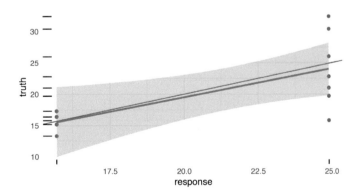

Figure 2.3: Comparing predicted and ground truth values for the mtcars dataset.

In the examples above, we made predictions by passing a task to `$predict()`. However, if you would rather pass a `data.frame` type object directly, then you can use `$predict_newdata()`. Note, the `truth` column values are all `NA`, as we did not include a target column in the generated data.

```
mtcars_new = data.table(cyl = c(5, 6), disp = c(100, 120),
  hp = c(100, 150), drat = c(4, 3.9), wt = c(3.8, 4.1),
  qsec = c(18, 19.5), vs = c(1, 0), am = c(1, 1),
  gear = c(6, 4), carb = c(3, 5))
prediction = lrn_rpart$predict_newdata(mtcars_new)
prediction
```

```
<PredictionRegr> for 2 observations:
 row_ids truth response
       1    NA    15.71
       2    NA    15.71
```

Changing the Prediction Type

While predicting a single numeric quantity is the most common prediction type in regression, it is not the only prediction type. Several regression models can also predict standard errors. To predict this, the $predict_type field of a LearnerRegr must be changed from "response" (the default) to "se" before training. The "rpart" learner we used above does not support predicting standard errors, so in the example below we will use a linear regression model (lrn("regr.lm")).

```
library(mlr3learners)
lrn_lm = lrn("regr.lm", predict_type = "se")
lrn_lm$train(tsk_mtcars, splits$train)
lrn_lm$predict(tsk_mtcars, splits$test)
```

```
<PredictionRegr> for 11 observations:
    row_ids truth response    se
          2  21.0   20.20 1.817
          9  22.8   33.32 4.246
         27  26.0   32.22 4.121
---
         29  15.8   28.26 4.182
         18  32.4   27.77 1.114
         28  30.4   27.57 2.735
```

Now the output includes an se column as desired. In Section 2.5.3, we will see prediction types playing an even bigger role in the context of classification.

Having covered the unified train/predict interface, we can now look at how to use hyperparameters to configure these methods for individual algorithms.

2.2.3   Hyperparameters

Learners encapsulate a machine learning algorithm and its hyperparameters, which affect *how* the algorithm is run and can be set by the user. Hyperparameters may affect how a model is trained or how it makes predictions and deciding how to set hyperparameters can require expert knowledge. Hyperparameters can be optimized automatically (Chapter 4), but in this chapter, we will focus on how to set them manually.

2.2.3.1 Paradox and Parameter Sets

We will continue our running example with a regression tree learner. To access the hyperparameters in the decision tree, we use `$param_set`:

```
lrn_rpart$param_set
```

```
<ParamSet>
                 id    class lower upper nlevels        default value
 1:              cp ParamDbl     0     1     Inf           0.01
 2:      keep_model ParamLgl    NA    NA       2          FALSE
 3:      maxcompete ParamInt     0   Inf     Inf              4
 4:        maxdepth ParamInt     1    30      30             30
 5:    maxsurrogate ParamInt     0   Inf     Inf              5
 6:       minbucket ParamInt     1   Inf     Inf <NoDefault[3]>
 7:        minsplit ParamInt     1   Inf     Inf             20
 8:   surrogatestyle ParamInt    0     1       2              0
 9:     usesurrogate ParamInt    0     2       3              2
10:            xval ParamInt     0   Inf     Inf             10     0
```

The output above is a `ParamSet` object, supplied by the paradox package. These objects provide information on hyperparameters including their name (`id`), data type (`class`), technically valid ranges for hyperparameter values (`lower`, `upper`), the number of levels possible if the data type is categorical (`nlevels`), the default value from the underlying package (`default`), and finally the set value (`value`). The second column references classes defined in paradox that determine the class of the parameter and the possible values it can take. Table 2.1 lists the possible hyperparameter types, all of which inherit from `Param`.

Table 2.1: Hyperparameter classes and the type of hyperparameter they represent.

Hyperparameter Class	Hyperparameter Type
ParamDbl	Real-valued (numeric)
ParamInt	Integer
ParamFct	Categorical (factor)
ParamLgl	Logical / Boolean
ParamUty	Untyped

In our decision tree example, we can infer from the `ParamSet` output that:

- cp must be a "double" (`ParamDbl`) taking values between 0 (`lower`) and 1 (`upper`) with a default of 0.01 (`default`).
- keep_model must be a "logical" (`ParamLgl`) taking values TRUE or FALSE with default FALSE
- xval must be an "integer" (`ParamInt`) taking values between 0 and Inf with a default of 10 and has a set value of 0.

In rare cases (we try to minimize it as much as possible), hyperparameters are initialized to values which deviate from the default in the underlying package. When this happens, the reason will always be given in the learner help page. In the case of `lrn("regr.rpart")`, the xval hyperparameter is initialized to 0 because xval controls internal cross-validations and if a user accidentally leaves this at the default 10, model training can take an unnecessarily long time.

2.2.3.2 Getting and Setting Hyperparameter Values

Now we have looked at how hyperparameter sets are stored, we can think about getting and setting them. Returning to our decision tree, say we are interested in growing a tree with depth 1, also known as a "decision stump", where data is split only once into two terminal nodes. From the parameter set output, we know that the `maxdepth` parameter has a default of 30 and that it takes integer values.

There are a few different ways we could change this hyperparameter. The simplest way is during the construction of the learner by passing the hyperparameter name and new value to `lrn()`:

```
lrn_rpart = lrn("regr.rpart", maxdepth = 1)
```

We can get a list of non-default hyperparameters (i.e., those that have been set) by using `$param_set$values`:

```
lrn_rpart$param_set$values
```

```
$xval
[1] 0

$maxdepth
[1] 1
```

Now we can see that `maxdepth = 1` (as we discussed above `xval = 0` is changed during construction) and the learned regression tree reflects this:

```
lrn_rpart$train(tsk("mtcars"))$model
```

```
n= 32

node), split, n, deviance, yval
      * denotes terminal node

1) root 32 1126.0 20.09
  2) cyl>=5 21   198.5 16.65 *
  3) cyl< 5 11   203.4 26.66 *
```

The `$values` field simply returns a `list` of set hyperparameters, so another way to update hyperparameters is by updating an element in the list:

```
lrn_rpart$param_set$values$maxdepth = 2
lrn_rpart$param_set$values
```

```
$xval
[1] 0

$maxdepth
[1] 2
```

```
# now with depth 2
lrn_rpart$train(tsk("mtcars"))$model
```

n= 32

node), split, n, deviance, yval
      * denotes terminal node

```
1) root 32 1126.00 20.09
  2) cyl>=5 21   198.50 16.65
    4) hp>=192.5 7    28.83 13.41 *
    5) hp< 192.5 14    59.87 18.26 *
  3) cyl< 5 11   203.40 26.66 *
```

To set multiple values at once, we recommend either setting these during construction or using `$set_values()`, which updates the given hyperparameters (argument names) with the respective values.

$set_values()

```
lrn_rpart = lrn("regr.rpart", maxdepth = 3, xval = 1)
lrn_rpart$param_set$values
```

```
$xval
[1] 1
```

```
$maxdepth
[1] 3
```

```
# or with set_values
lrn_rpart$param_set$set_values(xval = 2, cp = 0.5)
lrn_rpart$param_set$values
```

```
$xval
[1] 2
```

```
$maxdepth
[1] 3
```

```
$cp
[1] 0.5
```

 Setting Hyperparameters Using a `list`

As `lrn_rpart$param_set$values` returns a `list`, some users may be tempted to set hyperparameters by passing a new `list` to `$values` – this would work, but **we do not recommend it**. This is because passing a `list` will wipe any existing hyperparameter values if they are not included in the list. For example:

```
# set xval and cp
lrn_rpart_params = lrn("regr.rpart", xval = 0, cp = 1)
# passing maxdepth through a list, removing all other values
lrn_rpart_params$param_set$values = list(maxdepth = 1)
# we have removed xval and cp by mistake
lrn_rpart_params$param_set$values
```

```
$maxdepth
[1] 1
```

```
# now with set_values
lrn_rpart_params = lrn("regr.rpart", xval = 0, cp = 1)
lrn_rpart_params$param_set$set_values(maxdepth = 1)
lrn_rpart_params$param_set$values
```

```
$xval
[1] 0

$cp
[1] 1

$maxdepth
[1] 1
```

Whichever method you choose, all have safety checks to ensure your new values fall within the allowed parameter range:

```
lrn("regr.rpart", cp = 2, maxdepth = 2)
```

```
Error in self$assert(xs): Assertion on 'xs' failed: cp:
Element 1 is not <= 1.
```

2.2.3.3   Hyperparameter Dependencies

> **i**   **This section covers advanced ML or technical details.**

More complex hyperparameter spaces may include dependencies, which occur when setting a hyperparameter is conditional on the value of another hyperparameter; this is most important in the context of model tuning (Chapter 4). One such example is a support vector machine (`lrn("regr.svm")`). The field `$deps` returns a `data.table`, which lists the hyperparameter dependencies in the `Learner`. For example, we can see that the `cost` (`id`-column) parameter is dependent on the `type` (`on`-column) parameter.

```
lrn("regr.svm")$param_set$deps
```

```
        id      on            cond
1:    cost    type <CondAnyOf[9]>
2:      nu    type <CondEqual[9]>
3:  degree  kernel <CondEqual[9]>
4:   coef0  kernel <CondAnyOf[9]>
5:   gamma  kernel <CondAnyOf[9]>
6: epsilon    type <CondEqual[9]>
```

The cond column tells us what the condition is, which will either mean that id can be set if on equals a single value (CondEqual) or any value in the listed set (CondAnyOf).

```
lrn("regr.svm")$param_set$deps[[1, "cond"]]
```

CondAnyOf: x   {eps-regression, nu-regression}

```
lrn("regr.svm")$param_set$deps[[3, "cond"]]
```

CondEqual: x = polynomial

This tells us that the parameter cost should only be set if the type parameter is one of "eps-regression" or "nu-regression", and degree should only be set if kernel is equal to "polynomial".

The Learner will error if dependent hyperparameters are set when their conditions are not met:

```
# error as kernel is not polynomial
lrn("regr.svm", kernel = "linear", degree = 1)
```

Error in self$assert(xs): Assertion on 'xs' failed: The parameter 'degree' can only be set if the following condition is met 'kernel = polynomial'. Instead the current parameter value is: kernel=linear.

```
# works because kernel is polynomial
lrn("regr.svm", kernel = "polynomial", degree = 1)
```

```
<LearnerRegrSVM:regr.svm>
* Model: -
* Parameters: kernel=polynomial, degree=1
* Packages: mlr3, mlr3learners, e1071
* Predict Types:  [response]
* Feature Types: logical, integer, numeric
* Properties: -
```

2.2.4 Baseline Learners

Baselines

Before we move on to learner evaluation, we will highlight an important class of learners. These are extremely simple or "weak" learners known as baselines. Baselines are useful in model comparison (Chapter 3) and as fallback learners (Section 5.1.1, Section 10.2.2). For regression, we have implemented the baseline

lrn("regr.featureless"), which always predicts new values to be the mean (or
median, if the robust hyperparameter is set to TRUE) of the target in the training
data:

```
# generate data
df = as_task_regr(data.frame(x = runif(1000),
  y = rnorm(1000, 2, 1)), target = "y")
lrn("regr.featureless")$train(df, 1:995)$predict(df, 996:1000)
```

```
<PredictionRegr> for 5 observations:
 row_ids truth response
     996 2.996    1.976
     997 3.675    1.976
     998 3.651    1.976
     999 1.803    1.976
    1000 1.196    1.976
```

It is good practice to test all new models against a baseline, and also to include
baselines in experiments with multiple other models. In general, a model that does
not outperform a baseline is a "bad" model, on the other hand, a model is not
necessarily "good" if it outperforms the baseline.

2.3   Evaluation

Perhaps *the most* important step of the applied machine learning workflow is
evaluating model performance. Without this, we would have no way to know if our
trained model makes very accurate predictions, is worse than randomly guessing, or
somewhere in between. We will continue with our decision tree example to establish
if the quality of our predictions is "good", first we will rerun the above code so it is
easier to follow along.

```
lrn_rpart = lrn("regr.rpart")
tsk_mtcars = tsk("mtcars")
splits = partition(tsk_mtcars)
lrn_rpart$train(tsk_mtcars, splits$train)
prediction = lrn_rpart$predict(tsk_mtcars, splits$test)
```

2.3.1   Measures

The quality of predictions is evaluated using measures that compare them to the
ground truth data for supervised learning tasks. Similarly to Tasks and Learners,
the available measures in mlr3 are stored in a dictionary called mlr_measures and
can be accessed with msr():   msr()

```
as.data.table(msr())
```

```
            key                         label  task_type
 1:         aic   Akaike Information Criterion       <NA>
 2:         bic  Bayesian Information Criterion      <NA>
 3: classif.acc        Classification Accuracy    classif
 4: classif.auc        Area Under the ROC Curve   classif
 5: classif.bacc              Balanced Accuracy   classif
---
62: sim.jaccard        Jaccard Similarity Index       <NA>
63:     sim.phi      Phi Coefficient Similarity      <NA>
64:   time_both                   Elapsed Time        <NA>
65: time_predict                 Elapsed Time        <NA>
66:  time_train                  Elapsed Time        <NA>
3 variables not shown: [packages, predict_type, task_properties]
```

All measures implemented in `mlr3` are defined primarily by three components: 1) the function that defines the measure; 2) whether a lower or higher value is considered "good"; and 3) the range of possible values the measure can take. As well as these defining elements, other metadata are important to consider when selecting and using a `Measure`, including if the measure has any special properties (e.g., requires training data), the type of predictions the measure can evaluate, and whether the measure has any "control parameters". All this information is encapsulated in the

Measure

Measure object. By example, let us consider the mean absolute error (MAE):

```
measure = msr("regr.mae")
measure
```

```
<MeasureRegrSimple:regr.mae>: Mean Absolute Error
* Packages: mlr3, mlr3measures
* Range: [0, Inf]
* Minimize: TRUE
* Average: macro
* Parameters: list()
* Properties: -
* Predict type: response
```

This measure compares the absolute difference ("error") between true and predicted values: $f(y, \hat{y}) = |y - \hat{y}|$. Lower values are considered better (`Minimize: TRUE`), which is intuitive as we would like the true values, y, to be identical (or as close as possible) in value to the predicted values, $\hat{y}$. We can see that the range of possible values the learner can take is from 0 to ∞ (`Range: [0, Inf]`), it has no special properties (`Properties: -`), it evaluates `response` type predictions for regression models (`Predict type: response`), and it has no control parameters (`Parameters: list()`).

Now let us see how to use this measure for scoring our predictions.

2.3.2   Scoring Predictions

Usually, supervised learning measures compare the difference between predicted values and the ground truth. `mlr3` simplifies the process of bringing these quantities together by storing the predictions and true outcomes in the `Prediction` object as we have already seen.

```
prediction
```

```
<PredictionRegr> for 11 observations:
    row_ids truth response
          2  21.0   16.70
          8  24.4   26.81
         21  21.5   26.81
---
         31  15.0   16.70
         18  32.4   26.81
         26  27.3   26.81
```

To calculate model performance, we simply call the `$score()` method of a $score()
`Prediction` object and pass as a single argument the measure that we want to
compute:

```
prediction$score(measure)
```

```
regr.mae
   2.591
```

Note that all task types have default measures that are used if the argument to `$score()` is omitted, for regression this is the mean squared error
(`msr("regr.mse")`), which is the squared difference between true and predicted
values: $f(y, \hat{y}) = (y - \hat{y})^2$, averaged over the test set.

It is possible to calculate multiple measures at the same time by passing multiple
measures to `$score()`. For example, below we compute performance for mean
squared error (`"regr.mse"`) and mean absolute error (`"regr.mae"`) – note we use
`msrs()` to load multiple measures at once.   msrs()

```
measures = msrs(c("regr.mse", "regr.mae"))
prediction$score(measures)
```

```
regr.mse regr.mae
   9.567    2.591
```

2.3.3   Technical Measures

> **i**   **This section covers advanced ML or technical details.**

`mlr3` also provides measures that do not quantify the quality of the predictions of a
model, but instead provide "meta"-information about the model. These include:

- `msr("time_train")` – The time taken to train a model.
- `msr("time_predict")` – The time taken for the model to make predictions.
- `msr("time_both")` – The total time taken to train the model and then make
  predictions.
- `msr("selected_features")` – The number of features selected by a model, which
  can only be used if the model has the "selected_features" property.

For example, we could score our decision tree to see how many seconds it took to train the model and make predictions:

```
measures = msrs(c("time_train", "time_predict", "time_both"))
prediction$score(measures, learner = lrn_rpart)
```

```
time_train time_predict   time_both
     0.002        0.001       0.003
```

Notice a few key properties of these measures:

1) `time_both` is simply the sum of `time_train` and `time_predict`.
2) We had to pass `learner = lrn_rpart` to `$score()` as these measures have the `requires_learner` property:

```
msr("time_train")$properties
```

```
[1] "requires_learner"
```

3) These can be used after model training and predicting because we automatically store model run times whenever `$train()` and `$predict()` are called, so the measures above are equivalent to:

```
c(lrn_rpart$timings, both = sum(lrn_rpart$timings))
```

```
train predict    both
0.002   0.001   0.003
```

The `selected_features` measure calculates how many features were used in the fitted model.

```
msr_sf = msr("selected_features")
msr_sf
```

```
<MeasureSelectedFeatures:selected_features>: Absolute or Relative
Frequency of Selected Features
* Packages: mlr3
* Range: [0, Inf]
* Minimize: TRUE
* Average: macro
* Parameters: normalize=FALSE
* Properties: requires_task, requires_learner, requires_model
* Predict type: NA
```

Control Parameters   We can see that this measure contains control parameters (`Parameters: normalize=FALSE`), which control how the measure is computed. As with hyperparameters these can be accessed with `$param_set`:

```
msr_sf = msr("selected_features")
msr_sf$param_set
```

```
<ParamSet>
          id    class lower upper nlevels default value
1: normalize ParamLgl    NA    NA       2   FALSE FALSE
```

The `normalize` hyperparameter specifies whether the returned number of selected features should be normalized by the total number of features; this is useful if you are comparing this value across tasks with differing numbers of features. We would change this parameter in the exact same way as we did with the learner above:

```
msr_sf$param_set$values$normalize = TRUE
prediction$score(msr_sf, task = tsk_mtcars, learner = lrn_rpart)
```

```
selected_features
              0.1
```

Note that we passed the task and learner as the measure has the `requires_task` and `requires_learner` properties.

2.4   Our First Regression Experiment

We have now seen how to train a model, make predictions and score them. What we have not yet attempted is to ascertain if our predictions are any "good". So before look at how the building blocks of `mlr3` extend to classification, we will take a brief pause to put together everything above in a short experiment to assess the quality of our predictions. We will do this by comparing the performance of a featureless regression learner to a decision tree with changed hyperparameters.

```
library(mlr3)
set.seed(349)
# load and partition our task
tsk_mtcars = tsk("mtcars")
splits = partition(tsk_mtcars)
# load featureless learner
lrn_featureless = lrn("regr.featureless")
# load decision tree and set hyperparameters
lrn_rpart = lrn("regr.rpart", cp = 0.2, maxdepth = 5)
# load MSE and MAE measures
measures = msrs(c("regr.mse", "regr.mae"))
# train learners
lrn_featureless$train(tsk_mtcars, splits$train)
lrn_rpart$train(tsk_mtcars, splits$train)
# make and score predictions
lrn_featureless$predict(tsk_mtcars, splits$test)$score(measures)
```

```
regr.mse regr.mae
  26.727    4.513
```

```
lrn_rpart$predict(tsk_mtcars, splits$test)$score(measures)
```

```
regr.mse regr.mae
   6.933    2.206
```

Before starting the experiment, we load the `mlr3` library and set a seed. We loaded the `mtcars` task using `tsk()` and then split this using `partition` with the default 2/3 split. Next, we loaded a featureless baseline learner (`"regr.featureless"`) with the `lrn()` function. Then loaded a decision tree (`lrn("regr.rpart")`) but changed the complexity parameter and max tree depth from their defaults. We then used `msrs()` to load multiple measures at once, the mean squared error (MSE: `regr.mse`) and the mean absolute error (MAE: `regr.mae`). With all objects loaded, we trained our models, ensuring we passed the same training data to both. Finally, we made predictions from our trained models and scored these. For both MSE and MAE, lower values are "better" (`Minimize: TRUE`) and we can therefore conclude that our decision tree performs better than the featureless baseline. In Section 3.3, we will see how to formalize comparison between models in a more efficient way using `benchmark()`.

Now we have put everything together you may notice that our learners and measures both have the `"regr."` prefix, which is a handy way of reminding us that we are working with a regression task and therefore must make use of learners and measures built for regression. In the next section, we will extend the building blocks of `mlr3` to consider classification tasks, which make use of learners and measures with the `"classif."` prefix.

2.5 Classification

Classification problems are ones in which a model predicts a discrete, categorical target, as opposed to a continuous, numeric quantity. For example, predicting the species of penguin from its physical characteristics would be a classification problem as there is a defined set of species. `mlr3` ensures that the interface for all tasks is as similar as possible (if not identical) and therefore we will not repeat any content from the previous section but will just focus on differences that make classification a unique machine-learning problem. We will first demonstrate the similarities between regression and classification by performing an experiment very similar to the one in Section 2.4, using code that will now be familiar to you. We will then move to differences in tasks, learners and predictions, before looking at thresholding, which is a method specific to classification.

2.5.1 Our First Classification Experiment

The interface for classification tasks, learners, and measures, is identical to the regression setting, except the underlying objects inherit from `TaskClassif`, `LearnerClassif`, and `MeasureClassif`, respectively. We can therefore run a very similar experiment to the one above.

```
library(mlr3)
set.seed(349)
# load and partition our task
tsk_penguins = tsk("penguins")
splits = partition(tsk_penguins)
# load featureless learner
lrn_featureless = lrn("classif.featureless")
# load decision tree and set hyperparameters
lrn_rpart = lrn("classif.rpart", cp = 0.2, maxdepth = 5)
# load accuracy measure
measure = msr("classif.acc")
# train learners
lrn_featureless$train(tsk_penguins, splits$train)
lrn_rpart$train(tsk_penguins, splits$train)
# make and score predictions
lrn_featureless$predict(tsk_penguins, splits$test)$score(measure)
```

```
classif.acc
     0.4425
```

```
lrn_rpart$predict(tsk_penguins, splits$test)$score(measure)
```

```
classif.acc
     0.9469
```

In this experiment, we loaded the predefined task `penguins`, which is based on the penguins dataset, then partitioned the data into training and test splits. We loaded the featureless classification baseline (using the default which always predicts the most common class in the training data, but which also has the option of predicting (uniformly or weighted) random response values) and a classification decision tree, then the accuracy measure (number of correct predictions divided by the total number of predictions), trained our models and finally made and scored predictions. Once again we can be happy with our predictions, which are vastly more accurate than the baseline.

Now that we have seen the similarities between classification and regression, we can turn to some key differences.

2.5.2   TaskClassif

Classification tasks, objects inheriting from `TaskClassif`, are very similar to regression tasks, except that the target variable is of type factor and will have a limited number of possible classes/categories that observations can fall into.

You can view the predefined classification tasks in `mlr3` by filtering the `mlr_tasks` dictionary:

```
as.data.table(mlr_tasks)[task_type == "classif"]
```

	key	label	task_type
1:	breast_cancer	Wisconsin Breast Cancer	classif
2:	german_credit	German Credit	classif
3:	ilpd	Indian Liver Patient Data	classif
4:	iris	Iris Flowers	classif
5:	optdigits	Optical Recognition of Handwritten Digits	classif
---			
9:	sonar	Sonar: Mines vs. Rocks	classif
10:	spam	HP Spam Detection	classif
11:	titanic	Titanic	classif
12:	wine	Wine Regions	classif
13:	zoo	Zoo Animals	classif

10 variables not shown: [nrow, ncol, properties, lgl, int, dbl, chr, fct, ord, pxc]

You can create your own task with as_task_classif.

```
as_task_classif(palmerpenguins::penguins, target = "species")
```

```
<TaskClassif:palmerpenguins::penguins> (344 x 8)
* Target: species
* Properties: multiclass
* Features (7):
  - int (3): body_mass_g, flipper_length_mm, year
  - dbl (2): bill_depth_mm, bill_length_mm
  - fct (2): island, sex
```

There are two types of classification tasks supported in mlr3: binary classification, in which the outcome can be one of two categories, and multiclass classification, where the outcome can be one of three or more categories.

The sonar task is an example of a binary classification problem, as the target can only take two different values, in mlr3 terminology it has the "twoclass" property:

```
tsk_sonar = tsk("sonar")
tsk_sonar
```

```
<TaskClassif:sonar> (208 x 61): Sonar: Mines vs. Rocks
* Target: Class
* Properties: twoclass
* Features (60):
  - dbl (60): V1, V10, V11, V12, V13, V14, V15, V16, V17, V18,
    V19, V2, V20, V21, V22, V23, V24, V25, V26, V27, V28, V29,
    V3, V30, V31, V32, V33, V34, V35, V36, V37, V38, V39, V4,
    V40, V41, V42, V43, V44, V45, V46, V47, V48, V49, V5, V50,
    V51, V52, V53, V54, V55, V56, V57, V58, V59, V6, V60, V7,
    V8, V9
```

```
tsk_sonar$class_names
```

```
[1] "M" "R"
```

In contrast, `tsk("penguins")` is a multiclass problem as there are more than two species of penguins; it has the "multiclass" property:

```
tsk_penguins = tsk("penguins")
tsk_penguins$properties
```

```
[1] "multiclass"
```

```
tsk_penguins$class_names
```

```
[1] "Adelie"    "Chinstrap" "Gentoo"
```

A further difference between these tasks is that binary classification tasks have an extra field called `$positive`, which defines the "positive" class. In binary classi-   `$positive` fication, as there are only two possible class types, by convention one of these is known as the "positive" class, and the other as the "negative" class. It is arbitrary which is which, though often the more 'important' (and often smaller) class is set as the positive class. You can set the positive class during or after construction. If no positive class is specified then `mlr3` assumes the first level in the `target` column is the positive class, which can lead to misleading results.

```
# Load the "Sonar" dataset from the "mlbench" package as an
# example
data(Sonar, package = "mlbench")
# specifying the positive class:
tsk_classif = as_task_classif(Sonar, target = "Class",
        positive = "R")
tsk_classif$positive
```

```
[1] "R"
```

```
# changing after construction
tsk_classif$positive = "M"
tsk_classif$positive
```

```
[1] "M"
```

While the choice of positive and negative class is arbitrary, they are essential to ensuring results from models and performance measures are interpreted as expected – this is best demonstrated when we discuss thresholding (Section 2.5.4) and ROC metrics (Section 3.4).

Finally, plotting is possible with `autoplot.TaskClassif`, below we plot a comparison between the target column and features.

```
library(ggplot2)
autoplot(tsk("penguins"), type = "duo") +
  theme(strip.text.y = element_text(angle = -45, size = 8))
```

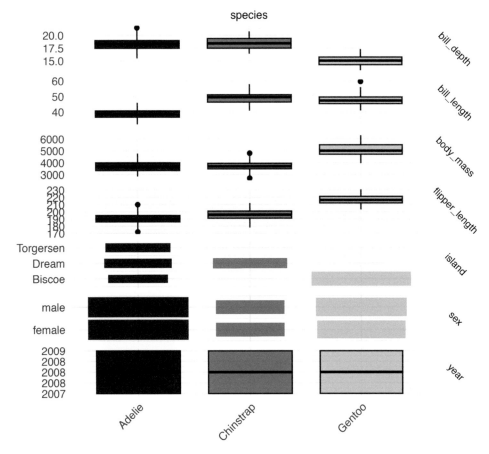

Figure 2.4: Overview of part of the penguins dataset.

2.5.3 LearnerClassif and MeasureClassif

Learner-
Classif

Classification learners, which inherit from `LearnerClassif`, have nearly the same interface as regression learners. However, a key difference is that the possible predictions in classification are either `"response"` – predicting an observation's class (a penguin's species in our example, this is sometimes called "hard labeling") – or `"prob"` – predicting a vector of probabilities, also called "posterior probabilities", of an observation belonging to each class. In classification, the latter can be more useful as it provides information about the confidence of the predictions:

```
lrn_rpart = lrn("classif.rpart", predict_type = "prob")
lrn_rpart$train(tsk_penguins, splits$train)
prediction = lrn_rpart$predict(tsk_penguins, splits$test)
prediction
```

```
<PredictionClassif> for 113 observations:
    row_ids       truth   response  prob.Adelie  prob.Chinstrap  prob.Gentoo
          2      Adelie     Adelie      0.97030          0.0297      0.00000
          4      Adelie     Adelie      0.97030          0.0297      0.00000
          7      Adelie     Adelie      0.97030          0.0297      0.00000
---
        338   Chinstrap  Chinstrap      0.04651          0.9302      0.02326
        341   Chinstrap     Adelie      0.97030          0.0297      0.00000
        344   Chinstrap  Chinstrap      0.04651          0.9302      0.02326
```

Notice how the predictions include the predicted probabilities for all three classes, as well as the response, which (by default) is the class with the highest predicted probability.

Also, the interface for classification measures, which are of class MeasureClassif, is identical to regression measures. The key difference in usage is that you will need to ensure your selected measure evaluates the prediction type of interest. To evaluate "response" predictions, you will need measures with predict_type = "response", or to evaluate probability predictions you will need predict_type = "prob". The easiest way to find these measures is by filtering the mlr_measures dictionary:

MeasureClas

```
as.data.table(msr())[
    task_type == "classif" & predict_type == "prob" &
    !sapply(task_properties, function(x) "twoclass" %in% x)]
```

```
               key                                    label
1:    classif.logloss                               Log Loss
2: classif.mauc_au1p     Weighted average 1 vs. 1 multiclass AUC
3: classif.mauc_au1u              Average 1 vs. 1 multiclass AUC
4: classif.mauc_aunp Weighted average 1 vs. rest multiclass AUC
5: classif.mauc_aunu           Average 1 vs. rest multiclass AUC
6:     classif.mbrier                 Multiclass Brier Score
4 variables not shown: [task_type, packages, predict_type,
   task_properties]
```

We also filtered to remove any measures that have the "twoclass" property as this would conflict with our "multiclass" task. We need to use sapply for this, the task_properties column is a list column. We can evaluate the quality of our probability predictions and response predictions simultaneously by providing multiple measures:

```
measures = msrs(c("classif.mbrier", "classif.logloss",
        "classif.acc"))
prediction$score(measures)
```

```
 classif.mbrier classif.logloss     classif.acc
         0.1017          0.2291          0.9469
```

The accuracy measure evaluates the "response" predictions whereas the Brier score ("classif.mbrier", squared difference between predicted probabilities and the truth) and logloss ("classif.logloss", negative logarithm of the predicted probability for the true class) are evaluating the probability predictions.

If no measure is passed to `$score()`, the default is the classification error (`msr("classif.ce")`), which is the number of misclassifications divided by the number of predictions, i.e., 1− `msr("classif.acc")`.

2.5.4 `PredictionClassif`, Confusion Matrix, and Thresholding

Prediction-
Classif

`PredictionClassif` objects have two important differences from their regression analog. Firstly, the added field `$confusion`, and secondly the added method `$set_threshold()`.

Confusion matrix

Confusion
Matrix

$confusion

A confusion matrix is a popular way to show the quality of classification (response) predictions in a more detailed fashion by seeing if a model is good at (mis)classifying observations in a particular class. For binary and multiclass classification, the confusion matrix is stored in the `$confusion` field of the `PredictionClassif` object:

```
prediction$confusion
```

```
           truth
response   Adelie Chinstrap Gentoo
  Adelie      49         3      0
  Chinstrap    1        18      1
  Gentoo       0         1     40
```

The rows in a confusion matrix are the predicted class and the columns are the true class. All off-diagonal entries are incorrectly classified observations, and all diagonal entries are correctly classified. In this case, the classifier does fairly well classifying all penguins, but we could have found that it only classifies the Adelie species well but often conflates Chinstrap and Gentoo, for example.

You can visualize the predicted class labels with `autoplot.PredictionClassif()`.

```
autoplot(prediction)
```

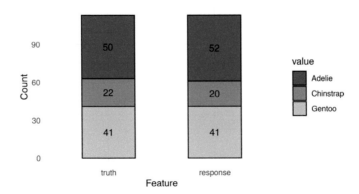

Figure 2.5: Counts of each class label in the ground truth data (left) and predictions (right).

In the binary classification case, the top left entry corresponds to true positives, the top right to false positives, the bottom left to false negatives and the bottom right to true negatives. Taking `tsk_sonar` as an example with M as the positive class:

```
splits = partition(tsk_sonar)
lrn_rpart$
  train(tsk_sonar, splits$train)$
  predict(tsk_sonar, splits$test)$
  confusion
```

```
        truth
response  M   R
       M 27  10
       R 10  22
```

We will return to the concept of binary (mis)classification in greater detail in Section 3.4.

Thresholding

The final big difference compared to regression we will discuss is thresholding. We saw previously that the default `response` prediction type is the class with the highest predicted probability. For k classes with predicted probabilities $p_1, \ldots, p_k$, this is the same as saying `response` = $\operatorname{argmax}\{p_1, \ldots, p_k\}$. If the maximum probability is not unique, i.e., multiple classes are predicted to have the highest probability, then the response is chosen randomly from these. In binary classification, this means that the positive class will be selected if the predicted class is greater than 50%, and the negative class otherwise.

This 50% value is known as the threshold and it can be useful to change this threshold if there is class imbalance (when one class is over- or under-represented in a dataset), or if there are different costs associated with classes, or simply if there is a preference to "over"-predict one class. As an example, let us take `tsk("german_credit")` in which 700 customers have good credit and 300 have bad. Now we could easily build a model with around "70%" accuracy simply by always predicting a customer will have good credit:

```
task_credit = tsk("german_credit")
lrn_featureless = lrn("classif.featureless",
        predict_type = "prob")
split = partition(task_credit)
lrn_featureless$train(task_credit, split$train)
prediction = lrn_featureless$predict(task_credit, split$test)
prediction$score(msr("classif.acc"))
```

```
classif.acc
        0.7
```

```
autoplot(prediction)
```

Threshold-
ing

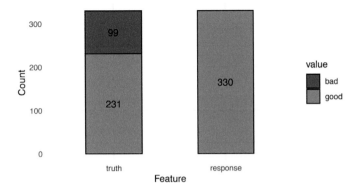

Figure 2.6: Class labels ground truth (left) and predictions (right). The learner completely ignores the "bad" class.

While this model may appear to have good performance on the surface, in fact, it just ignores all "bad" customers – this can create big problems in this finance example, as well as in healthcare tasks and other settings where false positives cost more than false negatives (see Section 13.1 for cost-sensitive classification).

Thresholding allows classes to be selected with a different probability threshold, so instead of predicting that a customer has bad credit if $P(good) < 50\%$, we might predict bad credit if $P(good) < 70\%$ – notice how we write this in terms of the positive class, which in this task is "good". Let us see this in practice:

```
prediction$set_threshold(0.7)
prediction$score(msr("classif.acc"))
```

```
classif.acc
     0.5394
```

```
lrn_rpart = lrn("classif.rpart", predict_type = "prob")
lrn_rpart$train(task_credit, split$train)
prediction = lrn_rpart$predict(task_credit, split$test)
prediction$score(msr("classif.acc"))
```

```
classif.acc
     0.6939
```

```
prediction$confusion
```

```
        truth
response good bad
    good  194  64
    bad    37  35
```

```
prediction$set_threshold(0.7)
prediction$score(msr("classif.acc"))
```

```
classif.acc
     0.6879
```

```
prediction$confusion
```

```
         truth
response good bad
    good  181  53
    bad    50  46
```

While our model performs "worse" overall, i.e., with lower accuracy, it is still a "better" model as it more accurately captures the relationship between classes.

In the binary classification setting, `$set_threshold()` only requires one numeric argument, which corresponds with the threshold for the positive class – hence it is essential to ensure the positive class is correctly set in your task.

In multiclass classification, thresholding works by first assigning a threshold to each of the n classes, dividing the predicted probabilities for each class by these thresholds to return n ratios, and then the class with the highest ratio is selected. For example, say we are predicting if a new observation will be of class A, B, C, or D and we have predicted $P(A = 0.2), P(B = 0.4), P(C = 0.1), P(D = 0.3)$. We will assume that the threshold for all classes is identical and 1:

```
probs = c(0.2, 0.4, 0.1, 0.3)
thresholds = c(A = 1, B = 1, C = 1, D = 1)
probs/thresholds
```

```
  A   B   C   D
0.2 0.4 0.1 0.3
```

We would therefore predict our observation is of class B as this is the highest ratio. However, we could change our thresholds so that D has the lowest threshold and is most likely to be predicted, A has the highest threshold, and B and C have equal thresholds:

```
thresholds = c(A = 0.5, B = 0.25, C = 0.25, D = 0.1)
probs/thresholds
```

```
  A   B   C   D
0.4 1.6 0.4 3.0
```

Now our observation will be predicted to be in class D.

In mlr3, this is achieved by passing a named list to `$set_threshold()`. This is demonstrated below with `tsk("zoo")`. Before changing the thresholds, some classes are never predicted and some are predicted more often than they occur.

```
library(ggplot2)
library(patchwork)

tsk_zoo = tsk("zoo")
```

```
splits = partition(tsk_zoo)
lrn_rpart = lrn("classif.rpart", predict_type = "prob")
lrn_rpart$train(tsk_zoo, splits$train)
prediction = lrn_rpart$predict(tsk_zoo, splits$test)
before = autoplot(prediction) + ggtitle("Default thresholds")
new_thresh = proportions(table(tsk_zoo$truth(splits$train)))
new_thresh
```

```
      mammal          bird        reptile          fish     amphibian
     0.40299       0.19403        0.04478       0.13433       0.04478
      insect mollusc.et.al
     0.07463       0.10448
```

```
prediction$set_threshold(new_thresh)
after = autoplot(prediction) +
        ggtitle("Inverse weighting thresholds")
before + after + plot_layout(guides = "collect")
```

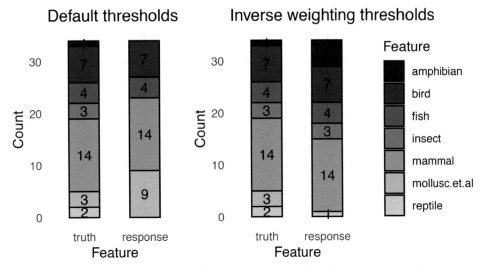

Figure 2.7: Comparing predicted and ground truth values for the zoo dataset.

Again we see that the model better represents all classes after thresholding. In this example, we set the new thresholds to be the proportions of each class in the training set. This is known as inverse weighting, as we divide the predicted probability by these class proportions before we select the label with the highest ratio.

In Section 13.1, we will look at cost-sensitive classification where each cell in the confusion matrix has a different associated cost.

2.6   Task Column Roles

> **i This section covers advanced ML or technical details.**

Now that we have covered regression and classification, we will briefly return to tasks and in particular to column roles, which are used to customize tasks further. Column roles are used by `Task` objects to define important metadata that can be used by learners and other objects to interact with the task. True to their name, they assign particular roles to columns in the data, we have already seen some of these in action with targets and features. There are seven column roles:

1. `"feature"`: Features used for prediction.
2. `"target"`: Target variable to predict.
3. `"name"`: Row names/observation labels, e.g., for `mtcars` this is the `"model"` column.
4. `"order"`: Variable(s) used to order data returned by `$data()`; must be sortable with `order()`.
5. `"group"`: Variable used to keep observations together during resampling.
6. `"stratum"`: Variable(s) to stratify during resampling.
7. `"weight"`: Observation weights. Only one numeric column may have this role.

We have already seen how features and targets work in Section 2.1, which are the only column roles that each task must have. In Section 3.2.5, we will have a look at the `stratum` and `group` column roles. So, for now, we will only look at `order`, and `weight`. We will not go into detail about `name`, which is primarily used in plotting and will almost always be the `rownames()` of the underlying data.

Column roles are updated using `$set_col_roles()`. When we set the `"order"` column role, the data is ordered according to that column(s). In the following example, we set the `"order"` column role and then order data by this column by including `ordered = TRUE`:

```
df = data.frame(mtcars[1:2, ], idx = 2:1)
tsk_mtcars_order = as_task_regr(df, target = "mpg")
# original order
tsk_mtcars_order$data(ordered = TRUE)
```

```
   mpg am carb cyl disp drat gear  hp idx  qsec vs    wt
1:  21  1    4   6  160  3.9    4 110   2 16.46  0 2.620
2:  21  1    4   6  160  3.9    4 110   1 17.02  0 2.875
```

```
# order by "idx" column
tsk_mtcars_order$set_col_roles("idx", roles = "order")
tsk_mtcars_order$data(ordered = TRUE)
```

```
   mpg am carb cyl disp drat gear  hp  qsec vs    wt
1:  21  1    4   6  160  3.9    4 110 17.02  0 2.875
2:  21  1    4   6  160  3.9    4 110 16.46  0 2.620
```

In this example, we can see that by setting "idx" to have the "order" column role, it is no longer used as a feature when we run $data() but instead is used to order the observations according to its value. This metadata is not passed to a learner.

The weights column role is used to weight data points differently. One example of why we would do this is in classification tasks with severe class imbalance, where weighting the minority class more heavily may improve the model's predictive performance for that class. For example in the breast_cancer dataset, there are more instances of benign tumors than malignant tumors, so if we want to better predict malignant tumors we could weight the data in favor of this class:

```
cancer_unweighted = tsk("breast_cancer")
summary(cancer_unweighted$data()$class)
```

```
malignant    benign
      239       444
```

```
# add column where weight is 2 if class "malignant", and 1 otherwise
df = cancer_unweighted$data()
df$weights = ifelse(df$class == "malignant", 2, 1)

# create new task and role
cancer_weighted = as_task_classif(df, target = "class")
cancer_weighted$set_col_roles("weights", roles = "weight")

# compare weighted and unweighted
# predictions
split = partition(cancer_unweighted)
lrn_rf = lrn("classif.ranger")
lrn_rf$train(cancer_unweighted, split$train)$
  predict(cancer_unweighted, split$test)$score()
```

```
classif.ce
    0.0177
```

```
lrn_rf$train(cancer_weighted, split$train)$
  predict(cancer_weighted, split$test)$score()
```

```
classif.ce
   0.00885
```

In this example, weighting improves the overall model performance (but see Chapter 3 for more thorough comparison methods). Not all models can handle weights in tasks so check a learner's properties to make sure this column role is being used as expected.

2.7   Supported Learning Algorithms

`mlr3` supports many learning algorithms (some with multiple implementations) as **Learners**. These are primarily provided by the `mlr3`, `mlr3learners` and `mlr3extralearners` packages. However, all packages that implement new tasks (Chapter 13) also include a handful of simple algorithms.

The list of learners included in `mlr3` is deliberately small to avoid large sets of dependencies:

- Featureless learners (`"regr.featureless"`/`"classif.featureless"`), which are baseline learners (Section 2.2.4).
- Debug learners (`"regr.debug"`/`"classif.debug"`), which are used to debug code (Section 10.2).
- Classification and regression trees (also known as CART: `"regr.rpart"`/`"classif.rpart"`).

The `mlr3learners` package contains a selection of algorithms (and select implementations) chosen by the mlr team that we recommend as a good starting point for most experiments:

- Linear (`"regr.lm"`) and logistic (`"classif.log_reg"`) regression.
- Penalized generalized linear models, where the penalization is either exposed as a hyperparameter (`"regr.glmnet"`/`"classif.glmnet"`) or where it is optimized automatically (`"regr.cv_glmnet"`/`"classif.cv_glmnet"`).
- Weighted k-Nearest Neighbors (`"regr.kknn"`/`"classif.kknn"`).
- Kriging / Gaussian process regression (`"regr.km"`).
- Linear (`"classif.lda"`) and quadratic (`"classif.qda"`) discriminant analysis.
- Naïve Bayes classification (`"classif.naive_bayes"`).
- Support-vector machines (`"regr.svm"`/`"classif.svm"`).
- Gradient boosting (`"regr.xgboost"`/`"classif.xgboost"`).
- Random forests for regression and classification (`"regr.ranger"`/ `"classif.ranger"`).

The majority of other supported learners are in `mlr3extralearners`. You can find an up-to-date list of learners at https://mlr-org.com/learners.html.

The dictionary `mlr_learners` contains learners that are supported in loaded packages:

```
learners_dt = as.data.table(mlr_learners)
learners_dt
```

```
                    key                         label  task_type
 1:   classif.AdaBoostM1            Adaptive Boosting     classif
 2:          classif.C50             Tree-based Model     classif
 3:          classif.IBk            Nearest Neighbour     classif
 4:          classif.J48             Tree-based Model     classif
 5:         classif.JRip  Propositional Rule Learner.     classif
---
```

```
134: surv.priority_lasso            Priority Lasso      surv
135:       surv.ranger              Random Forest       surv
136:        surv.rfsrc              Random Forest       surv
137:         surv.svm        Support Vector Machine     surv
138:     surv.xgboost           Gradient Boosting       surv
```
4 variables not shown: [feature_types, packages, properties, predict_types]

The resulting `data.table` contains a lot of metadata that is useful for identifying learners with particular properties. For example, we can list all learners that support classification problems:

```
learners_dt[task_type == "classif"]
```

```
                    key                      label task_type
 1: classif.AdaBoostM1          Adaptive Boosting    classif
 2:        classif.C50          Tree-based Model     classif
 3:        classif.IBk          Nearest Neighbour    classif
 4:        classif.J48          Tree-based Model     classif
 5:       classif.JRip Propositional Rule Learner.   classif
---
40:    classif.ranger                       <NA>    classif
41:     classif.rfsrc              Random Forest    classif
42:     classif.rpart          Classification Tree  classif
43:       classif.svm                       <NA>    classif
44:   classif.xgboost                       <NA>    classif
```
4 variables not shown: [feature_types, packages, properties, predict_types]

We can filter by multiple conditions, for example, to list all regression learners that can predict standard errors:

```
learners_dt[task_type == "regr" &
   sapply(predict_types, function(x) "se" %in% x)]
```

```
                key                                      label task_type
1:        regr.debug          Debug Learner for Regression          regr
2:        regr.earth Multivariate Adaptive Regression Splines     regr
3: regr.featureless          Featureless Regression Learner        regr
4:          regr.gam     Generalized Additive Regression Model     regr
5:          regr.glm          Generalized Linear Regression        regr
6:           regr.km                                     <NA>     regr
7:           regr.lm                                     <NA>     regr
8:          regr.mob       Model-based Recursive Partitioning      regr
9:       regr.ranger                                     <NA>     regr
```
4 variables not shown: [feature_types, packages, properties, predict_types]

2.8   Conclusion

In this chapter, we covered the building blocks of mlr3. We first introduced basic ML methodology and then showed how this is implemented in mlr3. We began by looking at the Task class, which is used to define machine learning tasks or problems to solve. We then looked at the Learner class, which encapsulates machine learning algorithms, hyperparameters, and other meta-information. Finally, we considered how to evaluate machine learning models with objects from the Measure class. After looking at regression implementations, we extended all the above to the classification setting, before finally looking at some extra details about tasks and the learning algorithms that are implemented across mlr3. The rest of this book will build on the basic elements seen in this chapter, starting with more advanced model comparison methods in Chapter 3 before moving on to improve model performance with automated hyperparameter tuning in Chapter 4.

Table 2.2: Important classes and functions covered in this chapter with underlying class (if applicable), class constructor or function, and important class fields and methods (if applicable).

Class	Constructor/Function	Fields/Methods
Task	tsk()/tsks()/as_task_X	$filter(); $select(); $data()
Learner	lrn()/lrns()	$train(); $predict();
		$predict_newdata(); $model()
Prediction	some_learner$predict()	$score(); $set_threshold();
		$confusion
Measure	msr()/msrs()	-

2.9   Exercises

1. Train a classification model with the `classif.rpart` learner on the "Pima Indians Diabetes" dataset. Do this without using `tsk("pima")`, and instead by constructing a task from the dataset in the `mlbench`-package: `data(PimaIndiansDiabetes2, package = "mlbench")`. Make sure to define the `pos` outcome as the positive class. Train the model on a random 80% subset of the given data and evaluate its performance with the classification error measure on the remaining data. (Note that the data set has NAs in its features. You can either rely on `rpart`'s capability to handle them internally ("surrogate splits") or remove them from the initial `data.frame` by using `na.omit`.

2. Calculate the true positive, false positive, true negative, and false negative rates of the predictions made by the model in Exercise 1. Try to solve this in two ways: (a) Using `mlr3measures`-predefined measure objects, and (b) without using `mlr3` tools by directly working on the ground truth and prediction vectors. Compare the results.

3. Change the threshold of the model from Exercise 1 such that the false negative rate is lower. What is one reason you might do this in practice?

3

Evaluation and Benchmarking

Giuseppe Casalicchio
Ludwig-Maximilians-Universität München, and Munich Center for Machine Learning (MCML), and Essential Data Science Training GmbH

Lukas Burk
Ludwig-Maximilians-Universität München, and Leibniz Institute for Prevention Research and Epidemiology – BIPS, and Munich Center for Machine Learning (MCML)

Generaliza-
tion
Performance

A supervised machine learning model can only be deployed in practice if it has a good generalization performance, which means it generalizes well to new, unseen data. Accurate estimation of the generalization performance is crucial for many aspects of machine learning application and research – whether we want to fairly compare a novel algorithm with established ones or to find the best algorithm for a particular task. The concept of performance estimation provides information on how well a model will generalize to new data and plays an important role in the context of model comparison (Section 3.3), model selection, and hyperparameter tuning (Chapter 4).

Assessing the generalization performance of a model begins with selecting a performance measure that is appropriate for our given task and evaluation goal. As we have seen in Section 2.3, performance measures typically compute a numeric score indicating how well the model predictions match the ground truth (though some technical measures were seen in Section 2.3.3). Once we have decided on a performance measure, the next step is to adopt a strategy that defines how to use the available data to estimate the generalization performance. Using the same data to train and test a model is a bad strategy as it would lead to an overly optimistic performance estimate. For example, a model that is overfitted (fit too closely to the data) could make perfect predictions on training data simply by memorizing it and then only make random guesses for new data. In Section 2.2.1.1, we introduced `partition()`, which splits a dataset into training data – data for training the model – and test data – data for testing the model and estimating the generalization performance, this is known as the holdout strategy (Section 3.1) and is where we will begin this chapter. We will then consider more advanced strategies for assessing the generalization performance (Section 3.2), look at robust methods for comparing models (Section 3.3), and finally will discuss specialized performance measures for binary classification (Section 3.4). For an in-depth overview of measures and performance estimation, we recommend Japkowicz and Shah (2011).

DOI: 10.1201/9781003402848-3

> ⚠ Resampling Does Not Avoid Model Overfitting
>
> A common **misunderstanding** is that holdout and other more advanced
> resampling strategies can prevent model overfitting. In fact, these methods just
> make overfitting visible as we can separately evaluate train/test performance.
> Resampling strategies also allow us to make (nearly) unbiased estimations of
> the generalization error.

3.1 Holdout and Scoring

An important goal of ML is to learn a model that can then be used to make
predictions about new data. For this model to be as accurate as possible, we would
ideally train it on as much data as is available. However, data is limited and as
we have discussed we cannot train and test a model on the same data. In practice,
one would usually create an intermediate model, which is trained on a subset of   Intermediate
the available data and then tested on the remainder of the data. The performance   Model
of this intermediate model, obtained by comparing the model predictions to the
ground truth is an estimate of the generalization performance of the final model,
which is the model fitted on all data.

The holdout strategy is a simple method to create this split between training and   Holdout
testing datasets, whereby the original data is split into two datasets using a defined
ratio. Ideally, the training dataset should be as large as possible so the intermediate
model represents the final model as well as possible. If the training data is too small,
the intermediate model is unlikely to perform as well as the final model, resulting in
a pessimistically biased performance estimate. On the other hand, if the training
data is too large, then we will not have a reliable estimate of the generalization
performance due to the high variance resulting from small test data. As a rule of
thumb, it is common to use 2/3 of the data for training and 1/3 for testing as
this provides a reasonable trade-off between bias and variance of the generalization
performance estimate (Kohavi 1995; Dobbin and Simon 2011).

In Chapter 2, we used `partition()` to apply the holdout method to a `Task` object.
To recap, let us split `tsk("penguins")` with a 2/3 holdout (default split):

```
tsk_penguins = tsk("penguins")
splits = partition(tsk_penguins)
lrn_rpart = lrn("classif.rpart")
lrn_rpart$train(tsk_penguins, splits$train)
prediction = lrn_rpart$predict(tsk_penguins, splits$test)
```

We can now estimate the generalization performance of a final model by evaluating
the quality of the predictions from our intermediate model. As we have seen in
Section 2.3, this is simply a case of choosing one or more measures and passing
them to the `$score()` function. So to estimate the accuracy of our final model we
would pass the accuracy measure to our intermediate model:

```
prediction$score(msr("classif.acc"))
```

```
classif.acc
    0.9558
```

> 💡 Permuting Observations for Performance Estimation
>
> When splitting data it is essential to permute observations before, to remove any information that is encoded in data ordering. The order of data is often informative in real-world datasets, for example, hospital data will likely be ordered by time of patient admission. In `tsk("penguins")`, the data is ordered such that the first 152 rows all have the label "Adelie", the next 68 have the label "Chinstrap", and the final 124 have the label "Gentoo"; so if we did not permute the data we could end up with a model that is only trained on one or two species.
>
> `partition()` and all resampling strategies discussed below automatically randomly split the data to prevent any biases (so do not forget to set a seed for reproducibility). Data *within* each set may still be ordered because of implementation details, but this is not a problem as long as the data is shuffled between sets.

Many performance measures are based on "decomposable" losses, which means they compute the differences between the predicted values and ground truth values first on an observation level and then aggregate the individual loss values over the test set into a single numeric score. For example, the classification accuracy compares whether the predicted values from the `response` column have the same value as the ground truth values from the `truth` column of the `Prediction` object. Hence, for each observation, the decomposable loss takes either value 1 (if `response` and `truth` have the same value) or 0 otherwise. The `$score()` method summarizes these individual loss values into a an average value – the percentage where our prediction was correct. Other performance measures that are not decomposable instead act on a set of observations, we will return to this in detail when we look at the AUC measure in Section 3.4. Figure 3.1 illustrates the input-output behavior of the `$score()` method, we will return to this when we turn to more complex evaluation strategies.

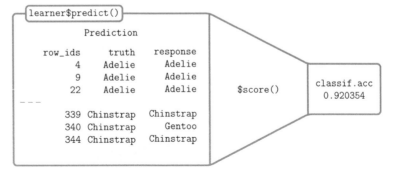

Figure 3.1: Illustration of the `$score()` method which aggregates predictions of multiple observations contained in a prediction object into a single numeric score.

3.2 Resampling

Resampling strategies repeatedly split all available data into multiple training and test sets, with one repetition corresponding to what is called a "resampling iteration" in mlr3. An intermediate model is then trained on each training set and the test set is used to measure the performance in each resampling iteration. The generalization performance is finally estimated by aggregating the performance scores over multiple resampling iterations (Figure 3.2). By repeating the data splitting process, data points are repeatedly used for both training and testing, allowing more efficient use of all available data for performance estimation. Furthermore, a high number of resampling iterations can reduce the variance in our scores and thus result in a more reliable performance estimate. This means that the performance estimate is less likely to be affected by an "unlucky" split (e.g., a split that does not reflect the original data distribution).

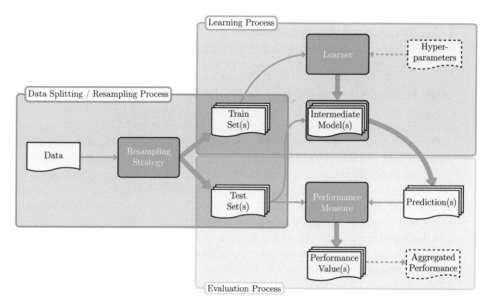

Figure 3.2: A general abstraction of the performance estimation process. The available data is (repeatedly) split into training data and test data (data splitting/resampling process). The learner is trained on each training dataset and produces intermediate models (learning process). Each intermediate model makes predictions based on the features in the test data. The performance measure compares these predictions with the ground truth from the test data and computes a performance value for each test dataset. All performance values are aggregated into a scalar value to estimate the generalization performance (evaluation process).

A variety of resampling strategies exist, each with its advantages and disadvantages, which depend on the number of available samples, the task complexity, and the type of model.

A very common strategy is k-fold cross-validation (CV), which randomly partitions the data into k non-overlapping subsets, called folds (Figure 3.3). The k models are always trained on $k - 1$ of the folds, with the remaining fold being used as

Cross-validation

test data; this process is repeated until each fold has acted exactly once as test set. Finally, the k performance estimates from each fold are aggregated, usually by averaging. CV guarantees that each observation will be used exactly once in a test set, making efficient use of the available data for performance estimation. Common values for k are 5 and 10, meaning each training set will consist of 4/5 or 9/10 of the original data, respectively. Several variations of CV exist, including repeated k-fold cross-validation where the k-fold process is repeated multiple times, and leave-one-out cross-validation (LOO-CV) where the number of folds is equal to the number of observations, leading to the test set in each fold consisting of only one observation.

Subsampling   Subsampling and bootstrapping are two related resampling strategies. Subsampling randomly selects a given ratio (4/5 and 9/10 are common) of the data for the training dataset where each observation in the dataset is drawn *without replacement* from the original dataset. The model is trained on this data and then tested on the remaining data, and this process is repeated k times. This differs from k-fold CV as the subsets of test data may be overlapping. Bootstrapping follows the same process as subsampling but data is drawn *with replacement* from the original dataset. Usually the number of bootstrap samples equals the size of the original dataset. This means an observation could be selected multiple times (and thus duplicated) in the training data (but never more than once per test dataset). On average, $1 - e^{-1} \approx 63.2\%$ of the data points will be contained in the training set during bootstrapping, referred to as "in-bag" samples (the other 36.8% are known as "out-of-bag" samples).

Note that terminology regarding resampling strategies is not consistent across the literature, for example, subsampling is sometimes referred to as "repeated holdout" or "Monte Carlo cross-validation".

The choice of the resampling strategy usually depends on the specific task at hand and the goals of the performance assessment, but some rules of thumb are available. If the available data is fairly small ($N \leq 500$), repeated cross-validation with a large number of repetitions can be used to keep the variance of the performance estimates low (10 folds and 10 repetitions is a good place to start). Traditionally, LOO-CV has also been recommended for these small sample size regimes, but this estimation scheme is quite expensive (except in special cases where computational shortcuts exist) and (counterintuitively) suffers from quite high variance. Furthermore, LOO-CV is also problematic in imbalanced binary classification tasks as concepts such as stratification (Section 3.2.5) cannot be applied. For the $500 \leq N \leq 50000$ range, 5- to 10-fold CV is generally recommended. In general, the larger the dataset, the fewer splits are required, yet sample-size issues can still occur, e.g., due to imbalanced data. For settings where one is more interested in proper inference (such as through statistical performance tests or confidence intervals) than bare point estimators of performance, bootstrapping and subsampling are often considered, usually with a higher number of iterations. Bootstrapping has become less common, as having repeated observations in training data can lead to problems in some machine learning setups, especially when combined with model selection methods and nested resampling (as duplicated observations can then end up simultaneously in training and test sets in nested schemes). Also note that in all of these common and simple schemes, resampling performance estimates are not independent, as models are fitted on overlapping training data, making proper inference less than trivial, but a proper treatment of these issues is out of scope for us here. For further details and critical discussion we refer to the literature, e.g., Molinaro, Simon, and Pfeiffer (2005), J.-H. Kim (2009), and Bischl et al. (2012).

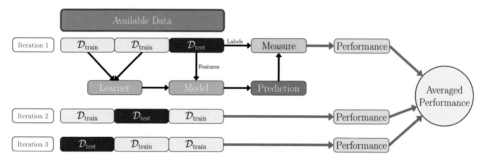

Figure 3.3: Illustration of a three-fold cross-validation.

In the rest of this section, we will go through querying and constructing resampling strategies in `mlr3`, instantiating train-test splits, and then performing resampling on learners.

3.2.1 Constructing a Resampling Strategy

All implemented resampling strategies are stored in the `mlr_resamplings` dictionary.

```
as.data.table(mlr_resamplings)
```

```
            key                              label           params iters
1:    bootstrap                          Bootstrap ratio,repeats    30
2:       custom                      Custom Splits                    NA
3:    custom_cv Custom Split Cross-Validation                       NA
4:           cv                   Cross-Validation            folds    10
5:      holdout                            Holdout            ratio     1
6:     insample                 Insample Resampling                   1
7:          loo                      Leave-One-Out                    NA
8:  repeated_cv      Repeated Cross-Validation folds,repeats   100
9:  subsampling                        Subsampling ratio,repeats    30
```

The `params` column shows the parameters of each resampling strategy (e.g., the train-test splitting `ratio` or the number of `repeats`) and `iters` displays the number of performed resampling iterations by default.

Resampling objects can be constructed by passing the strategy "key" to the sugar function `rsmp()`. For example, to construct the holdout strategy with a 4/5 split (2/3 by default):

`Resampling`
`rsmp()`

```
rsmp("holdout", ratio = 0.8)
```

```
<ResamplingHoldout>: Holdout
* Iterations: 1
* Instantiated: FALSE
* Parameters: ratio=0.8
```

Parameters for objects inheriting from `Resampling` work in the same way as measures and learners and can be set, retrieved, and updated accordingly:

```
# three-fold CV
cv3 = rsmp("cv", folds = 3)
# Subsampling with 3 repeats and 9/10 ratio
ss390 = rsmp("subsampling", repeats = 3, ratio = 0.9)
# 2-repeats 5-fold CV
rcv25 = rsmp("repeated_cv", repeats = 2, folds = 5)
```

When a "Resampling" object is constructed, it is simply a definition for how the data splitting process will be performed on the task when running the resampling strategy. However, it is possible to manually instantiate a resampling strategy, i.e., generate all train-test splits, by calling the $instantiate() method on a given task. So carrying on our tsk("penguins") example we can instantiate the three-fold CV object and then view the row indices of the data selected for training and testing each fold using $train_set() and $test_set() respectively:

$instantiate()

```
cv3$instantiate(tsk_penguins)
# first 5 observations in first training set
cv3$train_set(1)[1:5]
```

```
[1]  1  9 21 22 23
```

```
# first 5 observations in third test set
cv3$test_set(3)[1:5]
```

```
[1]  2  3  5 10 12
```

When the aim is to fairly compare multiple learners, best practice dictates that all learners being compared use the same training data to build a model and that they use the same test data to evaluate the model performance. Resampling strategies are instantiated automatically for you when using the resample() method, which we will discuss next. Therefore, manually instantiating resampling strategies is rarely required but might be useful for debugging or digging deeper into a model's performance.

3.2.2 Resampling Experiments

resample()

The resample() function takes a given Task, Learner, and Resampling object to run the given resampling strategy. resample() repeatedly fits a model on training sets, makes predictions on the corresponding test sets and stores them in a ResampleResult object, which contains all the information needed to estimate the generalization performance.

ResampleResult

```
rr = resample(tsk_penguins, lrn_rpart, cv3)
rr
```

```
<ResampleResult> with 3 resampling iterations
 task_id    learner_id resampling_id iteration warnings errors
 penguins classif.rpart           cv         1        0      0
 penguins classif.rpart           cv         2        0      0
 penguins classif.rpart           cv         3        0      0
```

Each row of the output corresponds to one of the three iterations/folds. As with
`Prediction` objects, we can calculate the score *for each iteration* with `$score()`:

```
acc = rr$score(msr("classif.ce"))
acc[, .(iteration, classif.ce)]
```

```
   iteration classif.ce
1:         1    0.06087
2:         2    0.04348
3:         3    0.06140
```

> 💡 Evaluating Train Sets
>
> By default, `$score()` evaluates the performance in the *test* sets in each
> iteration, however, you could evaluate the *train* set performance with
> `$score(predict_sets = "train")`.

While `$score()` returns the performance in each evaluation, `$aggregate()`, returns   `$aggregate()`
the aggregated score across all resampling iterations.

```
rr$aggregate(msr("classif.ce"))
```

```
classif.ce
   0.05525
```

By default, the majority of measures will aggregate scores using a macro average,
which first calculates the measure in each resampling iteration separately, and then
averages these scores across all iterations. However, it is also possible to aggregate
scores using a micro average, which pools predictions across resampling iterations
into one `Prediction` object and then computes the measure on this directly:

```
rr$aggregate(msr("classif.ce", average = "micro"))
```

```
classif.ce
   0.05523
```

We can see a *small* difference between the two methods. Classification error is a
decomposable loss (Section 3.1), in fact, if the test sets all had the same size then
the micro and macro methods would be identical (see box below). For errors like
AUC, which are defined across the set of observations, the difference between micro-
and macro-averaging will be larger. The default type of aggregation method can be
found by querying the `$average` field of a `Measure` object.

> 💡 Macro- and Micro-Averaging
>
> As a simple example to explain macro- and micro-averaging, consider the
> difference between taking the mean of a vector (micro) compared to the mean
> of two group-wise means (macro):

```
# macro
mean(mean(c(3, 5, 9)), mean(c(1, 5)))
```

```
[1] 5.667
```

```
# micro
mean(c(3, 5, 9, 1, 5))
```

```
[1] 4.6
```

In the example shown in the main text where we used `tsk("penguins")`, there is a difference in the classification error between micro and macro methods because the dataset has 344 rows, which is not divisible by three (the number of folds), hence the test sets are not of an equal size.

Note that the terms "macro-averaging" and "micro-averaging" are not used consistently in the literature, and sometimes refer to different concepts, e.g., the way in which the performance is aggregated across classes in a multi-class classification task.

The aggregated score returned by `$aggregate()` estimates the generalization performance of our selected learner on the given task using the resampling strategy defined in the `Resampling` object. While we are usually interested in this aggregated score, it can be useful to look at the individual performance values of each resampling iteration (as returned by the `$score()` method) as well, e.g., to see if any of the iterations lead to very different performance results. Figure 3.4 visualizes the relationship between `$score()` and `$aggregate()` for a small example based on the `"penguins"` task.

To visualize the resampling results, you can use the `autoplot.ResampleResult()` function to plot scores across folds as boxplots or histograms (Figure 3.5). Histograms can be useful to visually gauge the variance of the performance results across resampling iterations, whereas boxplots are often used when multiple learners are compared side-by-side (see Section 3.3).

```
rr = resample(tsk_penguins, lrn_rpart, rsmp("cv", folds = 10))
autoplot(rr, measure = msr("classif.acc"), type = "boxplot")
autoplot(rr, measure = msr("classif.acc"), type = "histogram")
```

3.2.3 ResampleResult Objects

As well as being useful for estimating the generalization performance, the `ResampleResult` object can also be used for model inspection. We can use the `$predictions()` method to obtain a list of `Prediction` objects corresponding to the predictions from each resampling iteration. This can be used to analyze the predictions of individual intermediate models from each resampling iteration. To understand the class better, we use it here to manually compute a macro-averaged performance estimate.

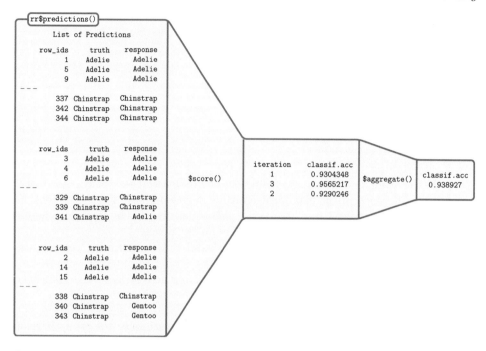

Figure 3.4: An example of the difference between `$score()` and `$aggregate()`: the former aggregates predictions to a single score within each resampling iteration, and the latter aggregates scores across all resampling iterations.

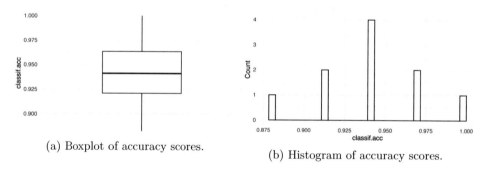

(a) Boxplot of accuracy scores.

(b) Histogram of accuracy scores.

Figure 3.5: Boxplot and Histogram of accuracy scores.

```
# list of prediction objects
rrp = rr$predictions()
# print first two
rrp[1:2]
```

```
[[1]]
<PredictionClassif> for 35 observations:
    row_ids      truth   response
         2      Adelie     Adelie
         4      Adelie     Adelie
        11      Adelie     Adelie
---
```

```
      333 Chinstrap Chinstrap
      334 Chinstrap Chinstrap
      337 Chinstrap Chinstrap

[[2]]
<PredictionClassif> for 35 observations:
    row_ids     truth response
          1    Adelie   Adelie
         21    Adelie   Adelie
         34    Adelie   Adelie
---
        309 Chinstrap   Adelie
        317 Chinstrap   Gentoo
        343 Chinstrap   Gentoo
```

```
# macro averaged performance
mean(sapply(rrp, function(.x) .x$score()))
```

```
[1] 0.05807
```

The `$prediction()` method can be used to extract a single `Prediction` object that combines the predictions of each intermediate model across all resampling iterations. The combined prediction object can, for example, be used to manually compute a micro-averaged performance estimate (see Section 3.2.2 for how you can micro-average more conveniently).

```
prediction = rr$prediction()
prediction
```

```
<PredictionClassif> for 344 observations:
    row_ids     truth  response
          2    Adelie    Adelie
          4    Adelie    Adelie
         11    Adelie    Adelie
---
        327 Chinstrap Chinstrap
        328 Chinstrap Chinstrap
        341 Chinstrap    Adelie
```

```
prediction$score()
```

```
classif.ce
   0.05814
```

By default, the intermediate models produced at each resampling iteration are discarded after the prediction step to reduce the memory consumption of the `ResampleResult` object (only the predictions are required to calculate most performance measures). However, it can sometimes be useful to inspect, compare, or extract information from these intermediate models. We can configure the `resample()` function to keep the fitted intermediate models by setting `store_models = TRUE`.

Each model trained in a specific resampling iteration can then be accessed via
`$learners[[i]]$model`, where i refers to the i-th resampling iteration:

```
rr = resample(tsk_penguins, lrn_rpart, cv3, store_models = TRUE)
# get the model from the first iteration
rr$learners[[1]]$model
```

n= 229

node), split, n, loss, yval, (yprob)
      * denotes terminal node

```
1) root 229 129 Adelie (0.436681 0.192140 0.371179)
  2) flipper_length< 207.5 141   42 Adelie (0.702128 0.290780 0.007092)
    4) bill_length< 44.65 100    3 Adelie (0.970000 0.030000 0.000000) *
    5) bill_length>=44.65 41     3 Chinstrap (0.048780 0.926829 0.024390) *
  3) flipper_length>=207.5 88    4 Gentoo (0.011364 0.034091 0.954545) *
```

In this example, we could then inspect the most important variables in each iteration
to help us learn more about the respective fitted models:

```
# print 2nd and 3rd iteration
lapply(rr$learners[2:3], function(x) x$model$variable.importance)
```

```
[[1]]
flipper_length    bill_length    bill_depth     body_mass
        87.23          81.27         66.26         59.46
        island
         51.22

[[2]]
   bill_length flipper_length    bill_depth     body_mass
         79.06          78.94         59.98         54.35
        island
         42.63
```

3.2.4   Custom Resampling

> **i**   **This section covers advanced ML or technical details.**

Sometimes it is necessary to perform resampling with custom splits, e.g., to reproduce
results reported in a study with pre-defined folds.

A custom holdout resampling strategy can be constructed using `rsmp("custom")`,
where the row IDs of the observations used for training and testing must be defined
manually when instantiated with a task. In the example below, we first construct a
custom holdout resampling strategy by manually assigning row IDs to the `$train`
and `$test` fields, then construct a resampling strategy with two iterations by passing
row IDs as list elements:

```
rsmp_custom = rsmp("custom")

# resampling strategy with two iterations
train_sets = c(1:5, 153:158, 277:280)
rsmp_custom$instantiate(tsk_penguins,
  train = list(train_sets, train_sets + 5),
  test = list(train_sets + 15, train_sets + 25)
)
resample(tsk_penguins, lrn_rpart, rsmp_custom)$prediction()
```

```
<PredictionClassif> for 30 observations:
   row_ids      truth response
        16     Adelie   Gentoo
        17     Adelie   Gentoo
        18     Adelie   Gentoo
---
       303 Chinstrap   Gentoo
       304 Chinstrap   Gentoo
       305 Chinstrap   Gentoo
```

A custom cross-validation strategy can be more efficiently constructed with `rsmp("custom_cv")`. In this case, we now have to specify either a custom `factor` variable or a `factor` column from the data to determine the folds. In the example below, we use a smaller version of `tsk("penguins")` and instantiate a custom two-fold CV strategy using a `factor` variable called `folds` where the first and third rows are used as the test set in Fold 1, and the second and fourth rows are used as the test set in Fold 2:

```
tsk_small = tsk("penguins")$filter(c(1, 100, 200, 300))
rsmp_customcv = rsmp("custom_cv")
folds = as.factor(c(1, 2, 1, 2))
rsmp_customcv$instantiate(tsk_small, f = folds)
resample(tsk_small, lrn_rpart, rsmp_customcv)$predictions()
```

```
[[1]]
<PredictionClassif> for 2 observations:
 row_ids  truth response
       1 Adelie   Adelie
     200 Gentoo   Adelie

[[2]]
<PredictionClassif> for 2 observations:
 row_ids      truth response
     100     Adelie   Adelie
     300 Chinstrap   Adelie
```

3.2.5   Stratification and Grouping

> ℹ **This section covers advanced ML or technical details.**

Using column roles (Section 2.6), it is possible to group or stratify observations according to a particular column in the data. We will look at each of these in turn.

Grouped Resampling

Keeping observations together when the data is split can be useful, and sometimes essential, during resampling – spatial analysis (Section 13.5) is a prominent example, as observations belong to natural groups (e.g., countries). When observations belong to groups, we need to ensure all observations of the same group belong to *either* the training set *or* the test set to prevent potential leakage of information between training and testing. For example, in a longitudinal study, measurements are taken from the same individual at multiple time points. If we do not group these, we might overestimate the model's generalization capability to unseen individuals, because observations of the same individuals might simultaneously be in the train and test set. In this context, the leave-one-out cross-validation strategy can be coarsened to the "leave-one-object-out" cross-validation strategy, where all observations associated with a certain group are left out (Figure 3.6).

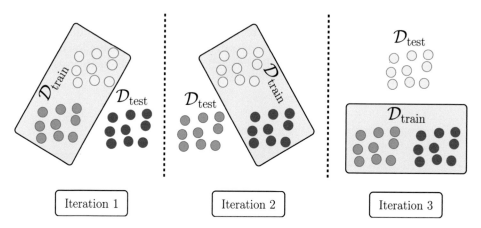

Figure 3.6: Illustration of the train-test splits of a leave-one-object-out cross-validation with 3 groups of observations (highlighted by different colors).

The `"group"` column role allows us to specify the column in the data that defines the group structure of the observations. In the following code, we construct a leave-one-out resampling strategy, assign the `"group"` role to the "year" column of `tsk("penguins")`, instantiate the resampling strategy, and finally show how the years are nicely separated in the first fold.

```
rsmp_loo = rsmp("loo")
tsk_grp = tsk("penguins")
tsk_grp$set_col_roles("year", "group")
rsmp_loo$instantiate(tsk_grp)
table(tsk_grp$data(rows = rsmp_loo$train_set(1), cols = "year"))
```

```
year
2008 2009
 114  120
```

```
  table(tsk_grp$data(rows = rsmp_loo$test_set(1), cols = "year"))
```

```
year
2007
 110
```

Other cross-validation techniques work in a similar way, where folds are determined at a group level (as opposed to an observation level).

Stratified Sampling

Stratified sampling ensures that one or more discrete features within the training and test sets will have a similar distribution as in the original task containing all observations. This is especially useful when a discrete feature is highly imbalanced and we want to make sure that the distribution of that feature is similar in each resampling iteration (Figure 3.7). We can also stratify the target feature to ensure that each intermediate model is fit on training data where the class distribution of the target is representative of the actual task, this is useful to ensure target classes are not strongly under-represented by random chance in individual resampling iterations, which would lead to degenerate estimations of the generalization performance.

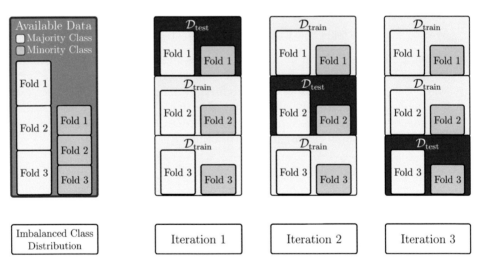

Figure 3.7: Illustration of a three-fold cross-validation with stratification for an imbalanced binary classification task with a majority class that is about twice as large as the minority class. In each resampling iteration, the class distribution from the available data is preserved (which is not necessarily the case for cross-validation without stratification).

Unlike grouping, it is possible to stratify by multiple discrete features using the `"stratum"` column role (Section 2.6). In this case, strata would be formed out of each combination of the stratified features, e.g., for two stratified features A and B with levels Aa, Ab; Ba; Bb respectively then the created stratum would have the levels AaBa, AaBb, AbBa, AbBb.

`tsk("penguins")` displays imbalance in the `species` column, as can be seen in the output below:

```
prop.table(table(tsk_penguins$data(cols = "species")))
```

```
species
   Adelie Chinstrap   Gentoo
   0.4419    0.1977   0.3605
```

Without specifying a `"stratum"` column role, the `species` column may have quite different class distributions across the CV folds, as can be seen in the example below.

```
rsmp_cv10 = rsmp("cv", folds = 10)
rsmp_cv10$instantiate(tsk_penguins)

fold1 = prop.table(table(tsk_penguins$data(rows =
    rsmp_cv10$test_set(1),cols = "species")))
fold2 = prop.table(table(tsk_penguins$data(rows =
    rsmp_cv10$test_set(2),cols = "species")))

rbind("Fold 1" = fold1, "Fold 2" = fold2)
```

```
        Adelie Chinstrap Gentoo
Fold 1 0.4286    0.1143 0.4571
Fold 2 0.4286    0.2286 0.3429
```

We can see across folds how Chinstrap is represented quite differently (0.11 vs. 0.23).

When the imbalance is severe, minority classes might not occur in the training sets entirely. Consequently, the intermediate models within these resampling iterations will never predict the missing class, resulting in a misleading performance estimate for any resampling strategy without stratification. The code below uses `species` as `"stratum"` column role to illustrate that the distribution of `species` in each test set will closely match the original distribution:

```
tsk_str = tsk("penguins")
# set species to have both the 'target' and 'stratum' column role
tsk_str$set_col_roles("species", c("target", "stratum"))
rsmp_cv10$instantiate(tsk_str)

fold1 = prop.table(table(tsk_str$data(rows =
    rsmp_cv10$test_set(1),cols = "species")))
fold2 = prop.table(table(tsk_str$data(rows =
    rsmp_cv10$test_set(2),cols = "species")))

rbind("Fold 1" = fold1, "Fold 2" = fold2)
```

```
        Adelie Chinstrap Gentoo
Fold 1 0.4444    0.1944 0.3611
Fold 2 0.4444    0.1944 0.3611
```

You can view the observations that fall into each stratum using the `$strata` field of a `Task` object, this can be particularly useful when we are interested in multiple strata:

```
tsk_str$set_col_roles("year", "stratum")
tsk_str$strata
```

```
        N                      row_id
1: 50                1,2,3,4,5,6,...
2: 50         51,52,53,54,55,56,...
3: 52 101,102,103,104,105,106,...
4: 34 153,154,155,156,157,158,...
5: 46 187,188,189,190,191,192,...
6: 44 233,234,235,236,237,238,...
7: 26 277,278,279,280,281,282,...
8: 18 303,304,305,306,307,308,...
9: 24 321,322,323,324,325,326,...
```

```
  # N above matches with numbers in table below
  table(tsk_penguins$data(cols = c("species", "year")))
```

```
            year
species     2007 2008 2009
   Adelie     50   50   52
   Chinstrap  26   18   24
   Gentoo     34   46   44
```

3.3 Benchmarking

Benchmarking in supervised machine learning refers to the comparison of different learners on one or more tasks. When comparing *multiple learners on a single task* or on a domain consisting of multiple similar tasks, the main aim is often to rank the learners according to a pre-defined performance measure and to identify the best-performing learner for the considered task or domain. When comparing *multiple learners on multiple tasks*, the main aim is often more of a scientific nature, e.g., to gain insights into how different learners perform in different data situations or whether there are certain data properties that heavily affect the performance of certain learners (or certain hyperparameters of learners). It is common (and good) practice for algorithm designers to analyze the generalization performance or runtime of a newly proposed learning algorithm in comparison to existing learners in a benchmark experiment. Since benchmarks usually consist of many evaluations that can run independently of each other, mlr3 offers the possibility of parallelizing them automatically, which we demonstrate in Section 10.1.2. In this section, we will focus on the basic setup of benchmark experiments that will be applicable in the majority of use cases. In Chapter 11, we will look at more complex, large-scale, benchmark experiments.

3.3.1 benchmark()

Benchmark experiments in `mlr3` are conducted with `benchmark()`, which simply runs `resample()` on each task and learner separately, then collects the results. The provided resampling strategy is automatically instantiated on each task to ensure that all learners are compared against the same training and test data.

To use the `benchmark()` function we first call `benchmark_grid()`, which constructs an exhaustive *design* to describe all combinations of the learners, tasks, and re-samplings to be used in a benchmark experiment, and instantiates the resampling strategies. For example, below we set up a design to see if a random forest, decision tree, or featureless baseline (Section 2.2.4), performs best across two classification tasks.

```
tasks = tsks(c("german_credit", "sonar"))
learners = lrns(c("classif.rpart", "classif.ranger",
  "classif.featureless"), predict_type = "prob")
rsmp_cv5 = rsmp("cv", folds = 5)

design = benchmark_grid(tasks, learners, rsmp_cv5)
head(design)
```

```
                task              learner resampling
1: german_credit        classif.rpart         cv
2: german_credit       classif.ranger         cv
3: german_credit classif.featureless         cv
4:         sonar        classif.rpart         cv
5:         sonar       classif.ranger         cv
6:         sonar classif.featureless         cv
```

The resulting design is essentially just a `data.table`, which can be modified if you want to remove particular combinations or could even be created from scratch without the `benchmark_grid()` function. Note that this `data.table` has list columns that contain R6 objects of tasks, learners, and resampling instances.

⚠ Reproducibility When Using `benchmark_grid()`

By default, `benchmark_grid()` instantiates the resamplings on the tasks, which means that concrete train-test splits are generated. Since this process is stochastic, it is necessary to set a seed **before** calling `benchmark_grid()` to ensure the reproducibility of the data splits.

The constructed benchmark design can then be passed to `benchmark()` to run the experiment and the result is a `BenchmarkResult` object:

```
bmr = benchmark(design)
bmr
```

```
<BenchmarkResult> of 30 rows with 6 resampling runs
 nr        task_id           learner_id resampling_id iters warnings
  1 german_credit        classif.rpart            cv     5        0
  2 german_credit       classif.ranger            cv     5        0
  3 german_credit classif.featureless            cv     5        0
  4         sonar         classif.rpart            cv     5        0
  5         sonar        classif.ranger            cv     5        0
  6         sonar classif.featureless            cv     5        0
1 variable not shown: [errors]
```

As benchmark() is just an extension of resample(), we can once again use $score(), or $aggregate() depending on your use-case, though note that in this case, $score() will return results over each fold of each learner/task/resampling combination.

```
    bmr$score()[c(1, 7, 13), .(iteration, task_id, learner_id,
           classif.ce)]

    iteration        task_id           learner_id classif.ce
1:          1 german_credit        classif.rpart      0.280
2:          2 german_credit       classif.ranger      0.235
3:          3 german_credit classif.featureless      0.275


    bmr$aggregate()[, .(task_id, learner_id, classif.ce)]

          task_id           learner_id classif.ce
1: german_credit        classif.rpart     0.2760
2: german_credit       classif.ranger     0.2490
3: german_credit classif.featureless     0.3000
4:         sonar         classif.rpart     0.2840
5:         sonar        classif.ranger     0.1535
6:         sonar classif.featureless     0.4661
```

This would conclude a basic benchmark experiment where you can draw tentative conclusions about model performance, in this case, we would possibly conclude that the random forest is the best of all three models on each task. We draw conclusions cautiously here as we have not run any statistical tests or included standard errors of measures, so we cannot definitively say if one model outperforms the other.

As the results of $score() and $aggregate() are returned in a data.table, you can post-process and analyze the results in any way you want. A common *mistake* is to average the learner performance across all tasks when the tasks vary significantly. This is a mistake as averaging the performance will miss out important insights into how learners compare on "easier" or more "difficult" predictive problems. A more robust alternative to compare the overall algorithm performance across multiple tasks is to compute the ranks of each learner on each task separately and then calculate the average ranks. This can provide a better comparison as task-specific "quirks" are taken into account by comparing learners within tasks before comparing them across tasks. However, using ranks will lose information about the numerical differences between the calculated performance scores. Analysis of benchmark experiments, including statistical tests, is covered in more detail in Section 11.3.

3.3.2   BenchmarkResult Objects

A `BenchmarkResult` object is a collection of multiple `ResampleResult` objects.

```
bmrdt = as.data.table(bmr)
bmrdt[1:2, .(task, learner, resampling, iteration)]
```

```
                   task                   learner           resampling
1: <TaskClassif[51]> <LearnerClassifRpart[38]> <ResamplingCV[20]>
2: <TaskClassif[51]> <LearnerClassifRpart[38]> <ResamplingCV[20]>
1 variable not shown: [iteration]
```

The contents of a `BenchmarkResult` and `ResampleResult` (Section 3.2.3) are almost identical and the stored `ResampleResults` can be extracted via the `$resample_result(i)` method, where i is the index of the performed resample experiment. This allows us to investigate the extracted `ResampleResult` and individual resampling iterations as shown in Section 3.2, as well as the predictions from each fold with `$resample_result(i)$predictions()`.

```
rr1 = bmr$resample_result(1)
rr1
```

```
<ResampleResult> with 5 resampling iterations
        task_id      learner_id resampling_id iteration warnings errors
 german_credit classif.rpart            cv         1        0      0
 german_credit classif.rpart            cv         2        0      0
 german_credit classif.rpart            cv         3        0      0
 german_credit classif.rpart            cv         4        0      0
 german_credit classif.rpart            cv         5        0      0
```

```
rr2 = bmr$resample_result(2)
```

In addition, `as_benchmark_result()` can be used to convert objects from `ResampleResult` to `BenchmarkResult`. The `c()`-method can be used to combine multiple `BenchmarkResult` objects, which can be useful when conducting experiments across multiple machines:

```
bmr1 = as_benchmark_result(rr1)
bmr2 = as_benchmark_result(rr2)
```

```
c(bmr1, bmr2)
```

```
<BenchmarkResult> of 10 rows with 2 resampling runs
 nr        task_id      learner_id resampling_id iters warnings errors
  1 german_credit   classif.rpart            cv     5        0      0
  2 german_credit  classif.ranger            cv     5        0      0
```

Boxplots are most commonly used to visualize benchmark experiments as they can intuitively summarize results across tasks and learners simultaneously.

```
autoplot(bmr, measure = msr("classif.acc"))
```

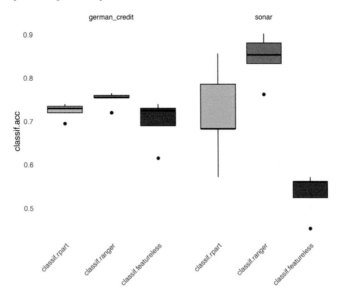

Figure 3.8: Boxplots of accuracy scores for each learner across resampling iterations and the three tasks. Random forests (`lrn("classif.ranger")`) consistently outperforms the other learners.

3.4 Evaluation of Binary Classifiers

In Section 2.5.3, we touched on the concept of a confusion matrix and how it can be used to break down classification errors in more detail. In this section, we will look at specialized performance measures for binary classification in more detail. We will first return to the confusion matrix and discuss measures that can be derived from it, and then will look at ROC analysis, which builds on these measures. See Chapters 7 and 8 of Provost and Fawcett (2013) for a more detailed introduction to ROC measures.

3.4.1 Confusion Matrix

To recap, a confusion matrix summarizes the following quantities in a two-dimensional contingency table (see also Figure 3.9):

- True positives (TPs): Positive instances that are correctly classified as positive.
- True negatives (TNs): Negative instances that are correctly classified as negative.
- False positives (FPs): Negative instances that are incorrectly classified as positive.
- False negatives (FNs): Positive instances that are incorrectly classified as negative.

Different applications may have a particular interest in one (or multiple) of the aforementioned quantities. For example, the `tsk("spam")` classification task is concerned with classifying if mail is spam (positive class) or not (negative class). In this case, we are likely to accept FNs (some spam classified as genuine mail) as long as we have a low number of FPs (genuine and possibly important mail classified as spam). In another example, say we are predicting if a travel bag contains a weapon (positive class) or not (negative class) at an airport. This classifier must have a very

		True Class y		
		$+$	$-$	
Predicted Class $\hat{y}$	$+$	TP	FP	$PPV = \frac{TP}{TP+FP}$
	$-$	FN	TN	$NPV = \frac{TN}{FN+TN}$
		$TPR = \frac{TP}{TP+FN}$	$TNR = \frac{TN}{FP+TN}$	$ACC = \frac{TP+TN}{TP+FP+FN+TN}$

Figure 3.9: Binary confusion matrix of ground truth class vs. predicted class.

high number of TPs (as FNs are not acceptable at all), even if this comes at the expense of more FPs (false alarms).

As we saw in Section 2.5.3, it is possible for a classifier to have a good classification accuracy but to overlook the nuances provided by a full confusion matrix, as in the following `tsk("german_credit")` example:

```
tsk_german = tsk("german_credit")
lrn_ranger = lrn("classif.ranger", predict_type = "prob")
splits = partition(tsk_german, ratio = 0.8)

lrn_ranger$train(tsk_german, splits$train)
prediction = lrn_ranger$predict(tsk_german, splits$test)
prediction$score(msr("classif.acc"))
```

```
classif.acc
     0.72
```

```
prediction$confusion
```

```
        truth
response good bad
    good  123  39
     bad   17  21
```

The classification accuracy only takes into account the TPs and TNs, whereas the confusion matrix provides a more holistic picture of the classifier's performance.

On their own, the absolute numbers in a confusion matrix can be less useful when there is a class imbalance. Instead, several normalized measures can be derived (Figure 3.9):

- **True Positive Rate (TPR)**, **Sensitivity** or **Recall**: How many of the true positives did we predict as positive?
- **True Negative Rate (TNR)** or **Specificity**: How many of the true negatives did we predict as negative?

- **False Positive Rate (FPR)**, or 1 – **Specificity**: How many of the true negatives did we predict as positive?
- **Positive Predictive Value (PPV)** or **Precision**: If we predict positive how likely is it a true positive?
- **Negative Predictive Value (NPV)**: If we predict negative how likely is it a true negative?
- **Accuracy (ACC)**: The proportion of correctly classified instances out of the total number of instances.
- **F1-score**: The harmonic mean of precision and recall, which balances the trade-off between precision and recall. It is calculated as $2 \times \frac{Precision \times Recall}{Precision + Recall}$.

The `mlr3measures` package allows you to compute several common confusion matrix-based measures using the `confusion_matrix()` function:

```
mlr3measures::confusion_matrix(truth = prediction$truth,
    response = prediction$response, positive = tsk_german$positive)
```

```
         truth
response good bad
    good  123  39
    bad    17  21
acc :  0.7200; ce  :  0.2800; dor :  3.8959; f1  :  0.8146
fdr :  0.2407; fnr :  0.1214; fomr:  0.4474; fpr :  0.6500
mcc :  0.2670; npv :  0.5526; ppv :  0.7593; tnr :  0.3500
tpr :  0.8786
```

We now have a better idea of the random forest predictions on `tsk("german_credit")`, in particular, the false positive rate is quite high. It is generally difficult to achieve a high TPR and a low FPR simultaneously because there is often a trade-off between the two rates. When a binary classifier predicts probabilities instead of discrete classes (`predict_type = "prob"`), we could set a threshold to cut off the probabilities to change how we assign observations to the positive/negative class (see Section 2.5.4). Increasing the threshold for identifying the positive cases, leads to a higher number of negative predictions, fewer positive predictions, and therefore a lower (and better) FPR but a lower (and worse) TPR – the reverse holds if we lower the threshold. Instead of arbitrarily changing a threshold to "game" these two numbers, a more robust way to tradeoff between TPR and FPR is to use ROC analysis, discussed next.

3.4.2 ROC Analysis

ROC (Receiver Operating Characteristic) analysis is widely used to evaluate binary classifiers by visualizing the trade-off between the TPR and the FPR.

The ROC curve is a line graph with TPR on the y-axis and the FPR on the x-axis. To understand the usefulness of this curve, first consider the simple case of a hard labeling classifier (`predict_type = "response"`) that classifies observations as either positive or negative. This classifier would be represented as a single point in the ROC space (see Figure 3.10, panel (a)). The best classifier would lie in the top-left corner where the TPR is 1 and the FPR is 0. Classifiers on the diagonal predict class labels randomly (with different class proportions). For example, if each positive instance will be randomly classified (ignoring features) with 25% as the

positive class, we would obtain a TPR of 0.25. If we assign each negative instance
randomly to the positive class, we would have an FPR of 0.25. In practice, we should
never obtain a classifier below the diagonal and a point in the ROC space below
the diagonal might indicate that the positive and negative class labels have been
switched by the classifier.

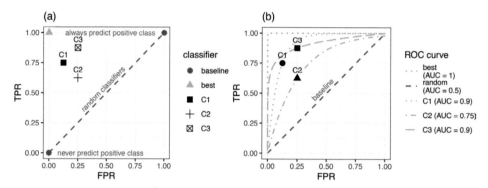

Figure 3.10: Panel (a): ROC space with best discrete classifier, two baseline classifiers
– one that always predicts the positive class and one that never predicts the positive
class – and three "real" classifiers C1, C2, and C3. We cannot say if C1 or C3 is
better than the other as both are better in one metric. C2 is clearly worse than
C1 and C3, which are better in at least one metric than C2 while not being worse
in any other metric. Panel (b): ROC curves of the best classifier (AUC = 1), of a
random guessing classifier (AUC = 0.5), and the classifiers C1, C3, and C2.

Now consider classifiers that predict probabilities instead of discrete classes. Using different thresholds to cut off predicted probabilities and assign them to the
positive and negative class will lead to different TPRs and FPRs and by plotting these values across different thresholds we can characterize the behavior of
a binary classifier – this is the ROC curve. For example, we can use the previous
`Prediction` object to compute all possible TPR and FPR combinations by thresholding the predicted probabilities across all possible thresholds, which is exactly what
`mlr3viz::autoplot.PredictionClassif` will do when `type = "roc"` is selected:

```
autoplot(prediction, type = "roc")
```

A natural performance measure that can be derived from the ROC curve is the area
under the curve (AUC), implemented in `msr("classif.auc")`. The AUC can be   Area Under
interpreted as the probability that a randomly chosen positive instance has a higher   the Curve
predicted probability of belonging to the positive class than a randomly chosen
negative instance. Therefore, higher values (closer to 1) indicate better performance.
Random classifiers (such as the featureless baseline) will always have an AUC of
(approximately, when evaluated empirically) 0.5 (see Figure 3.10, panel (b)).

```
prediction$score(msr("classif.auc"))
```

```
classif.auc
     0.7475
```

Evaluating our random forest on `tsk("german_credit")` results in an AUC of
around 0.75, which is acceptable but could be better.

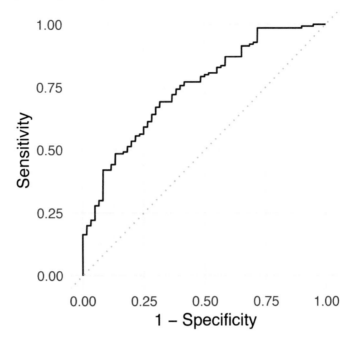

Figure 3.11: ROC-curve based on the `german_credit` dataset and the `classif.ranger` random forest learner. Recall FPR = 1− Specificity and TPR = Sensitivity.

> 💡 Multiclass ROC and AUC
>
> Extensions of ROC analysis for multiclass classifiers exist (see e.g., Hand and Till 2001) but we only cover the more common binary classification case in this book. Generalizations of the AUC measure to multiclass classification are implemented in `mlr3`, see `msr("classif.mauc_au1p")`.

Precision-recall Curve We can also plot the precision-recall curve (PRC) which visualizes the PPV/precision vs. TPR/recall. The main difference between ROC curves and PR curves is that the number of true-negatives are ignored in the latter. This can be useful in imbalanced populations where the positive class is rare, and where a classifier with high TPR may still not be very informative and have low PPV. See Davis and Goadrich (2006) for a detailed discussion about the relationship between the PRC and ROC curves.

```
autoplot(prediction, type = "prc")
```

Another useful way to think about the performance of a classifier is to visualize the relationship of a performance metric over varying thresholds, for example, see Figure 3.13 to inspect the FPR and accuracy across all possible thresholds:

```
autoplot(prediction, type = "threshold",
         measure = msr("classif.fpr"))
autoplot(prediction, type = "threshold",
         measure = msr("classif.acc"))
```

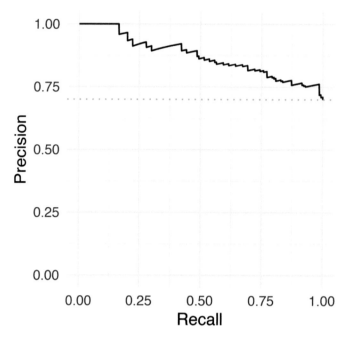

Figure 3.12: Precision-Recall curve based on `tsk("german_credit")` and `lrn("classif.ranger")`.

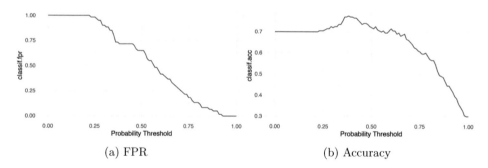

(a) FPR   (b) Accuracy

Figure 3.13: Comparing threshold and FPR (left) with threshold and accuracy (right) for the random forest trained on `tsk("german_credit")`.

This visualization would show us that changing the threshold from the default 0.5 to a higher value like 0.7 would greatly reduce the FPR while reducing accuracy by only a few percentage points. Depending on the problem at hand, this might be a perfectly desirable trade-off.

These visualizations are also available for `ResampleResult` objects. In this case, the predictions of individual resampling iterations are merged before calculating an ROC or PR curve (micro-averaged):

```
rr = resample(
   task = tsk("german_credit"),
   learner = lrn("classif.ranger", predict_type = "prob"),
   resampling = rsmp("cv", folds = 5)
```

```
)
autoplot(rr, type = "roc")
autoplot(rr, type = "prc")
```

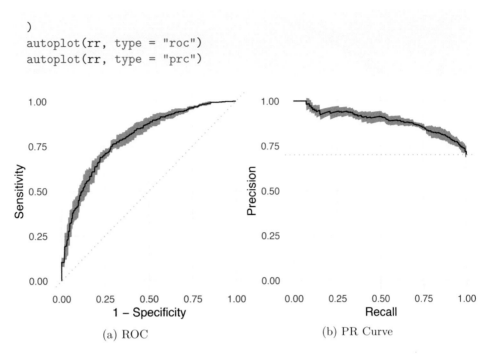

(a) ROC   (b) PR Curve

Figure 3.14: Comparing ROC (left) and PR curve (right) for a random forest trained on `tsk("german_credit")`.

Finally, we can visualize ROC/PR curves for a `BenchmarkResult` to compare multiple learners on the same `Task`:

```
library(patchwork)

design = benchmark_grid(
  tasks = tsk("german_credit"),
  learners = lrns(c("classif.rpart", "classif.ranger"),
    predict_type = "prob"),
  resamplings = rsmp("cv", folds = 5)
)
bmr = benchmark(design)
autoplot(bmr, type = "roc") + autoplot(bmr, type = "prc") +
  plot_layout(guides = "collect")
```

3.5   Conclusion

In this chapter, we learned how to estimate the generalization performance of a model via resampling strategies, from holdout to cross-validation and bootstrap, and how to automate the comparison of multiple learners in benchmark experiments. We also covered the basics of performance measures for binary classification, including the confusion matrix, ROC analysis, and precision-recall curves. These topics are fundamental in supervised learning and will continue to be built upon throughout

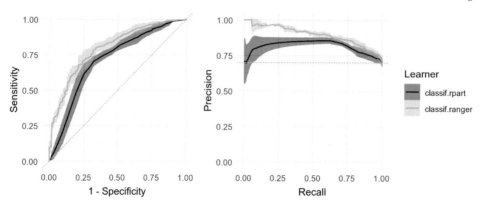

Figure 3.15: Comparing random forest (green) and decision tree (purple) using ROC and PR Curves.

this book. In particular, Chapter 4 utilizes evaluation in automated model tuning to improve performance, in Chapter 11 we look at large benchmarks and their statistical analysis, and in Chapter 13 we will take a look at specialized tasks that require different resampling strategies.

Table 3.1: Important classes and functions covered in this chapter with underlying class (if applicable), class constructor or function, and important class fields and methods (if applicable).

Class	Constructor/Function	Fields/Methods
PredictionClassif	classif_lrn$predict()	confusion_matrix(); autoplot(some_prediction_classif, type = "roc")
-	partition()	-
Resampling	rsmp()	$instantiate()
ResampleResult	resample()	$score(); $aggregate(); $predictions(); as_benchmark_result(); autoplot(some_resample_result, type = "roc")
-	benchmark_grid()	-
BenchmarkResult	benchmark()	$aggregate(); $resample_result(); $score(); autoplot(some_benchmark_result, type = "roc")

3.6 Exercises

1. Apply a repeated cross-validation resampling strategy on tsk("mtcars") and evaluate the performance of lrn("regr.rpart"). Use five repeats of three folds each. Calculate the MSE for each iteration and visualize the result. Finally, calculate the aggregated performance score.

2. Use `tsk("spam")` and five-fold CV to benchmark `lrn("classif. ranger")`, `lrn("classif.log_reg")`, and `lrn("classif.xgboost", nrounds = 100)` with respect to AUC. Which learner appears to perform best? How confident are you in your conclusion? Think about the stability of results and investigate this by re-rerunning the experiment with different seeds. What can be done to improve this?

3. A colleague reports a 93.1% classification accuracy using `lrn("classif.rpart")` on `tsk("penguins_simple")`. You want to reproduce their results and ask them about their resampling strategy. They said they used a custom three-fold CV with folds assigned as `factor(task$row_ids %% 3)`. See if you can reproduce their results.

4. (*) Program your own ROC plotting function without using mlr3's `autoplot()` function. The signature of your function should be `my_roc_plot(task, learner, train_indices, test_indices)`. Your function should use the `$set_threshold()` method of `Prediction`, as well as `mlr3measures`.

Part II

Tuning and Feature Selection

4

Hyperparameter Optimization

Marc Becker
Ludwig-Maximilians-Universität München, and Munich Center for Machine Learning (MCML)

Lennart Schneider
Ludwig-Maximilians-Universität München, and Munich Center for Machine Learning (MCML)

Sebastian Fischer
Ludwig-Maximilians-Universität München, and Munich Center for Machine Learning (MCML)

Hyperparam-
eters
Machine learning algorithms usually include parameters and hyperparameters. Parameters are the model coefficients or weights or other information that are determined by the learning algorithm based on the training data. In contrast, hyperparameters, are configured by the user and determine how the model will fit its parameters, i.e., how the model is built. Examples include setting the number of trees in a random forest, penalty settings in support vector machines, or the learning rate in a neural network.

Hyperparam-
eter
Optimiza-
tion
The goal of hyperparameter optimization (HPO) or model tuning is to find the optimal configuration of hyperparameters of a machine learning algorithm for a given task. There is no closed-form mathematical representation (nor analytic gradient information) for model-agnostic HPO. Instead, we follow a black box optimization approach: a machine learning algorithm is configured with values chosen for one or more hyperparameters, this algorithm is then evaluated (using a resampling method) and its performance is measured. This process is repeated with multiple configurations and finally, the configuration with the best performance is selected (Figure 4.1). HPO closely relates to model evaluation (Chapter 3) as the objective is to find a hyperparameter configuration that optimizes the generalization performance. Broadly speaking, we could think of finding the optimal model configuration in the same way as selecting a model from a benchmark experiment, where in this case each model in the experiment is the same algorithm but with different hyperparameter configurations. For example, we could benchmark three support vector machines (SVMs) with three different `cost` values. However, human trial-and-error is time-consuming, subjective and often biased, error-prone, and computationally inefficient. Instead, many sophisticated hyperparameter optimization methods (or "tuners", see Section 4.1.4) have been developed over the past few decades for robust and efficient HPO. Besides simple approaches such as a random search or grid search, most hyperparameter optimization methods employ iterative techniques that propose different configurations over time, often exhibiting adaptive behavior guided towards potentially optimal hyperparameter configurations. These methods continually propose new configurations until a termination criterion is met, at which point the best configuration so far is returned (Figure 4.1). For more general details on HPO

DOI: 10.1201/9781003402848-4

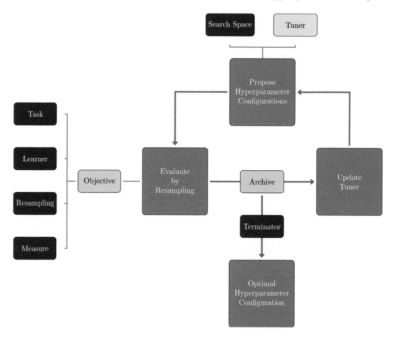

Figure 4.1: Representation of the hyperparameter optimization loop in mlr3tuning. Purple – Hyperparameter optimization loop. Blue – Objects of the tuning instance supplied by the user. Green – Optimization Algorithm.

and a more theoretical background, we recommend Bischl et al. (2023) and Feurer and Hutter (2019).

4.1 Model Tuning

mlr3tuning is the hyperparameter optimization package of the mlr3 ecosystem. At the heart of the package are the R6 classes

- TuningInstanceSingleCrit, a tuning "instance" that describes the optimization problem and store the results; and
- Tuner which is used to configure and run optimization algorithms.

In this section, we will cover these classes as well as other supporting functions and classes. Throughout this section, we will look at optimizing an SVM classifier from e1071 on tsk("sonar") as a running example.

4.1.1 Learner and Search Space

The tuning process begins by deciding which hyperparameters to tune and what range to tune them over. The first place to start is therefore picking a learner and looking at the possible hyperparameters to tune with $param_set:

```
as.data.table(lrn("classif.svm")$param_set)[,
  .(id, class, lower, upper, nlevels)]
```

```
                 id       class lower upper nlevels
 1:       cachesize ParamDbl  -Inf   Inf     Inf
 2: class.weights   ParamUty    NA    NA     Inf
 3:           coef0 ParamDbl  -Inf   Inf     Inf
 4:            cost ParamDbl     0   Inf     Inf
 5:           cross ParamInt     0   Inf     Inf
---
12:              nu ParamDbl  -Inf   Inf     Inf
13:           scale ParamUty    NA    NA     Inf
14:       shrinking ParamLgl    NA    NA       2
15:       tolerance ParamDbl     0   Inf     Inf
16:            type ParamFct    NA    NA       2
```

Given infinite resources, we could tune all hyperparameters jointly, but in reality that is not possible (or maybe necessary), so usually only a subset of hyperparameters can be tuned. This subset of possible hyperparameter values to tune over is referred to as the search space or tuning space. In this example we will tune the numeric regularization and kernel width hyperparameters, `cost` and `gamma`; see the help page for `svm()` for details. In practice, search spaces are usually more complex and can require expert knowledge to define them. Section 4.4 provides more detailed insight into the creation of tuning spaces, including using `mlr3tuningspaces` to load predefined search spaces.

Search Space

> 💡 Untunable Hyperparameters
>
> In rare cases, parameter sets may include hyperparameters that should not be tuned. These will usually be "technical" (or "control") parameters that *provide information* about how the model is being fit but do not control the training process itself, for example, the `verbose` hyperparameter in `lrn("classif.ranger")` controls how much information is displayed to the user during training.

For numeric hyperparameters (we will explore others later) one must specify the bounds to tune over. We do this by constructing a learner and using `to_tune()` to set the lower and upper limits for the parameters we want to tune. This function allows us to *mark* the hyperparameter as requiring tuning in the specified range.

```
learner = lrn("classif.svm",
  type   = "C-classification",
  kernel = "radial",
  cost   = to_tune(1e-1, 1e5),
  gamma  = to_tune(1e-1, 1)
)
learner
```

```
<LearnerClassifSVM:classif.svm>
* Model: -
* Parameters: type=C-classification, kernel=radial,
  cost=<RangeTuneToken>, gamma=<RangeTuneToken>
* Packages: mlr3, mlr3learners, e1071
* Predict Types:  [response], prob
* Feature Types: logical, integer, numeric
* Properties: multiclass, twoclass
```

Here we have constructed a classification SVM, `lrn("classif.svm")`, selected the type of model as `"C-classification"`, set the kernel to `"radial"`, and specified that we plan to tune the `cost` and `gamma` parameters over the range $[0.1, 10^5]$ and $[0.1, 1]$ respectively (though these are usually tuned on a log scale, see Section 4.1.5). Note that calling `$train()` on a learner with a tune token (e.g., `cost=<RangeTuneToken>`) will throw an error.

Now we have decided which hyperparameters to tune, we specify when to stop the tuning process.

4.1.2   Terminator

`mlr3tuning` includes many methods to specify when to terminate an algorithm (Table 4.1), which are implemented in `Terminator` classes. Terminators are stored in the `mlr_terminators` dictionary and are constructed with the sugar function `trm()`.

Terminator

trm()

Table 4.1: Terminators available in `mlr3tuning` at the time of publication, their function call and default parameters. A complete and up-to-date list can be found at https://mlr-org.com/terminators.html.

Terminator	Function call and default parameters
Clock Time	`trm("clock_time")`
Combo	`trm("combo", any = TRUE)`
None	`trm("none")`
Number of Evaluations	`trm("evals", n_evals = 100, k = 0)`
Performance Level	`trm("perf_reached", level = 0.1)`
Run Time	`trm("run_time", secs = 30)`
Stagnation	`trm("stagnation", iters = 10, threshold = 0)`

The most commonly used terminators are those that stop the tuning after a certain time (`trm("run_time")`) or a given number of evaluations (`trm("evals")`). Choosing a runtime is often based on practical considerations and intuition. Using a time limit can be important on compute clusters where a maximum runtime for a compute job may need to be specified. `trm("perf_reached")` stops the tuning when a specified performance level is reached, which can be helpful if a certain performance is seen as sufficient for the practical use of the model, however, if this is set too optimistically the tuning may never terminate. `trm("stagnation")` stops when no progress greater than the `threshold` has been made for a set number of `iterations`. The threshold can be difficult to select as the optimization could stop too soon for complex search spaces despite room for (possibly significant) improvement. `trm("none")` is used for tuners that control termination themselves and so this terminator does nothing. Finally, any of these terminators can be freely combined by using `trm("combo")`, which can be used to specify if HPO finishes when any (`any = TRUE`) terminator is triggered or when all (`any = FALSE`) are triggered.

4.1.3 Tuning Instance with `ti`

The tuning instance collects the tuner-agnostic information required to optimize a model, i.e., all information about the tuning process, except for the tuning algorithm itself. This includes the task to tune over, the learner to tune, the resampling method and measure used to analytically compare hyperparameter optimization configurations, and the terminator to determine when the measure has been optimized "enough". This implicitly defines a "black box" objective function, mapping hyperparameter configurations to (stochastic) performance values, to be optimized. This concept will be revisited in Chapter 5.

A tuning instance can be constructed explicitly with the `ti()` function, or we can tune a learner with the `tune()` function, which implicitly creates a tuning instance, as shown in Section 4.2. We cover the `ti()` approach first as this allows finer control of tuning and a more nuanced discussion about the design and use of `mlr3tuning`.

Continuing our example, we will construct a single-objective tuning problem (i.e., tuning over *one* measure) by using the `ti()` function to create a `TuningInstanceSingleCrit`, we will return to multi-objective tuning in Section 5.2.

For this example, we will use three-fold CV and optimize the classification error measure. Note that in the next section, we will continue our example with a grid search tuner, so we select `trm("none")` below as we will want to iterate over the full grid without stopping too soon.

```
tsk_sonar = tsk("sonar")

learner = lrn("classif.svm",
  cost  = to_tune(1e-1, 1e5),
  gamma = to_tune(1e-1, 1),
  kernel = "radial",
  type = "C-classification"
)

instance = ti(
  task = tsk_sonar,
  learner = learner,
  resampling = rsmp("cv", folds = 3),
  measures = msr("classif.ce"),
  terminator = trm("none")
)

instance
```

```
<TuningInstanceSingleCrit>
* State:  Not optimized
* Objective: <ObjectiveTuning:classif.svm_on_sonar>
* Search Space:
        id    class lower upper nlevels
1:    cost ParamDbl   0.1 1e+05     Inf
2:   gamma ParamDbl   0.1 1e+00     Inf
* Terminator: <TerminatorNone>
```

4.1.4   Tuner

With all the pieces of our tuning problem assembled, we can now decide *how* to
tune our model. There are multiple Tuner classes in `mlr3tuning`, which implement   **Tuner**
different HPO (or more generally speaking black box optimization) algorithms
(Table 4.2).

Table 4.2: Tuning algorithms available in `mlr3tuning`, their function call and the
package in which the algorithm is implemented. A complete and up-to-date list can
be found at https://mlr-org.com/tuners.html.

Tuner	Function call	Package
Random Search	`tnr("random_search")`	`mlr3tuning`
Grid Search	`tnr("grid_search")`	`mlr3tuning`
Bayesian Optimization	`tnr("mbo")`	`mlr3mbo`
CMA-ES	`tnr("cmaes")`	`adagio`
Iterated Racing	`tnr("irace")`	`irace`
Hyperband	`tnr("hyperband")`	`mlr3hyperband`
Generalized Simulated Annealing	`tnr("gensa")`	`GenSA`
Nonlinear Optimization	`tnr("nloptr")`	`nloptr`

Search strategies

Grid search and random search (Bergstra and Bengio 2012) are the most basic
algorithms and are often selected first in initial experiments. The idea of grid search is
to exhaustively evaluate every possible combination of given hyperparameter values.
Categorical hyperparameters are usually evaluated over all possible values they can
take. Numeric and integer hyperparameter values are then spaced equidistantly in
their box constraints (upper and lower bounds) according to a given resolution,
which is the number of distinct values to try per hyperparameter. Random search
involves randomly selecting values for each hyperparameter independently from a pre-
specified distribution, usually uniform. Both methods are non-adaptive, which means
each proposed configuration ignores the performance of previous configurations. Due
to their simplicity, both grid search and random search can handle mixed search
spaces (i.e., hyperparameters can be numeric, integer, or categorical) as well as
hierarchical search spaces (Section 4.4).

Adaptive algorithms

Adaptive algorithms learn from previously evaluated configurations to find good
configurations quickly, examples in `mlr3` include Bayesian optimization (also called

model-based optimization), Covariance Matrix Adaptation Evolution Strategy (CMA-ES), Iterated Racing, and Hyperband.

Bayesian optimization (e.g., Snoek, Larochelle, and Adams 2012) describes a family of iterative optimization algorithms that use a surrogate model to approximate the unknown function that is to be optimized – in HPO this would be the mapping from a hyperparameter configuration to the estimated generalization performance. If a suitable surrogate model is chosen, e.g. a random forest, Bayesian optimization can be quite flexible and even handle mixed and hierarchical search spaces. Bayesian optimization is discussed in full detail in Section 5.4.

CMA-ES (Hansen and Auger 2011) is an evolutionary strategy that maintains a probability distribution over candidate points, with the distribution represented by a mean vector and covariance matrix. A new set of candidate points is generated by sampling from this distribution, with the probability of each candidate being proportional to its performance. The covariance matrix is adapted over time to reflect the performance landscape. Further evolutionary strategies are available in `mlr3` via the `miesmuschel` package, however, these will not be covered in this book.

Racing algorithms work by iteratively discarding configurations that show poor performance, as determined by statistical tests. Iterated Racing (López-Ibáñez et al. 2016) starts by "racing" down an initial population of randomly sampled configurations from a parameterized density and then uses the surviving configurations of the race to stochastically update the density of the subsequent race to focus on promising regions of the search space, and so on.

Multi-fidelity HPO is an adaptive method that leverages the predictive power of computationally cheap lower fidelity evaluations (i.e., poorer quality predictions such as those arising from neural networks with a small number of epochs) to improve the overall optimization efficiency. This concept is used in Hyperband (Li et al. 2018), a popular multi-fidelity hyperparameter optimization algorithm that dynamically allocates increasingly more resources to promising configurations and terminates low-performing ones. Hyperband is discussed in full detail in Section 5.3.

Other implemented algorithms for numeric search spaces are Generalized Simulated Annealing (Xiang et al. 2013; Tsallis and Stariolo 1996) and various nonlinear optimization algorithms.

Choosing strategies

As a rule of thumb, if the search space is small or does not have a complex structure, grid search may be able to exhaustively evaluate the entire search space in a reasonable time. However, grid search is generally not recommended due to the curse of dimensionality – the grid size "blows up" very quickly as the number of parameters to tune increases – and insufficient coverage of numeric search spaces. By construction, grid search cannot evaluate a large number of unique values per hyperparameter, which is suboptimal when some hyperparameters have minimal impact on performance while others do. In such scenarios, random search is often a better choice as it considers more unique values per hyperparameter compared to grid search.

For higher-dimensional search spaces or search spaces with more complex structures, more guided optimization algorithms such as evolutionary strategies or Bayesian optimization tend to perform better and are more likely to result in peak performance. When choosing between evolutionary strategies and Bayesian optimization, the cost of function evaluation is highly relevant. If hyperparameter configurations can be evaluated quickly, evolutionary strategies often work well. On the other hand, if model evaluations are time-consuming and the optimization budget is limited, Bayesian optimization is usually preferred, as it is quite sample efficient compared to other algorithms, i.e., less function evaluations are needed to find good configurations. Hence, Bayesian optimization is usually recommended for HPO. While the optimization overhead of Bayesian optimization is comparably large (e.g., in each iteration, training of the surrogate model and optimizing the acquisition function), this has less of an impact in the context of relatively costly function evaluations such as resampling of ML models.

Finally, in cases where the hyperparameter optimization problem involves a meaningful fidelity parameter (e.g., number of epochs, number of trees, number of boosting rounds) and where the optimization budget needs to be spent efficiently, multi-fidelity hyperparameter optimization algorithms like Hyperband may be worth considering. For further details on different tuners and practical recommendations, we refer to Bischl et al. (2023).

💡 **$param_classes and $properties**

The $param_classes and $properties fields of a Tuner respectively provide information about which classes of hyperparameters can be handled and what properties the tuner can handle (e.g., hyperparameter dependencies, which are shown in Section 4.4, or multicriteria optimization, which is presented in Section 5.2):

```
tnr("random_search")$param_classes
```

```
[1] "ParamLgl" "ParamInt" "ParamDbl" "ParamFct"
```

```
tnr("random_search")$properties
```

```
[1] "dependencies" "single-crit"  "multi-crit"
```

For our SVM example, we will use a grid search with a resolution of five for runtime reasons here (in practice a larger resolution would be preferred). The resolution is the number of distinct values to try *per hyperparameter*, which means in our example the tuner will construct a 5x5 grid of 25 configurations of equally spaced points between the specified upper and lower bounds. All configurations will be tried by the tuner (in random order) until either all configurations are evaluated or the terminator (Section 4.1.2) signals that the budget is exhausted. For grid and random search tuners, the batch_size parameter controls how many configurations are evaluated at the same time when parallelization is enabled (see Section 10.1.3), and also determines how many configurations should be applied before the terminator should check if the termination criterion has been reached.

```
tuner = tnr("grid_search", resolution = 5, batch_size = 10)
tuner
```

```
<TunerGridSearch>: Grid Search
* Parameters: resolution=5, batch_size=10
* Parameter classes: ParamLgl, ParamInt, ParamDbl, ParamFct
* Properties: dependencies, single-crit, multi-crit
* Packages: mlr3tuning
```

Control
Parameters
The `resolution` and `batch_size` parameters are termed control parameters of the tuner, and other tuners will have other control parameters that can be set, as with learners these are accessible with `$param_set`.

```
tuner$param_set
```

```
<ParamSet>
                    id    class lower upper nlevels        default value
1:        batch_size ParamInt     1   Inf     Inf <NoDefault[3]>    10
2:        resolution ParamInt     1   Inf     Inf <NoDefault[3]>     5
3: param_resolutions ParamUty    NA    NA     Inf <NoDefault[3]>
```

While changing the control parameters of the tuner can improve optimal performance, we have to take care that is likely the default settings will fit most needs. While it is not possible to cover all application cases, `mlr3tuning`'s defaults were chosen to work well in most cases. However, some control parameters like `batch_size` often interact with the parallelization setup (further described in Section 10.1.3) and may need to be adjusted accordingly.

Triggering the tuning process

Now that we have introduced all our components, we can start the tuning process. To do this we simply pass the constructed `TuningInstanceSingleCrit` to the `$optimize()` method of the initialized Tuner, which triggers the hyperparameter optimization loop (Figure 4.1).

```
tuner$optimize(instance)
```

```
    cost gamma learner_param_vals  x_domain classif.ce
1: 25000   0.1          <list[4]> <list[2]>     0.2449
```

The optimizer returns the best hyperparameter configuration and the corresponding performance, this information is also stored in `instance$result`. The first columns (here `cost` and `gamma`) will be named after the tuned hyperparameters and show the optimal values from the searched tuning spaces. The `$learner_param_vals` field of the `$result` lists the optimal hyperparameters from tuning, as well as the values of any other hyperparameters that were set, this is useful for onward model use (Section 4.1.6).

```
instance$result$learner_param_vals
```

```
[[1]]
[[1]]$kernel
[1] "radial"

[[1]]$type
[1] "C-classification"

[[1]]$cost
[1] 25000

[[1]]$gamma
[1] 0.1
```

The `$x_domain` field is most useful in the context of hyperparameter transformations, which we will briefly turn to next.

> ⚠ Overconfident Performance Estimates
>
> A common mistake when tuning is to report the performance estimated on the resampling sets on which the tuning was performed (`instance$result$classif.ce`) as an unbiased estimate of the model's performance and to ignore its optimistic bias. The correct method is to test the model on more unseen data, which can be efficiently performed with nested resampling, we will discuss this in Section 4.3.2.

4.1.5 Logarithmic Transformations

For many non-negative hyperparameters that have a large upper bound, tuning on a logarithmic scale can be more efficient than tuning on a linear scale. For example, consider sampling uniformly in the interval $[\log(1e-5), \log(1e5)]$ and then exponentiating the outcome, the histograms in Figure 4.2 show how we are initially sampling within a narrow range $([-11.5, 11.5])$ but then exponentiating results in the majority of points being relatively small but a few being very large.

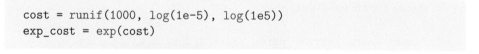

```
cost = runif(1000, log(1e-5), log(1e5))
exp_cost = exp(cost)
```

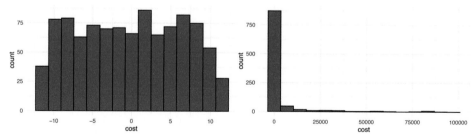

(a) Linear scale sampled by the tuner.   (b) Logarithmic scale seen by the learner.

Figure 4.2: Histograms of uniformly sampled values from the interval $[\log(1e-5), \log(1e5)]$ before (left) and after (right) exponentiation.

To add this transformation to a hyperparameter we simply pass `logscale = TRUE` to `to_tune()`.

```
learner = lrn("classif.svm",
  cost  = to_tune(1e-5, 1e5, logscale = TRUE),
  gamma = to_tune(1e-5, 1e5, logscale = TRUE),
  kernel = "radial",
  type = "C-classification"
)

instance = ti(
  task = tsk_sonar,
  learner = learner,
  resampling = rsmp("cv", folds = 3),
  measures = msr("classif.ce"),
  terminator = trm("none")
)

tuner$optimize(instance)
```

```
    cost  gamma learner_param_vals  x_domain classif.ce
1: 5.756 -5.756          <list[4]> <list[2]>     0.1394
```

We can see from this example that using the log transformation improved the hyperparameter search, as `classif.ce` is smaller.

Note that the fields `cost` and `gamma` show the optimal values *before* transformation, whereas `x_domain` and `learner_param_vals` contain optimal values *after* transformation, it is these latter fields you would take forward for future model use.

```
instance$result$x_domain
```

```
[[1]]
[[1]]$cost
[1] 316.2

[[1]]$gamma
[1] 0.003162
```

In Section 4.4, we will look at how to implement more complex, custom transformations for any hyperparameter or combination of hyperparameters. Now we will look at how to put everything into practice so we can make use of the tuned model (and the transformed hyperparameters).

4.1.6 Analyzing and Using the Result

Independently of whether you use `ti()` or `tune()`, or if you include transformations or not, the created objects and the output are structurally the same and the instance's archive lists all evaluated hyperparameter configurations:

```
as.data.table(instance$archive)[1:3, .(cost, gamma, classif.ce)]
```

```
      cost   gamma classif.ce
1: -11.513  -5.756     0.4665
2:  -5.756 -11.513     0.4665
3:  -5.756  11.513     0.4665
```

Each row of the archive is a different evaluated configuration. The columns show the tested configurations (before transformation) and the chosen performance measure. We can also manually inspect the archive to determine other important features such as time of evaluation, model runtime, and any errors or warnings that occurred during tuning.

```
as.data.table(instance$archive)[1:3,
   .(timestamp, runtime_learners, errors, warnings)]
```

```
            timestamp runtime_learners errors warnings
1: 2023-07-04 15:21:47            0.031      0        0
2: 2023-07-04 15:21:47            0.028      0        0
3: 2023-07-04 15:21:47            0.031      0        0
```

Another powerful feature of the instance is that we can score the internal `ResampleResults` on a different performance measure, for example, by looking at false negative rate and false positive rate as well as classification error:

```
as.data.table(instance$archive,
   measures = msrs(c("classif.fpr", "classif.fnr")))[1:5 ,
   .(cost, gamma, classif.ce, classif.fpr, classif.fnr)]
```

```
      cost   gamma classif.ce classif.fpr classif.fnr
1: -11.513  -5.756     0.4665      1.0000     0.00000
2:  -5.756 -11.513     0.4665      1.0000     0.00000
3:  -5.756  11.513     0.4665      1.0000     0.00000
4:   0.000  -5.756     0.2308      0.3186     0.14997
5:   5.756  -5.756     0.1394      0.2089     0.08056
```

You can access all the resamplings combined in a `BenchmarkResult` object with `instance$archive$benchmark_result`.

Finally, to visualize the results, you can use `autoplot.TuningInstanceSingleCrit` (Figure 4.3). In this example we can observe one of the flaws (by design) in grid search, despite testing 25 configurations, we only saw five unique values for each hyperparameter.

```
autoplot(instance, type = "surface")
```

Training an optimized model

Once we found good hyperparameters for our learner through tuning, we can use them to train a final model on the whole data. To do this we simply construct a new learner with the same underlying algorithm and set the learner hyperparameters to the optimal configuration:

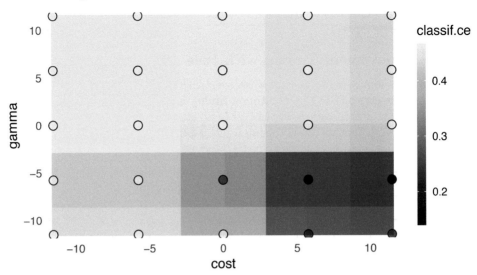

Figure 4.3: Model performance with different configurations for `cost` and `gamma`. Bright yellow regions represent the model performing worse and dark blue performing better. We can see that high `cost` values and low `gamma` values achieve the best performance. Note that we should not directly infer the performance of new unseen values from the heatmap since it is only an interpolation based on a surrogate model (`regr.ranger`). However, we can see the general interaction between the hyperparameters.

```
lrn_svm_tuned = lrn("classif.svm")
lrn_svm_tuned$param_set$values =
        instance$result_learner_param_vals
```

Now we can train the learner on the full dataset and we are ready to make predictions.

```
lrn_svm_tuned$train(tsk_sonar)$model
```

```
Call:
svm.default(x = data, y = task$truth(), type = "C-classification",
    kernel = "radial", gamma = 0.00316227766016838,
    cost = 316.227766016838,
    probability = (self$predict_type == "prob"))


Parameters:
   SVM-Type:  C-classification
 SVM-Kernel:  radial
       cost:  316.2

Number of Support Vectors:  93
```

4.2   Convenient Tuning with `tune` and `auto_tuner`

In the previous section, we looked at constructing and manually putting together
the components of HPO by creating a tuning instance using `ti()`, passing this to
the tuner, and then calling `$optimize()` to start the tuning process. `mlr3tuning`
includes two helper methods to simplify this process further.

The first helper function is `tune()`, which creates the tuning instance and calls
`$optimize()` for you. You may prefer the manual method with `ti()` if you want to
view and make changes to the instance before tuning.

```
tnr_grid_search = tnr("grid_search", resolution = 5,
        batch_size = 5)
lrn_svm = lrn("classif.svm",
  cost  = to_tune(1e-5, 1e5, logscale = TRUE),
  gamma = to_tune(1e-5, 1e5, logscale = TRUE),
  kernel = "radial",
  type = "C-classification"
)
rsmp_cv3 = rsmp("cv", folds = 3)
msr_ce = msr("classif.ce")

instance = tune(
  tuner = tnr_grid_search,
  task = tsk_sonar,
  learner = lrn_svm,
  resampling = rsmp_cv3,
  measures = msr_ce
)
instance$result
```

```
   cost  gamma learner_param_vals  x_domain classif.ce
1: 5.756 -5.756         <list[4]> <list[2]>     0.1444
```

The other helper function is `auto_tuner`, which creates an object of class `AutoTuner`
(Figure 4.4). The `AutoTuner` inherits from the `Learner` class and wraps all the
information needed for tuning, which means you can treat a learner waiting to be
optimized just like any other learner. Under the hood, the `AutoTuner` essentially
runs `tune()` on the data that is passed to the model when `$train()` is called and
then sets the learner parameters to the optimal configuration.

```
at = auto_tuner(
  tuner = tnr_grid_search,
  learner = lrn_svm,
  resampling = rsmp_cv3,
  measure = msr_ce
)

at
```

```
<AutoTuner:classif.svm.tuned>
* Model: list
* Search Space:
<ParamSet>
      id    class  lower upper nlevels        default value
1:  cost ParamDbl -11.51 11.51     Inf <NoDefault[3]>
2: gamma ParamDbl -11.51 11.51     Inf <NoDefault[3]>
Trafo is set.
* Packages: mlr3, mlr3tuning, mlr3learners, e1071
* Predict Type: response
* Feature Types: logical, integer, numeric
* Properties: multiclass, twoclass
```

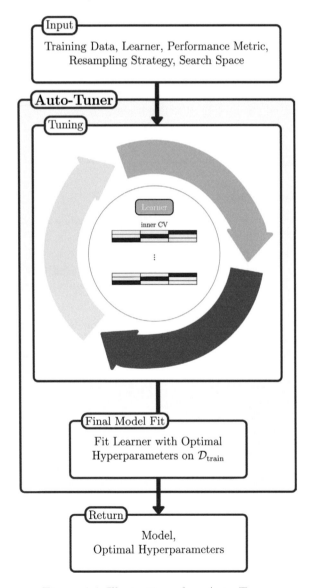

Figure 4.4: Illustration of an Auto-Tuner.

And we can now call `$train()`, which will first tune the hyperparameters in the search space listed above before fitting the optimal model.

```
split = partition(tsk_sonar)
at$train(tsk_sonar, row_ids = split$train)
at$predict(tsk_sonar, row_ids = split$test)$score()
```

```
classif.ce
   0.2029
```

The `AutoTuner` contains a tuning instance that can be analyzed like any other instance.

```
at$tuning_instance$result
```

```
    cost  gamma learner_param_vals  x_domain classif.ce
1: 5.756 -5.756            <list[4]> <list[2]>     0.1727
```

We could also pass the `AutoTuner` to `resample()` and `benchmark()`, which would result in a nested resampling, discussed next.

4.3   Nested Resampling

HPO requires additional resampling to reduce bias when estimating the performance of a model. If the same data is used for determining the optimal configuration and the evaluation of the resulting model itself, the actual performance estimate might be biased (Simon 2007). This is analogous to optimism of the training error described in James et al. (2014), which occurs when training error is taken as an estimate of out-of-sample performance.

Nested resampling separates model optimization from the process of estimating the performance of the tuned model by adding an additional resampling, i.e., while model performance is estimated using a resampling method in the "usual way", tuning is then performed by resampling the resampled data (Figure 4.5). For more details and a formal introduction to nested resampling, the reader is referred to Bischl et al. (2023) and Simon (2007).

Figure 4.5 represents the following example of nested resampling:

1. Outer resampling start – Instantiate three-fold CV to create different testing and training datasets.
2. Inner resampling – Within the outer training data instantiate four-fold CV to create different inner testing and training datasets.
3. HPO – Tune the hyperparameters on the outer training set (large, light blue blocks) using the inner data splits.
4. Training – Fit the learner on the outer training dataset using the optimal hyperparameter configuration obtained from the inner resampling (small blocks).

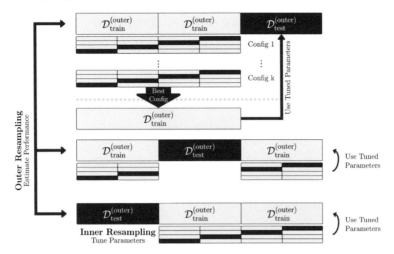

Figure 4.5: An illustration of nested resampling. The large blocks represent three-fold CV for the outer resampling for model evaluation and the small blocks represent four-fold CV for the inner resampling for HPO. The light blue blocks are the training sets and the dark blue blocks are the test sets.

5. Evaluation – Evaluate the performance of the learner on the outer testing data (large, dark blue block).
6. Outer resampling repeats – Repeat (2)–(5) for each of the three outer folds.
7. Aggregation – Take the sample mean of the three performance values for an unbiased performance estimate.

The inner resampling produces generalization performance estimates for each configuration and selects the optimal configuration to be evaluated on the outer resampling. The outer resampling then produces generalization estimates for these optimal configurations. The result from the outer resampling can be used for comparison to other models trained and tested on the same outer folds.

> 💡 Nested Resampling and Parallelization
>
> Nested resampling is computationally expensive. For example, an experiment with three outer folds and four inner folds and a grid search of resolution five used to tune two parameters, would result in $3 * 5 * 5^2 = 300$ iterations of model training/testing. If you have the resources we recommend utilizing parallelization when tuning (Section 10.1).

A common mistake is to think of nested resampling as a method to select optimal model configurations. Nested resampling is a method to compare models and to estimate the generalization performance of a tuned model, however, this is the performance based on multiple different configurations (one from each outer fold) and not performance based on a *single* configuration (Section 4.3.2). If you are interested in identifying optimal configurations, then use `tune()`/`ti()` or `auto_tuner()` with `$train()` on the complete dataset.

4.3.1   Nested Resampling with an `AutoTuner`

While the theory of nested resampling may seem complicated, it is all automated in `mlr3tuning` by simply passing an `AutoTuner` to `resample()` or `benchmark()`. Continuing with our previous example, we will use the auto-tuner to resample a support vector classifier with three-fold CV in the outer resampling and four-fold CV in the inner resampling.

```
at = auto_tuner(
  tuner = tnr_grid_search,
  learner = lrn_svm,
  resampling = rsmp("cv", folds = 4),
  measure = msr_ce,
)


rr = resample(tsk_sonar, at, rsmp_cv3, store_models = TRUE)


rr
```

```
<ResampleResult> with 3 resampling iterations
 task_id       learner_id resampling_id iteration warnings errors
   sonar classif.svm.tuned            cv         1        0      0
   sonar classif.svm.tuned            cv         2        0      0
   sonar classif.svm.tuned            cv         3        0      0
```

Note that we set `store_models = TRUE` so that the `AutoTuner` models (fitted on the outer training data) are stored, which also enables investigation of the inner tuning instances. While we used k-fold CV for both the inner and outer resampling strategy, you could use different resampling strategies (Section 3.2) and different parallelization methods (Section 10.1.4).

The estimated performance of a tuned model is reported as the aggregated performance of all outer resampling iterations, which is a less biased estimate of future model performance.

```
rr$aggregate()
```

```
classif.ce
    0.1589
```

In addition to the methods described in Section 3.2, `extract_inner_tuning_results()` and `extract_inner_tuning_archives()` return the optimal configurations (across all outer folds) and full tuning archives, respectively.

```
extract_inner_tuning_results(rr)[,
  .(iteration, cost, gamma, classif.ce)]
```

```
   iteration  cost  gamma classif.ce
1:         1 11.51 -5.756     0.2174
2:         2 11.51 -5.756     0.2086
3:         3 11.51 -5.756     0.1796
```

```
extract_inner_tuning_archives(rr)[1:3,
   .(iteration, cost, gamma, classif.ce)]
```

```
   iteration    cost    gamma  classif.ce
1:         1  -11.51  -11.513      0.5286
2:         1  -11.51   11.513      0.5286
3:         1    0.00   -5.756      0.2981
```

4.3.2 The Right (and Wrong) Way to Estimate Performance

> **i** This section covers advanced ML or technical details.

In this short section, we will empirically demonstrate that directly reporting tuning performance without nested resampling results in optimistically biased performance estimates. In this experiment, we tune several parameters from `lrn("classif.xgboost")`. To best estimate the generalization performance, we make use of the `"moons"` TaskGenerator. The TaskGenerator class is used when TaskGenerator you want to simulate data for use in experiments, these are very useful in cases such as this experiment when you need access to an infinite number of data points to estimate quantities such as the generalization error.

We begin by loading our learner, task generator, and generating 100 training data points and 1,000,000 testing data points.

```
lrn_xgboost = lrn("classif.xgboost",
  eta                = to_tune(1e-4, 1, logscale = TRUE),
  max_depth          = to_tune(1, 20),
  colsample_bytree   = to_tune(1e-1, 1),
  colsample_bylevel  = to_tune(1e-1, 1),
  lambda             = to_tune(1e-3, 1e3, logscale = TRUE),
  alpha              = to_tune(1e-3, 1e3, logscale = TRUE),
  subsample          = to_tune(1e-1, 1)
)
tsk_moons = tgen("moons")
tsk_moons_train = tsk_moons$generate(100)
tsk_moons_test = tsk_moons$generate(1000000)
```

Now, we will tune the learner with respect to the classification error, using holdout resampling and random search with 700 evaluations. We then report the tuning performance without nested resampling.

```
tnr_random = tnr("random_search")
rsmp_holdout = rsmp("holdout")
trm_evals700 = trm("evals", n_evals = 700)

instance = tune(
  tuner = tnr_random,
  task = tsk_moons_train,
```

```
  learner = lrn_xgboost,
  resampling = rsmp_holdout,
  measures = msr_ce,
  terminator = trm_evals700
)


insample = instance$result_y
```

Next, we estimate generalization error by nested resampling (below we use an outer five-fold CV), using an `AutoTuner`:

```
# same setup as above
at = auto_tuner(
  tuner = tnr_random,
  learner = lrn_xgboost,
  resampling = rsmp_holdout,
  measure = msr_ce,
  terminator = trm_evals700
)


rsmp_cv5 = rsmp("cv", folds = 5)


outsample = resample(tsk_moons_train, at, rsmp_cv5)$aggregate()
```

And finally, we estimate the generalization error by training the tuned learner (i.e., using the values from the `instance` above) on the full training data again and predicting on the test data.

```
lrn_xgboost_tuned = lrn("classif.xgboost")
lrn_xgboost_tuned$param_set$set_values(
  .values = instance$result_learner_param_vals)
generalization = lrn_xgboost_tuned$train(tsk_moons_train)$
  predict(tsk_moons_test)$score()
```

Now, we can compare these three values:

```
round(c(true_generalization = as.numeric(generalization),
  without_nested_resampling = as.numeric(insample),
  with_nested_resampling = as.numeric(outsample)), 2)

    true_generalization without_nested_resampling
               0.19                      0.06
with_nested_resampling
               0.22
```

We find that the performance estimate from unnested tuning optimistically overestimates the true performance (which could indicate "meta-overfitting" to the specific inner holdout-splits), while the outer estimate from nested resampling works much better.

4.4   More Advanced Search Spaces

Up until now, we have only considered tuning simple search spaces limited to a few numeric hyperparameters. In this section, we will first look at how to tune different scalar parameter classes with to_tune(), and then how to define your own search space with ParamSet to create more advanced search spaces that may include tuning over vectors, transformations, and handling parameter dependencies. Finally, we will consider how to access a database of standardized search spaces from the literature.

4.4.1   Scalar Parameter Tuning

The to_tune() function can be used to tune parameters of any class, whether they are scalar or vectors. To best understand this function, we will consider what is happening behind the scenes. When to_tune() is used in a learner, implicitly a ParamSet is created just for the tuning search space:

```
learner = lrn("classif.svm",
  cost  = to_tune(1e-1, 1e5),
  gamma = to_tune(1e-1, 1),
  kernel = "radial",
  type = "C-classification"
)

learner$param_set$search_space()
```

```
<ParamSet>
      id     class lower upper nlevels      default value
1:  cost ParamDbl   0.1 1e+05     Inf <NoDefault[3]>
2: gamma ParamDbl   0.1 1e+00     Inf <NoDefault[3]>
```

Recall from Section 2.2.3 that the class field corresponds to the hyperparameter class as defined in paradox. In this example, we can see that the gamma hyperparameter has the class ParamDbl, with lower = 0.1 and upper = 1, which was automatically created by to_tune() as we passed two numeric values to this function. If we wanted to tune over a non-numeric hyperparameter, we can still use to_tune(), which will infer the correct class to construct in the resulting parameter set. For example, say we wanted to tune the numeric cost, factor kernel, and logical scale hyperparameter in our SVM:

```
learner = lrn("classif.svm",
  cost  = to_tune(1e-1, 1e5),
  kernel = to_tune(c("radial", "linear")),
  shrinking = to_tune(),
  type = "C-classification"
)

learner$param_set$search_space()
```

```
<ParamSet>
           id    class lower upper nlevels     default value
1:       cost ParamDbl   0.1 1e+05     Inf <NoDefault[3]>
2:     kernel ParamFct    NA    NA       2 <NoDefault[3]>
3: shrinking ParamLgl    NA    NA       2          TRUE
```

Here the `kernel` hyperparameter is a factor, so we simply pass in a vector corresponding to the levels we want to tune over. The `shrinking` hyperparameter is a logical, there are only two possible values this could take so we do not need to pass anything to `to_tune()`, it will automatically recognize this is a logical from `learner$param_set` and passes this detail to `learner$param_set$search_space()`. Similarly, for factor parameters, we could also use `to_tune()` without any arguments if we want to tune over all possible values. Finally, we can use `to_tune()` to treat numeric parameters as factors if we want to discretize them over a small subset of possible values. For example, if we wanted to find the optimal number of trees in a random forest, we might only consider three scenarios: 100, 200, or 400 trees:

```
lrn("classif.ranger", num.trees = to_tune(c(100, 200, 400)))
```

Before we look at tuning over vectors, we must first learn how to create parameter sets from scratch.

> ⚠ **Ordered Hyperparameters**
>
> Treating an integer as a factor for tuning results in "unordered" hyperparameters. Therefore, algorithms that make use of ordering information will perform worse when ordering is ignored. For these algorithms, it would make more sense to define a `ParamDbl` or `ParamInt` (Section 4.4.2) with a custom transformation (Section 4.4.3).

4.4.2   Defining Search Spaces with `ps`

As we have seen, `to_tune()` is a helper function that creates a parameter set that will go on to be used by `tune()`, `ti()`, or `auto_tuner()` during the tuning process. However, there will be use cases where you will need to create a parameter set manually using `ps()`. This function takes named arguments of class `Param`, which can be created using the sugar functions in Table 4.3.

Table 4.3: `Domain` Constructors and their resulting `Param`.

Constructor	Description	Underlying Class
`p_dbl`	Real valued parameter ("double")	`ParamDbl`
`p_int`	Integer parameter	`ParamInt`
`p_fct`	Discrete valued parameter ("factor")	`ParamFct`
`p_lgl`	Logical/Boolean parameter	`ParamLgl`
`p_uty`	Untyped parameter	`ParamUty`

As a simple example, let us look at how to create a search space to tune `cost` and `gamma` again:

```
search_space = ps(
  cost   = p_dbl(lower = 1e-1, upper = 1e5),
  kernel = p_fct(c("radial", "linear")),
  shrinking = p_lgl()
)
```

This search space would then be passed to the `search_space` argument in `auto_tuner()`:

```
ti(tsk_sonar, lrn("classif.svm", type = "C-classification"),
  rsmp_cv3,  msr_ce, trm("none"), search_space = search_space)
```

```
<TuningInstanceSingleCrit>
* State:  Not optimized
* Objective: <ObjectiveTuning:classif.svm_on_sonar>
* Search Space:
          id    class lower upper nlevels
1:      cost ParamDbl   0.1 1e+05     Inf
2:    kernel ParamFct    NA    NA       2
3: shrinking ParamLgl    NA    NA       2
* Terminator: <TerminatorNone>
```

> ⚠ Bounded Search Spaces
>
> When manually creating search spaces, make sure all numeric hyperparameters in your search space are bounded, e.g., if you are trying to tune a hyperparameter that could take any value in $(-\infty, \infty)$ then the tuning process will throw an error for nearly all tuners if you do not pass lower and upper limits to `p_dbl()` or `p_int()`. You can use `$is_bounded` on the constructed `ParamSet` if you are unsure:
>
> ```
> ps(cost = p_dbl(lower = 0.1, upper = 1))$is_bounded
> ```
>
> ```
> [1] TRUE
> ```
>
> ```
> ps(cost = p_dbl(lower = 0.1, upper = Inf))$is_bounded
> ```
>
> ```
> [1] FALSE
> ```

4.4.3   Transformations and Tuning Over Vectors

> ℹ **This section covers advanced ML or technical details.**

In Section 4.1.5, we saw how to quickly apply log transformations with `to_tune()`. As you now know, `to_tune()` is just a wrapper that creates `ParamSet` objects, so let us look at what is taking place when we set `logscale = TRUE`:

```
lrn("classif.svm", cost = to_tune(1e-5, 1e5, logscale = TRUE))$
  param_set$search_space()
```

```
<ParamSet>
     id    class  lower upper nlevels        default value
1: cost ParamDbl -11.51 11.51     Inf <NoDefault[3]>
Trafo is set.
```

Notice that now the `lower` and `upper` fields correspond to the transformed bounds, i.e. $[\log(1e-5), \log(1e5)]$. To manually create the same transformation, we can pass the transformation to the `trafo` argument in `p_dbl()` and set the bounds:

```
search_space = ps(cost = p_dbl(log(1e-5), log(1e5),
  trafo = function(x) exp(x))) # alternatively: 'trafo = exp'
search_space
```

```
<ParamSet>
     id     class  lower upper nlevels        default value
1: cost ParamDbl -11.51 11.51     Inf <NoDefault[3]>
Trafo is set.
```

We can confirm it is correctly set by making use of the `$trafo()` method, which takes a named list and applies the specified transformations

```
search_space$trafo(list(cost = 1))
```

```
$cost
[1] 2.718
```

Where transformations become the most powerful is in the ability to pass arbitrary functions that can act on single parameters or even the entire parameter set. As an example, consider a simple transformation to add "2" to our range:

```
search_space = ps(cost = p_dbl(0, 3, trafo = function(x) x + 2))
search_space$trafo(list(cost = 1))
```

```
$cost
[1] 3
```

Simple transformations such as this can even be added directly to a learner by passing a `Param` object to `to_tune()`:

```
lrn("classif.svm",
  cost = to_tune(p_dbl(0, 3, trafo = function(x) x + 2)))
```

More complex transformations that require multiple arguments should be passed to the `.extra_trafo` parameter in `ps()`. `.extra_trafo` takes a function with parameters `x` and `param_set` where, during tuning, `x` will be a list containing the configuration being tested, and `param_set` is the whole parameter set. Below we first exponentiate the value of `cost` and then add "2" if the `kernel` is `"polynomial"`.

```
search_space = ps(
  cost = p_dbl(-1, 1, trafo = function(x) exp(x)),
  kernel = p_fct(c("polynomial", "radial")),
  .extra_trafo = function(x, param_set) {
    if (x$kernel == "polynomial") {
      x$cost = x$cost + 2
    }
    x
  }
)
search_space$trafo(list(cost = 1, kernel = "radial"))
```

```
$cost
[1] 2.718

$kernel
[1] "radial"
```

```
  search_space$trafo(list(cost = 1, kernel = "polynomial"))
```

```
$cost
[1] 4.718

$kernel
[1] "polynomial"
```

Vector transformations

Any function can be passed to `trafo` and `.extra_trafo`, which enables tuning of "untyped" parameters of class `ParamUty` that could be vectors, functions, or any non-atomic class. By example, consider the `class.weights` parameter of the SVM, which takes a named vector of class weights with one entry for each target class. To tune this parameter we could tune a scalar and then transform this to a vector. The code below would result in a value, x, between 0.1 and 0.9 being sampled, the result is then transformed to $(x, 1 - x)$ and is then passed to the `Learner`.

```
search_space = ps(
  class.weights = p_dbl(lower = 0.1, upper = 0.9,
    trafo = function(x) c(M = x, R = 1 - x))
)
```

In other cases, we may need to tune two or more "pseudoparameters" that do not exist in our learner's parameter set but are required to tune a vector parameter. For example, say we want to tune the architecture of a neural network, in which we need to decide the number of layers and the number of nodes in each layer, this is the case in the `num_nodes` hyperparameter in `lrn("surv.coxtime")` (we use this learner as it provides a useful template for this sort of transformation, interested readers can read about survival analysis in Section 13.2). In this case, the learner expects a vector where each element of the vector corresponds to the number of nodes in a layer and the length of the vector is the number of layers. We could then tune this as follows:

```
search_space = ps(
  num_layers = p_int(lower = 1, upper = 20),
  num_nodes_per_layer = p_int(4, 64),
  .extra_trafo = function(x, param_set) {
    x$num_nodes = rep(x$num_nodes_per_layer, x$num_layers)
    x$num_layers = NULL
    x$num_nodes_per_layer = NULL
    x
  }
)
```

Here we are tuning the pseudo-parameter `num_layers` between 1 and 20, then tuning the pseudo-parameter `num_nodes_per_layer` between 4 and 64, then combining these into a vector called `num_nodes` (the real hyperparameter) and removing the pseudo-parameters.

```
search_space$trafo(list(num_layers = 4, num_nodes_per_layer = 12))
```

```
$num_nodes
[1] 12 12 12 12
```

Even though this transformation looks complex, it only affects one of the hyperparameters (and does not need access to others), so we could include it in the learner using `to_tune()` by passing the whole `ParamSet` object:

```
learner = lrn("surv.coxtime")
learner$param_set$set_values(num_nodes = to_tune(search_space))
learner$param_set$search_space()
```

```
<ParamSet>
                     id    class lower upper nlevels        default
1:            num_layers ParamInt     1    20      20 <NoDefault[3]>
2: num_nodes_per_layer ParamInt     4    64      61 <NoDefault[3]>
1 variable not shown: [value]
Trafo is set.
```

4.4.4 Hyperparameter Dependencies

> **i** **This section covers advanced ML or technical details.**

Hyperparameter dependencies occur when a hyperparameter should only be set if another hyperparameter has a particular value. For example, the `degree` parameter in SVM is only valid when `kernel` is `"polynomial"`. In the `ps()` function, we specify this using the `depends` argument, which takes a named argument of the form `<param> == value` or `<param> %in% <vector>`:

```
ps(
  kernel = p_fct(c("polynomial", "radial")),
  degree = p_int(1, 3, depends = (kernel == "polynomial")),
  gamma = p_dbl(1e-5, 1e5,
    depends = (kernel %in% c("polynomial", "radial"))))
)
```

```
<ParamSet>
       id    class lower upper nlevels        default parents value
1: degree ParamInt 1e+00 3e+00       3 <NoDefault[3]>  kernel
2:  gamma ParamDbl 1e-05 1e+05     Inf <NoDefault[3]>  kernel
3: kernel ParamFct    NA    NA       2 <NoDefault[3]>
```

Above we have said that **degree** should only be set if **kernel** is (==) "polynomial", and **gamma** should only be set if **kernel** is one of (%in%) "polynomial" or "radial". In practice, some underlying implementations ignore unused parameters and others throw errors, either way, this is problematic during tuning if, for example, we were wasting time trying to tune **degree** when the kernel was not polynomial. Hence setting the dependency tells the tuning process to tune **degree** if **kernel** is "polynomial" and to ignore it otherwise.

Dependencies can also be passed straight into a learner using to_tune():

```
lrn("classif.svm",
  kernel = to_tune(c("polynomial", "radial")),
  degree = to_tune(p_int(1, 3, depends = (kernel == "polynomial")))
)$param_set$search_space()
```

```
<ParamSet>
       id    class lower upper nlevels        default        parents
1: degree ParamInt     1     3       3 <NoDefault[3]> kernel,kernel
2: kernel ParamFct    NA    NA       2 <NoDefault[3]>
1 variable not shown: [value]
```

4.4.5 Recommended Search Spaces with `mlr3tuningspaces`

> **i** This section covers advanced ML or technical details.

Selected search spaces can require a lot of background knowledge or expertise. The package `mlr3tuningspaces` tries to make HPO more accessible by providing implementations of published search spaces for many popular machine learning algorithms, the hope is that these search spaces are applicable to a wide range of datasets. The search spaces are stored in the dictionary `mlr_tuning_spaces`.

```
library(mlr3tuningspaces)
as.data.table(mlr_tuning_spaces)[1:3, .(key, label)]
```

```
                        key                                  label
1:  classif.glmnet.default    Classification GLM with Default
2:      classif.glmnet.rbv1 Classification GLM with RandomBot
3:      classif.glmnet.rbv2 Classification GLM with RandomBot
```

The tuning spaces are named according to the scheme
`{learner-id}.{tuning-space-id}`. The `default` tuning spaces are published in Bischl et al. (2023), and other tuning spaces are part of the random bot
experiments `rbv1` and `rbv2` published in Kuehn et al. (2018) and Binder, Pfisterer,
and Bischl (2020). The sugar function `lts()` (learner tuning space) is used to
retrieve a TuningSpace.

```
lts_rpart = lts("classif.rpart.default")
lts_rpart
```

```
<TuningSpace:classif.rpart.default>: Classification Rpart with Default
           id lower upper levels logscale
1:   minsplit 2e+00 128.0            TRUE
2: minbucket 1e+00  64.0            TRUE
3:         cp 1e-04   0.1            TRUE
```

A tuning space can be passed to `ti()` or `auto_tuner()` as the `search_space`.

```
instance = ti(
  task = tsk_sonar,
  learner = lrn("classif.rpart"),
  resampling = rsmp("cv", folds = 3),
  measures = msr("classif.ce"),
  terminator = trm("evals", n_evals = 20),
  search_space = lts_rpart
)
```

Alternatively, as loaded search spaces are just a collection of tune tokens, we could
also pass these straight to a learner:

```
vals = lts_rpart$values
vals
```

```
$minsplit
Tuning over:
range [2, 128] (log scale)


$minbucket
Tuning over:
range [1, 64] (log scale)


$cp
Tuning over:
range [1e-04, 0.1] (log scale)
```

```
learner = lrn("classif.rpart")
learner$param_set$set_values(.values = vals)
learner$param_set
```

```
<ParamSet>
             id        class lower upper nlevels         default
 1:          cp  ParamDbl     0     1     Inf            0.01
 2:  keep_model  ParamLgl    NA    NA       2           FALSE
 3:  maxcompete  ParamInt     0   Inf     Inf               4
 4:    maxdepth  ParamInt     1    30      30              30
 5: maxsurrogate ParamInt     0   Inf     Inf               5
 6:   minbucket  ParamInt     1   Inf     Inf <NoDefault[3]>
 7:    minsplit  ParamInt     1   Inf     Inf              20
 8: surrogatestyle ParamInt   0     1       2               0
 9:  usesurrogate ParamInt    0     2       3               2
10:        xval  ParamInt     0   Inf     Inf              10
1 variable not shown: [value]
```

Note how we used the `.values` parameter of `$set_values()`, which allows us to safely pass a list to the `ParamSet` without accidentally overwriting any other hyperparameter values (Section 2.2.3).

We could also apply the default search spaces from Bischl et al. (2023) by passing the learner to `lts()`:

```
lts(lrn("classif.rpart"))
```

```
<LearnerClassifRpart:classif.rpart>: Classification Tree
* Model: -
* Parameters: xval=0, minsplit=<RangeTuneToken>,
  minbucket=<RangeTuneToken>, cp=<RangeTuneToken>
* Packages: mlr3, rpart
* Predict Types:  [response], prob
* Feature Types: logical, integer, numeric, factor, ordered
* Properties: importance, missings, multiclass,
  selected_features, twoclass, weights
```

Finally, it is possible to overwrite a predefined tuning space in construction, for example, changing the range of the `maxdepth` hyperparameter in a decision tree:

```
lts("classif.rpart.rbv2", maxdepth = to_tune(1, 20))
```

```
<TuningSpace:classif.rpart.rbv2>: Classification Rpart with
RandomBot
          id lower upper levels logscale
1:        cp 1e-04     1            TRUE
2:  maxdepth 1e+00    20           FALSE
3: minbucket 1e+00   100           FALSE
4:  minsplit 1e+00   100           FALSE
```

4.5   Conclusion

In this chapter, we learned how to optimize a model using tuning instances, about different tuners and terminators, search spaces and transformations, how to make use of convenience methods for quicker implementation in larger experiments, and the importance of nested resampling.

Table 4.4: Important classes and functions covered in this chapter with underlying class (if applicable), class constructor or function, and important class fields and methods (if applicable).

Class	Constructor/Function	Fields/Methods
Terminator	`trm()`	-
TuningInstanceSingleCritti or TuningInstanceMultiCrit	`i()`/`tune()`	`$result`; `$archive`
Tuner	`tnr()`	`$optimize()`
TuneToken	`to_tune()`	-
AutoTuner	`auto_tuner()`	`$train()`; `$predict()`; `$tuning_instance`
-	`extract_inner_tuning_results()`	
-	`extract_inner_tuning_archives()`	
ParamSet	`ps()`	-
TuningSpace	`lts()`	`$values`

4.6   Exercises

1. Tune the `mtry`, `sample.fraction`, and `num.trees` hyperparameters of `lrn("regr.ranger")` on `tsk("mtcars")`. Use a simple random search with 50 evaluations. Evaluate with a three-fold CV and the root mean squared error. Visualize the effects that each hyperparameter has on the performance via simple marginal plots, which plot a single hyperparameter versus the cross-validated MSE.

2. Evaluate the performance of the model created in Exercise 1 with nested resampling. Use a holdout validation for the inner resampling and a three-fold CV for the outer resampling.

3. Tune and benchmark an XGBoost model against a logistic regression (without tuning the latter) and determine which has the best Brier score. Use `mlr3tuningspaces` and nested resampling, and try to pick appropriate inner and outer resampling strategies that balance computational efficiency vs. stability of the results.

4. (*) Write a function that implements an iterated random search procedure that drills down on the optimal configuration by applying random search to iteratively smaller search spaces. Your function should have seven inputs: `task`, `learner`, `search_space`, `resampling`, `measure`,

random_search_stages, and random_search_size. You should only worry about programming this for fully numeric and bounded search spaces that have no dependencies. In pseudo-code:

(1) Create a random design of size random_search_size from the given search space and evaluate the learner on it.

(2) Identify the best configuration.

(3) Create a smaller search space around this best config, where you define the new range for each parameter as: new_range[i] = (best_conf[i] - 0.25 * current_range[i], best_conf[i] + 0.25*current_range[i]). Ensure that this new_range respects the initial bound of the original search_space by taking the max() of the new and old lower bound, and the min() of the new and the old upper bound ("clipping").

(4) Iterate the previous steps random_search_stages times and at the end return the best configuration you have ever evaluated. As a stretch goal, look into mlr3tuning's internal source code and turn your function into an R6 class inheriting from the Tuner class – test it out on a learner of your choice.

5

Advanced Tuning Methods and Black Box Optimization

Lennart Schneider
Ludwig-Maximilians-Universität München, and Munich Center for Machine Learning (MCML)

Marc Becker
Ludwig-Maximilians-Universität München, and Munich Center for Machine Learning (MCML)

Having looked at the basic usage of `mlr3tuning`, we will now turn to more advanced methods. We will begin in Section 5.1 by continuing to look at single-objective tuning but will consider what happens when experiments go wrong and how to prevent fatal errors. We will then extend the methodology from Chapter 4 to enable multi-objective tuning, where learners are optimized to multiple measures simultaneously, in Section 5.2, we will demonstrate how this is handled relatively simply in `mlr3` by making use of the same classes and methods we have already used. The final two sections focus on specific optimization methods. Section 5.3 looks in detail at multi-fidelity tuning and the Hyperband tuner, and then demonstrates it in practice with `mlr3hyperband`. Finally, Section 5.4 takes a deep dive into black box Bayesian optimization. This is a more theory-heavy section to motivate the design of the classes and methods in `mlr3mbo`.

5.1 Error Handling and Memory Management

In this section, we will look at how to use `mlr3` to ensure that tuning workflows are efficient and robust. In particular, we will consider how to enable features that prevent fatal errors leading to irrecoverable data loss in the middle of an experiment, and then how to manage tuning experiments that may use up a lot of computer memory.

5.1.1 Encapsulation and Fallback Learner

Error handling is discussed in detail in Section 10.2, however, it is very important in the context of tuning so here we will just practically demonstrate how to make use of encapsulation and fallback learners and explain why they are essential during HPO.

Even in simple machine learning problems, there is a lot of potential for things to go wrong. For example, when learners do not converge, run out of memory, or terminate with an error due to issues in the underlying data. As a common issue, learners can fail if there are factor levels present in the test data that were not in the training data, models fail in this case as there have been no weights/coefficients trained for these new factor levels:

```
tsk_pen = tsk("penguins")
# remove rows with missing values
tsk_pen$filter(tsk_pen$row_ids[complete.cases(tsk_pen$data())])
# create custom resampling with new factors in test data
rsmp_custom = rsmp("custom")
rsmp_custom$instantiate(tsk_pen,
  list(tsk_pen$row_ids[tsk_pen$data()$island != "Torgersen"]),
  list(tsk_pen$row_ids[tsk_pen$data()$island == "Torgersen"])
)
msr_ce = msr("classif.ce")
tnr_random = tnr("random_search")
learner = lrn("classif.lda", method = "t", nu = to_tune(3, 10))


tune(tnr_random, tsk_pen, learner, rsmp_custom, msr_ce, 10)
```

```
Error in lda.default(x, grouping, ...): variable 6 appears to be
constant within groups
```

In the above example, we can see the tuning process breaks and we lose all information about the hyperparameter optimization process. This is even worse in nested resampling or benchmarking when errors could cause us to lose all progress across multiple configurations or even learners and tasks.

Encapsulation (Section 10.2.1) allows errors to be isolated and handled, without disrupting the tuning process. We can tell a learner to encapsulate an error by setting the $encapsulate field as follows:

```
learner$encapsulate = c(train = "evaluate", predict = "evaluate")
```

Note by passing "evaluate" to both train and predict, we are telling the learner to set up encapsulation in both the training and prediction stages (see Section 10.2 for other encapsulation options).

Another common issue that cannot be easily solved during HPO is learners not converging and the process running indefinitely. We can prevent this from happening by setting the timeout field in a learner, which signals the learner to stop if it has been running for that much time (in seconds), again this can be set for training and prediction individually:

```
learner$timeout = c(train = 30, predict = 30)
```

Now if either an error occurs, or the model timeout threshold is reached, then instead of breaking, the learner will simply not make predictions when errors are found and the result is NA for resampling iterations with errors. When this happens, our hyperparameter optimization experiment will fail as we cannot aggregate results across resampling iterations. Therefore it is essential to select a fallback learner (Section 10.2.2), which is a learner that will be fitted if the learner of interest fails.

A common approach is to use a featureless baseline (`lrn("regr.featureless")` or `lrn("classif.featureless")`). Below we set `lrn("classif.featureless")`, which always predicts the majority class, by passing this learner to the `$fallback` field.

```
learner$fallback = lrn("classif.featureless")
```

We can now run our experiment and see errors that occurred during tuning in the archive.

```
instance = tune(tnr_random, tsk_pen, learner, rsmp_custom, msr_ce,
    10)

as.data.table(instance$archive)[1:3, .(df, classif.ce, errors)]
```

```
              df classif.ce errors
1: <function[1]>          1      1
2: <function[1]>          1      1
3: <function[1]>          1      1
```

```
# Reading the error in the first resample result
instance$archive$resample_result(1)$errors
```

```
   iteration                                              msg
1:         1 variable 6 appears to be constant within groups
```

The learner was tuned without breaking because the errors were encapsulated and logged before the fallback learners were used for fitting and predicting:

```
instance$result
```

```
   nu learner_param_vals  x_domain classif.ce
1:  5          <list[2]> <list[1]>          1
```

5.1.2 Memory Management

Running a large tuning experiment can use a lot of memory, especially when using nested resampling. Most of the memory is consumed by the models since each resampling iteration creates one new model. Storing the models is therefore disabled by default and in most cases is not required. The option `store_models` in the functions `ti()` and `auto_tuner()` allows us to enable the storage of the models.

The archive stores a `ResampleResult` for each evaluated hyperparameter configuration. The contained `Prediction` objects can also take up a lot of memory, especially with large datasets and many resampling iterations. We can disable the storage of the resample results by setting `store_benchmark_result = FALSE` in the functions `ti()` and `auto_tuner()`. Note that without the resample results, it is no longer possible to score the configurations with another measure.

When we run nested resampling with many outer resampling iterations, additional memory can be saved if we set `store_tuning_instance = FALSE` in the `auto_tuner()` function. However, the functions `extract_inner_tuning_results()` and `extract_inner_tuning_archives()` will then no longer work.

The option `store_models = TRUE` sets `store_benchmark_result` and `store_tuning_instance` to TRUE because the models are stored in the benchmark results which in turn is part of the instance. This also means that `store_benchmark_result = TRUE` sets `store_tuning_instance` to TRUE.

Finally, we can set `store_models = FALSE` in the `resample()` or `benchmark()` functions to disable the storage of the auto tuners when running nested resampling. This way we can still access the aggregated performance (`rr$aggregate()`) but lose information about the inner resampling.

5.2 Multi-Objective Tuning

So far we have considered optimizing a model with respect to one metric, but multi-criteria, or multi-objective optimization, is also possible. A simple example of multi-objective optimization might be optimizing a classifier to simultaneously maximize true positive predictions and minimize false negative predictions. In another example, consider the single-objective problem of tuning a neural network to minimize classification error. The best-performing model is likely to be quite complex, possibly with many layers that will have drawbacks like being harder to deploy on devices with limited resources. In this case, we might want to simultaneously minimize the classification error and model complexity.

Multi-objective

By definition, optimization of multiple metrics means these will be in competition (otherwise we would only optimize one of them) and therefore in general no *single* configuration exists that optimizes all metrics. Therefore, we instead focus on the concept of Pareto optimality. One hyperparameter configuration is said to *Pareto-dominate* another if the resulting model is equal or better in all metrics and strictly better in at least one metric. For example, say we are minimizing classification error, CE, and complexity, CP, for configurations A and B with CE of CE_A and CE_B respectively and CP of CP_A and CP_B respectively. Then, A pareto-dominates B if: 1) $CE_A \leq CE_B$ and $CP_A < CP_B$ or; 2) $CE_A < CE_B$ and $CP_A \leq CP_B$. All configurations that are not Pareto-dominated by any other configuration are called *Pareto-efficient* and the set of all these configurations is the *Pareto set*. The metric values corresponding to the Pareto set are referred to as the Pareto front.

Pareto front

The goal of multi-objective hyperparameter optimization is to find a set of non-dominated solutions so that their corresponding metric values approximate the Pareto front. We will now demonstrate multi-objective hyperparameter optimization by tuning a decision tree on `tsk("sonar")` with respect to the classification error, as a measure of model performance, and the number of selected features, as a measure of model complexity (in a decision tree the number of selected features is straightforward to obtain by simply counting the number of unique splitting variables). Methodological details on multi-objective hyperparameter optimization can be found in Karl et al. (2022) and Morales-Hernández, Van Nieuwenhuyse, and Rojas Gonzalez (2022).

We will tune `cp`, `minsplit`, and `maxdepth`:

```
learner = lrn("classif.rpart", cp = to_tune(1e-04, 1e-1),
  minsplit = to_tune(2, 64), maxdepth = to_tune(1, 30))

measures = msrs(c("classif.ce", "selected_features"))
```

As we are tuning with respect to multiple measures, the function `ti()` automatically creates a TuningInstanceMultiCrit instead of a TuningInstanceSingleCrit. Below we set `store_models = TRUE` as this is required by the selected features measure.

```
instance = ti(
  task = tsk("sonar"),
  learner = learner,
  resampling = rsmp("cv", folds = 3),
  measures = measures,
  terminator = trm("evals", n_evals = 30),
  store_models = TRUE
)
instance
```

```
<TuningInstanceMultiCrit>
* State:  Not optimized
* Objective: <ObjectiveTuning:classif.rpart_on_sonar>
* Search Space:
         id    class lower upper nlevels
1:       cp ParamDbl 1e-04   0.1     Inf
2: minsplit ParamInt 2e+00  64.0      63
3: maxdepth ParamInt 1e+00  30.0      30
* Terminator: <TerminatorEvals>
```

We can then select and tune a tuning algorithm as usual:

```
tuner = tnr("random_search")
tuner$optimize(instance)
```

Finally, we inspect the best-performing configurations, i.e., the Pareto set, and visualize the corresponding estimated Pareto front (Figure 5.1). Note that the `selected_features` measure is averaged across the folds, so the values in the archive may not always be integers.

```
instance$archive$best()[, .(cp, minsplit, maxdepth, classif.ce,
  selected_features)]
```

```
        cp minsplit maxdepth classif.ce selected_features
1: 0.06881       60        4     0.2596             1.000
2: 0.09451       23       14     0.2596             1.000
3: 0.09891       21       29     0.2596             1.000
4: 0.05475       57       17     0.2596             1.000
5: 0.09774       38       16     0.2596             1.000
6: 0.08944        4        8     0.2547             2.333
```

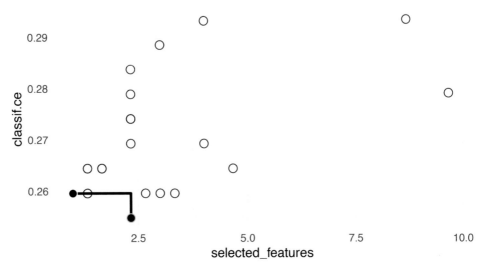

Figure 5.1: Pareto front of selected features and classification error. White dots represent tested configurations, each black dot individually represents a Pareto-optimal configuration and all black dots together represent the approximated Pareto front.

Determining which configuration to deploy from the Pareto front is up to you. By definition, there is no optimal configuration so this may depend on your use case, for example, if you would prefer lower complexity at the cost of higher error then you might prefer a configuration where `selected_features = 1`.

You can select one configuration and pass it to a learner for training using `$result_learner_param_vals`, so if we want to select the second configuration we would run:

```
learner = lrn("classif.rpart")
learner$param_set$values = instance$result_learner_param_vals[[2]]
```

As multi-objective tuning requires manual intervention to select a configuration, it is currently not possible to use `auto_tuner()`.

5.3 Multi-Fidelity Tuning via Hyperband

Increasingly large datasets and search spaces and increasingly complex models make hyperparameter optimization a time-consuming and computationally expensive task. To tackle this, some HPO methods make use of evaluating a configuration at multiple fidelity levels. Multi-fidelity HPO is motivated by the idea that the performance of a lower-fidelity model is indicative of the full-fidelity model, which can be used to make HPO more efficient (as we will soon see with Hyperband).

Multi-fidelity HPO

To unpack what these terms mean and to motivate multi-fidelity tuning, say that we think a gradient-boosting algorithm with up to 1000 rounds will be a very good fit to our training data. However, we are concerned this model will take too long to tune and train. Therefore, we want to gauge the performance of this model using a similar model that is quicker to train by setting a smaller number of rounds. In this example, the hyperparameter controlling the number of rounds is a *fidelity parameter*, as it controls the tradeoff between model performance and speed. The different configurations of this parameter are known as *fidelity levels*. We refer to the model with 1000 rounds as the model at *full-fidelity* and we want to approximate this model's performance using models at different fidelity levels. Lower fidelity levels result in low-fidelity models that are quicker to train but may poorly predict the full-fidelity model's performance. On the other hand, higher fidelity levels result in high-fidelity models that are slower to train but may better indicate the full-fidelity model's performance.

Other common models that have natural fidelity parameters include neural networks (number of epochs) and random forests (number of trees). The proportion of data to subsample before running any algorithm can also be viewed as a model-agnostic fidelity parameter, we will return to this in Section 8.4.4.

5.3.1   Hyperband and Successive Halving

A popular multi-fidelity HPO algorithm is *Hyperband* (Li et al. 2018). After having evaluated randomly sampled configurations on low fidelities, Hyperband iteratively allocates more resources to promising configurations and terminates low-performing ones early. Hyperband builds upon the Successive Halving algorithm by Jamieson and Talwalkar (2016). Successive Halving is initialized with a number of starting configurations m_0, the proportion of configurations discarded in each stage η, and the minimum, r_0, and maximum, r_{max}, budget (fidelity) of a single evaluation. The algorithm starts by sampling m_0 random configurations and allocating the minimum budget r_0 to them. The configurations are evaluated and $\frac{\eta-1}{\eta}$ of the worst-performing configurations are discarded. The remaining configurations are promoted to the next stage, or "bracket", and evaluated on a larger budget. This continues until one or more configurations are evaluated on the maximum budget r_{max} and the best-performing configuration is selected. The total number of stages is calculated so that each stage consumes approximately the same overall budget. A big disadvantage of this method is that it is unclear if it is better to start with many configurations (large m_0) and a small budget or fewer configurations (small m_0) but a larger budget.

Hyperband solves this problem by running Successive Halving with different numbers of starting configurations, each at different budget levels r_0. The algorithm is initialized with the same η and r_{max} parameters (but not m_0). Each bracket starts with a different budget, r_0, where smaller values mean that more configurations can be evaluated and so the most exploratory bracket (i.e., the one with the most number of stages) is allocated the global minimum budget r_{min}. In each bracket, the starting budget increases by a factor of η until the last bracket essentially performs a random search with the full budget r_{max}. The total number of brackets, $s_{max} + 1$, is calculated as $s_{max} = \log_\eta \frac{r_{max}}{r_{min}}$. The number of starting configurations m_0 of each bracket are calculated so that each bracket uses approximately the same amount of budget. The optimal hyperparameter configuration in each bracket is the configuration with the best performance in the final stage. The optimal hyperparameter configuration at the end of tuning is the configuration with the best performance across all brackets.

An example Hyperband schedule is given in Table 5.1 where $s = 3$ is the most exploratory bracket and $s = 0$ essentially performs a random search using the full budget. Table 5.2 demonstrates how this schedule may look if we were to tune 20 different hyperparameter configurations; note that each entry in the table is a unique ID referring to a possible configuration of multiple hyperparameters to tune.

Table 5.1: Hyperband schedule with the number of configurations, m_i, and resources, r_i, for each bracket, s, and stage, i, when $\eta = 2$, $r_{min} = 1$, and $r_{max} = 8$.

	$s = 3$		$s = 2$		$s = 1$		$s = 0$	
i	m_i	r_i	m_i	r_i	m_i	r_i	m_i	r_i
0	8	1	6	2	4	4	4	8
1	4	2	3	4	2	8		
2	2	4	1	8				
3	1	8						

Table 5.2: Hyperparameter configurations in each stage and bracket from the schedule in Table 5.1. Entries are unique identifiers for tested hyperparameter configurations (HPCs). HPC_s^* is the optimal hyperparameter configuration in bracket s and HPC^* is the optimal hyperparameter configuration across all brackets.

	$s = 3$	$s = 2$	$s = 1$	$s = 0$
$i = 0$	$\{1, 2, 3, 4, 5, 6, 7, 8\}$	$\{9, 10, 11, 12, 13, 14\}$	$\{15, 16, 17, 18\}$	$\{19, 20, 21, 22\}$
$i = 1$	$\{1, 2, 7, 8\}$	$\{9, 14, 15\}$	$\{20, 21\}$	
$i = 2$	$\{1, 8\}$	$\{15\}$		
$i = 3$	$\{1\}$			
HPC_s^*	$\{1\}$	$\{15\}$	$\{21\}$	$\{22\}$
HPC^*	$\{15\}$			

5.3.2 mlr3hyperband

The Successive Halving and Hyperband algorithms are implemented in mlr3hyperband as `tnr("successive_halving")` and `tnr("hyperband")` respectively; in this section, we will only showcase the Hyperband method.

By example, we will optimize `lrn("classif.xgboost")` on `tsk("sonar")` and use the number of boosting iterations (`nrounds`) as the fidelity parameter, this is a suitable choice as increasing iterations increases model training time but generally also improves performance. Hyperband will allocate increasingly more boosting iterations to well-performing hyperparameter configurations.

We will load the learner and define the search space. We specify a range from 16 (r_{min}) to 128 (r_{max}) boosting iterations and tag the parameter with `"budget"` to identify it as a fidelity parameter. For the other hyperparameters, we take the search space for XGBoost from Bischl et al. (2023), which usually works well for a wide range of datasets.

```
library(mlr3hyperband)

learner = lrn("classif.xgboost")
learner$param_set$set_values(
  nrounds            = to_tune(p_int(16, 128, tags = "budget")),
  eta                = to_tune(1e-4, 1, logscale = TRUE),
  max_depth          = to_tune(1, 20),
  colsample_bytree   = to_tune(1e-1, 1),
  colsample_bylevel  = to_tune(1e-1, 1),
  lambda             = to_tune(1e-3, 1e3, logscale = TRUE),
  alpha              = to_tune(1e-3, 1e3, logscale = TRUE),
  subsample          = to_tune(1e-1, 1)
)
```

We now construct the tuning instance and a hyperband tuner with `eta = 2`. We use `trm("none")` and set the `repetitions` control parameter to 1 so that Hyperband can terminate itself after all brackets have been evaluated a single time. Note that setting `repetition = Inf` can be useful if you want a terminator to stop the optimization, for example, based on runtime. The `hyperband_schedule()` function can be used to display the schedule across the given fidelity levels and budget increase factor.

```
instance = ti(
  task = tsk("sonar"),
  learner = learner,
  resampling = rsmp("holdout"),
  measures = msr("classif.ce"),
  terminator = trm("none")
)

tuner = tnr("hyperband", eta = 2, repetitions = 1)

hyperband_schedule(r_min = 16, r_max = 128, eta = 2)
```

	bracket	stage	budget	n
1:	3	0	16	8
2:	3	1	32	4
3:	3	2	64	2
4:	3	3	128	1
5:	2	0	32	6
6:	2	1	64	3
7:	2	2	128	1
8:	1	0	64	4
9:	1	1	128	2
10:	0	0	128	4

Finally, we can tune as normal and print the result and archive. Note that the archive resulting from a Hyperband run contains the additional columns `bracket` and `stage` which break down the results by the corresponding bracket and stage.

```
tuner$optimize(instance)
```

```
      nrounds    eta max_depth colsample_bytree colsample_bylevel lambda
1:        64 -2.618         3            0.666            0.4722 -5.816
5 variables not shown: [alpha, subsample, learner_param_vals, x_domain,
classif.ce]
```

```
instance$result[, .(classif.ce, nrounds)]
```

```
   classif.ce nrounds
1:     0.1304      64
```

```
as.data.table(instance$archive)[,
   .(bracket, stage, classif.ce, eta, max_depth, colsample_bytree)]
```

	bracket	stage	classif.ce	eta	max_depth	colsample_bytree
1:	3	0	0.4203	-6.664	8	0.8640
2:	3	0	0.5942	-3.139	2	0.8902
3:	3	0	0.2899	-6.968	15	0.8204
4:	3	0	0.2609	-6.555	12	0.6761
5:	3	0	0.2174	-2.618	3	0.6660

31:	0	0	0.2609	-8.070	1	0.9717
32:	3	3	0.1594	-2.618	3	0.6660
33:	2	2	0.1739	-6.455	5	0.9380
34:	1	1	0.2029	-4.509	10	0.7219
35:	1	1	0.2464	-5.749	3	0.2345

5.4 Bayesian Optimization

In this section, we will take a deep dive into Bayesian optimization (BO), also known as Model Based Optimization (MBO). The design of BO is more complex than what we have seen so far in other tuning methods so to help motivate this we will spend a little more time in this section on theory and methodology.

In hyperparameter optimization (Chapter 4), learners are passed a hyperparameter configuration and evaluated on a given task via a resampling technique to estimate its generalization performance with the goal to find the optimal hyperparameter configuration. In general, no analytical description for the mapping from hyperparameter configuration to performance exists and gradient information is also not available. HPO is, therefore, a prime example for black box optimization, which considers the optimization of a function whose mathematical structure and analytical description is unknown or unexploitable. As a result, the only observable information is the output value (i.e., generalization performance) of the function given an input value (i.e., hyperparameter configuration). In fact, as evaluating the performance of a learner can take a substantial amount of time, HPO is quite an expensive black box optimization problem. Black box optimization problems occur in the real-world, for example they are encountered quite often in engineering such as in modeling experiments like crash tests or chemical reactions.

Black Box Optimization

Many optimization algorithm classes exist that can be used for black box optimization, which differ in how they tackle this problem; for example, we saw in Chapter 4 methods including grid/random search and briefly discussed evolutionary strategies. Bayesian optimization refers to a class of sample-efficient iterative global black box optimization algorithms that rely on a "surrogate model" trained on observed data to model the black box function. This surrogate model is typically a non-linear regression model that tries to capture the unknown function using limited observed data. During each iteration, BO algorithms employ an "acquisition function" to determine the next candidate point for evaluation. This function measures the expected "utility" of each point within the search space based on the prediction of the surrogate model. The algorithm then selects the candidate point with the best acquisition function value and evaluates the black box function at that point to then update the surrogate model. This iterative process continues until a termination criterion is met, such as reaching a pre-specified maximum number of evaluations or achieving a desired level of performance. BO is a powerful method that often results in good optimization performance, especially if the cost of the black box evaluation becomes expensive and the optimization budget is tight.

In the rest of this section, we will first provide an introduction to black box optimization with the `bbotk` package and then introduce the building blocks of BO algorithms and examine their interplay and interaction during the optimization process before we assemble these building blocks in a ready to use black box optimizer with `mlr3mbo`. Readers who are primarily interested in how to utilize BO for HPO without delving deep into the underlying building blocks may want to skip to Section 5.4.4. Detailed introductions to black box optimization and BO are given in Bischl et al. (2023), Feurer and Hutter (2019), and Garnett (2022).

As a running example throughout this section, we will optimize the sinusoidal function $f : [0, 1] \to \mathbb{R}, x \mapsto 2x + \sin(14x)$ (Figure 5.2), which is characterized by two local minima and one global minimum.

5.4.1   Black Box Optimization

The `bbotk` (black box optimization toolkit) package is the workhorse package for general black box optimization within the `mlr3` ecosystem. At the heart of the package are the R6 classes:

- `OptimInstanceSingleCrit` and `OptimInstanceMultiCrit`, which are used to construct an optimization instance that describes the optimization problem and stores the results   *Optimization Instance*
- `Optimizer` which is used to construct and configure optimization algorithms.

These classes might look familiar after reading Chapter 4, and in fact `TuningInstanceSingleCrit` and `TuningInstanceMultiCrit` inherit from `OptimInstanceSingle/MultiCrit` and `Tuner` is closely based on `Optimizer`.

`OptimInstanceSingleCrit` relies on an `Objective` function that wraps the actual   *Objective* mapping from a domain (all possible function inputs) to a codomain (all possible function outputs).

Objective functions can be created using different classes, all of which inherit from `Objective`. These classes provide different ways to define and evaluate objective functions and picking the right one will reduce type conversion overhead:

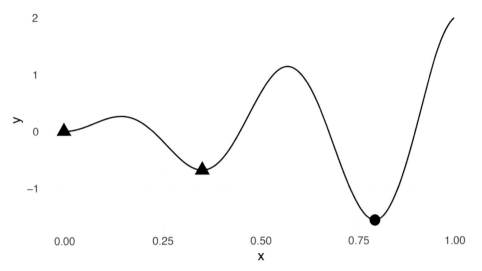

Figure 5.2: Visualization of the sinusoidal function. Local minima in triangles and global minimum in the circle.

- `ObjectiveRFun` wraps a function that takes a list describing a *single configuration* as input where elements can be of any type. It is suitable when the underlying function evaluation mechanism is given by evaluating a single configuration at a time.
- `ObjectiveRFunMany` wraps a function that takes a list of *multiple configurations* as input where elements can be of any type and even mixed types. It is useful when the function evaluation of multiple configurations can be parallelized.
- `ObjectiveRFunDt` wraps a function that operates on a `data.table`. It allows for efficient vectorized or batched evaluations directly on the `data.table` object, avoiding unnecessary data type conversions.

To start translating our problem, to code, we will use the `ObjectiveRFun` class to take a single configuration as input. The `Objective` requires specification of the function to optimize its domain and codomain. By tagging the codomain with `"minimize"` or `"maximize"`, we specify the optimization direction. Note how below our optimization function takes a `list` as an input with one element called `x`.

```
library(bbotk)
sinus_1D = function(xs) 2 * xs$x * sin(14 * xs$x)

domain = ps(x = p_dbl(lower = 0, upper = 1))
codomain = ps(y = p_dbl(tags = "minimize"))
objective = ObjectiveRFun$new(sinus_1D,
  domain = domain, codomain = codomain)
```

We can visualize our objective by generating a grid of points on which we evaluate the function (Figure 5.2), this will help us identify its local minima and global minimum.

```
library(ggplot2)

xydt = generate_design_grid(domain, resolution = 1001)$data
xydt[, y := objective$eval_dt(xydt)$y]
optima = data.table(x = c(0, 0.3509406, 0.7918238))
optima[, y := objective$eval_dt(optima)$y]
optima[, type := c("local", "local", "global")]

ggplot(aes(x = x, y = y), data = xydt) +
  geom_line() +
  geom_point(aes(pch = type), color = "black", size = 4,
        data = optima) +
  theme_minimal() +
  theme(legend.position = "none")
```

The global minimizer, 0.792, corresponds to the point of the domain with the lowest function value:

```
xydt[y == min(y), ]
```

```
       x       y
1: 0.792 -1.577
```

With the objective function defined, we can proceed to optimize it using `OptimInstanceSingleCrit`. This class allows us to wrap the objective function and explicitly specify a search space. The search space defines the set of input values we want to optimize over, and it is typically a subset or transformation of the domain, though by default the entire domain is taken as the search space. In black box optimization, it is common for the domain, and hence also the search space, to have finite box constraints. Similarly to HPO, transformations can sometimes be used to more efficiently search the space (Section 4.1.5).

In the following, we use a simple random search to optimize the sinusoidal function over the whole domain and inspect the result from the `instance` in the usual way (Section 4.1.4). Analogously to tuners, `Optimizers` in `bbotk` are stored in the `mlr_optimizers` dictionary and can be constructed with `opt()`.   `opt()`

```
instance = OptimInstanceSingleCrit$new(objective,
  search_space = domain,
  terminator = trm("evals", n_evals = 20))
optimizer = opt("random_search", batch_size = 20)
optimizer$optimize(instance)
```

Similarly to how we can use `tune()` to construct a tuning instance, here we can use `bb_optimize()`, which returns a list with elements `"par"` (best found parameters), `"val"` (optimal outcome), and `"instance"` (the optimization instance); the values given as `"par"` and `"val"` are the same as the values found in `instance$result`:

```
optimal = bb_optimize(objective, method = "random_search",
  max_evals = 20)
optimal$instance$result
```

```
      x  x_domain       y
1: 0.7377 <list[1]> -1.158
```

Now we have introduced the basic black box optimization setup, we can introduce the building blocks of any Bayesian optimization algorithm.

5.4.2   Building Blocks of Bayesian Optimization

Bayesian optimization (BO) is a global optimization algorithm that usually follows the following process (Figure 5.3):

1. Generate and evaluate an initial design
2. Loop:
   a. Fit a surrogate model on the archive of all observations made so far to model the unknown black box function.
   b. Optimize an acquisition function to determine, which points of the search space are promising candidate(s) that should be evaluated next.
   c. Evaluate the next candidate(s) and update the archive of all observations made so far.
   d. Check if a given termination criterion is met, if not go back to (a).

The acquisition function relies on the mean and standard deviation prediction of the surrogate model and requires no evaluation of the true black box function, making it comparably cheap to optimize. A good acquisition function will balance *exploiting* knowledge about regions where we observed that performance is good and the surrogate model has low uncertainty, with *exploring* regions where it has not yet evaluated points and as a result the uncertainty of the surrogate model is high.

We refer to these elements as the "building blocks" of BO as it is a highly modular algorithm; as long as the above structure is in place, then the surrogate models, acquisition functions, and acquisition function optimizers are all interchangeable to a certain extent. The design of `mlr3mbo` reflects this modularity, with the base class for `OptimizerMbo` holding all the key elements: the BO algorithm loop structure (`loop_function`), *surrogate* model (`Surrogate`), *acquisition function* (`AcqFunction`), and *acquisition function optimizer* (`AcqOptimizer`). In this section,

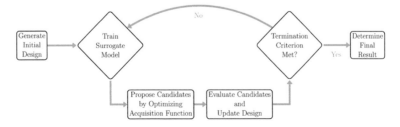

Figure 5.3: Bayesian optimization loop.

we will provide a more detailed explanation of these building blocks and explore their interplay and interaction during optimization.

5.4.2.1   The Initial Design

Before we can fit a surrogate model to model the unknown black box function, we need data. The initial set of points that is evaluated before a surrogate model can be fit is referred to as the initial design.

`mlr3mbo` allows you to either construct the initial design manually or let a `loop_function` construct and evaluate this for you. In this section, we will demonstrate the first method, which requires more user input but therefore allows more control over the initial design.

A simple method to construct an initial design is to use one of the four design generators in `paradox`:

- `generate_design_random()`: Generate points uniformly at random
- `generate_design_grid()`: Generate points in a uniform-sized grid
- `generate_design_lhs()`: Latin hypercube sampling (Stein 1987)
- `generate_design_sobol()`: Sobol sequence (Niederreiter 1988)

Figure 5.4 illustrates the difference in generated designs from these four methods assuming an initial design of size nine and a domain of two numeric variables from 0 to 1. We already covered the difference between grid and random designs in Section 4.1.4. An LHS design divides each input variable into equally sized intervals (indicated by the horizontal and vertical dotted lines in Figure 5.4) and ensures that each interval is represented by exactly one sample point, resulting in uniform marginal distributions. Furthermore, in LHS designs the minimal distance between two points is usually maximized, resulting in its space-filling coverage of the space. The Sobol design works similarly to LHS but can provide better coverage than LHS

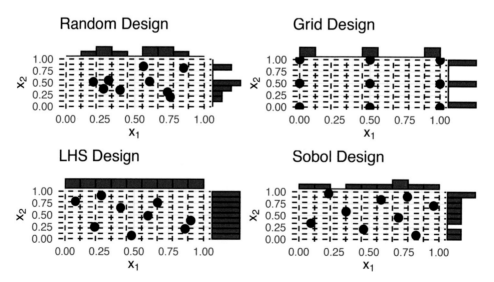

Figure 5.4: Comparing different samplers for constructing an initial design of nine points on a domain of two numeric variables ranging from 0 to 1. Dotted horizontal and vertical lines partition the domain into equally sized bins. Histograms on the top and right visualize the marginal distributions of the generated sample.

when the number of dimensions is large. For this reason, LHS or Sobol designs are usually recommended for BO, but usually the influence of the initial design will be smaller compared to other design choices of BO. A random design might work well-enough, but grid designs are usually discouraged.

Whichever of these methods you choose, the result is a `Design` object, which is mostly just a wrapper around a `data.table`:

```
sample_domain = ps(x1 = p_dbl(0, 1), x2 = p_dbl(0, 1))
generate_design_random(sample_domain, n = 3)$data
```

```
        x1      x2
1: 0.9930 0.3773
2: 0.6782 0.4612
3: 0.4355 0.8019
```

Therefore, you could also specify a completely custom initial design by defining your own `data.table`. Either way, when manually constructing an initial design (as opposed to letting `loop_function` automate this), it needs to be evaluated on the `OptimInstance` before optimizing it. Returning to our running example of minimizing the sinusoidal function, we will evaluate a custom initial design with `$eval_batch()`:

```
instance = OptimInstanceSingleCrit$new(objective,
  terminator = trm("evals", n_evals = 20))
design = data.table(x = c(0.1, 0.34, 0.65, 1))
instance$eval_batch(design)
instance$archive$data
```

```
      x       y  x_domain            timestamp batch_nr
1: 0.10  0.1971 <list[1]> 2023-07-04 15:25:40        1
2: 0.34 -0.6792 <list[1]> 2023-07-04 15:25:40        1
3: 0.65  0.4148 <list[1]> 2023-07-04 15:25:40        1
4: 1.00  1.9812 <list[1]> 2023-07-04 15:25:40        1
```

We can see how each point in our design was evaluated by the sinusoidal function, giving us data we can now use to start the iterative BO algorithm by fitting the surrogate model on that data.

5.4.2.2 Surrogate Model

A surrogate model wraps a regression learner that models the unknown black box function based on observed data. In `mlr3mbo`, the `SurrogateLearner` is a higher-level R6 class inheriting from the base `Surrogate` class, designed to construct and manage the surrogate model, including automatic construction of the `TaskRegr` that the learner should be trained on at each iteration of the BO loop.

Any regression learner in `mlr3` can be used. However, most acquisition functions depend on both mean and standard deviation predictions from the surrogate model, the latter of which requires the `"se"` `predict_type` to be supported. Therefore, not all learners are suitable for all scenarios. Typical choices of regression learners used as surrogate models include Gaussian processes (`lrn("regr.km")`) for low to medium dimensional numeric search spaces and random forests (e.g., `lrn("regr.ranger")`) for higher dimensional mixed (and/or hierarchical) search spaces. A detailed introduction to Gaussian processes can be found in Williams and Rasmussen (2006) and an in-depth focus on Gaussian processes in the context of surrogate models in BO is given in Garnett (2022). In this example, we use a Gaussian process with Matérn 5/2 kernel, which uses `BFGS` as an optimizer to find the optimal kernel parameters and set `trace = FALSE` to prevent too much output during fitting.

```
lrn_gp = lrn("regr.km", covtype = "matern5_2",
   optim.method = "BFGS",control = list(trace = FALSE))
```

A `SurrogateLearner` can be constructed by passing a `LearnerRegr` object to the sugar function `srlrn()`, alongside the `archive` of the instance: srlrn()

```
library(mlr3mbo)
surrogate = srlrn(lrn_gp, archive = instance$archive)
```

Internally, the regression learner is fit on a `TaskRegr` where features are the variables of the domain and the target is the codomain, the data is from the `Archive` of the `OptimInstance` that is to be optimized.

In our running example, we have already initialized our archive with the initial design, so we can update our surrogate model, which essentially fits the Gaussian process, note how we use `$learner` to access the wrapped model:

```
surrogate$update()
surrogate$learner$model
```

```
Call:
DiceKriging::km(design = data, response = task$truth(),
    covtype = "matern5_2", optim.method = "BFGS",
    control = pv$control)

Trend  coeff.:
              Estimate
 (Intercept)    0.7899

Covar. type   : matern5_2
Covar. coeff.:
              Estimate
    theta(x)    0.3014

Variance estimate: 1.07
```

Having introduced the concept of a surrogate model, we can now move on to the acquisition function, which makes use of the surrogate model predictions to decide which candidate to evaluate next.

5.4.2.3 Acquisition Function

Roughly speaking, an acquisition function relies on the prediction of a surrogate model and quantifies the expected "utility" of each point of the search space if it were to be evaluated in the next iteration.

A popular example is the expected improvement (Jones, Schonlau, and Welch 1998), which tells us how much we can expect a candidate point to improve over the best function value observed so far (the performance of the "incumbent"), given the performance prediction of the surrogate model:

$$\alpha_{\mathrm{EI}}(\mathbf{x}) = \mathbb{E}\left[\max\left(f_{\min} - Y(\mathbf{x}), 0\right)\right]$$

Here, $Y(\mathbf{x})$ is the surrogate model prediction (a random variable) for a given point $\mathbf{x}$ (which when using a Gaussian process follows a normal distribution) and $f_{\min}$ is the best function value observed so far (assuming minimization). Calculating the expected improvement requires mean and standard deviation predictions from the model.

acqf()

In mlr3mbo, acquisition functions (of class AcqFunction) are stored in the mlr_acqfunctions dictionary and can be constructed with acqf(), passing the key of the method you want to use and our surrogate learner. In our running example, we will use the expected improvement (acqf("ei")) to choose the next candidate for evaluation. Before we can do that, we have to update ($update()) the AcqFunction's view of the incumbent, to ensure it is still using the best value observed so far.

```
acq_function = acqf("ei", surrogate = surrogate)
acq_function$update()
acq_function$y_best
```

```
[1] -0.6792
```

You can use $eval_dt() to evaluate the acquisition function for the domain given as a data.table. In Figure 5.5, we evaluated the expected improvement on a uniform grid of points between 0 and 1 using the predicted mean and standard deviation from the Gaussian process. We can see that the expected improvement is high in regions where the mean prediction (gray dashed lines) of the Gaussian process is low, or where the uncertainty is high.

```
xydt = generate_design_grid(domain, resolution = 1001)$data
# evaluate our sinusoidal function
xydt[, y := objective$eval_dt(xydt)$y]
# evaluate expected improvement
xydt[, ei :=  acq_function$eval_dt(xydt[, "x"])]
# make predictions from our data
xydt[, c("mean", "se") :=  surrogate$predict(xydt[, "x"])]
xydt[1:3]
```

```
        x        y       ei    mean      se
1: 0.000 0.000000 4.642e-05 0.5191 0.3632
2: 0.001 0.000028 4.171e-05 0.5166 0.3597
3: 0.002 0.000112 3.738e-05 0.5142 0.3562
```

```
ggplot(xydt, mapping = aes(x = x, y = y)) +
  geom_point(size = 2, data = instance$archive$data) +
  geom_line() +
  geom_line(aes(y = mean), colour = "gray", linetype = 2) +
  geom_ribbon(aes(min = mean - se, max = mean + se),
    fill = "gray", alpha = 0.3) +
  geom_line(aes(y = ei * 40), linewidth = 1,
  colour = "darkgray") +
  scale_y_continuous("y",
    sec.axis = sec_axis(~ . * 0.025, name = "EI",
      breaks = c(0, 0.025, 0.05))) +
  theme_minimal()
```

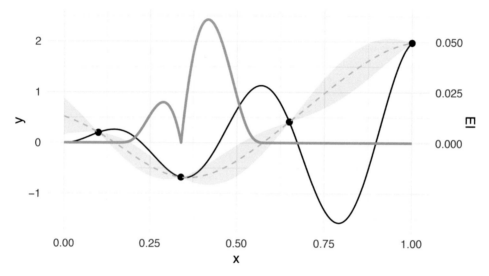

Figure 5.5: Expected improvement (solid dark gray line) based on the mean and uncertainty prediction (dashed gray line) of the Gaussian process surrogate model trained on an initial design of four points (black). Ribbons represent the mean plus minus the standard deviation prediction.

We will now proceed to optimize the acquisition function itself to find the candidate with the largest expected improvement.

5.4.2.4 Acquisition Function Optimizer

An acquisition function optimizer of class `AcqOptimizer` is used to optimize the acquisition function by efficiently searching the space of potential candidates within a limited computational budget.

`AcqOptimizer`

Due to the non-convex nature of most commonly used acquisition functions (Garnett 2022) it is typical to employ global optimization techniques for acquisition function optimization. Widely used approaches for optimizing acquisition functions include derivative-free global optimization methods like branch and bound algorithms, such as the DIRECT algorithm (Jones, Perttunen, and Stuckman 1993), as well as multi-start local optimization methods, such as running the L-BFGS-B algorithm (Byrd et al. 1995) or a local search multiple times from various starting points (Kim and Choi 2021). Consequently the optimization problem of the acquisition function can be handled as a black box optimization problem itself, but a much cheaper one than the original.

acqo()   AcqOptimizer objects are constructed with acqo(), which takes as input a Optimizer, a Terminator, and the acquisition function. Optimizers are stored in the mlr_optimizers dictionary and can be constructed with the sugar function opt()   opt(). The terminators are the same as those introduced in Section 4.1.2.

Below we use the DIRECT algorithm and we terminate the acquisition function optimization if there is no improvement of at least 1e-5 for 100 iterations. The $optimize() method optimizes the acquisition function and returns the next candidate.

```
acq_optimizer = acqo(
  optimizer = opt("nloptr", algorithm = "NLOPT_GN_ORIG_DIRECT"),
  terminator = trm("stagnation", iters = 100, threshold = 1e-5),
  acq_function = acq_function
)

candidate = acq_optimizer$optimize()
candidate
```

```
       x  x_domain  acq_ei .already_evaluated
1: 0.4173 <list[1]> 0.06074             FALSE
```

We have now shown how to run a single iteration of the BO algorithm loop manually. In practice, one would use OptimizerMbo to put all these pieces together to automate the process. Before demonstrating this class, we will first take a step back and introduce the loop_function which tells the algorithm how it should be run.

5.4.2.5   Using and Building Loop Functions

The loop_function determines the behavior of the BO algorithm on a global level, i.e., how to define the subroutine that is performed at each iteration to generate new candidates for evaluation. Loop functions are relatively simple functions that take as input the classes that we have just discussed and define the BO loop. Loop functions are stored in the mlr_loop_functions dictionary. As these are S3 (not R6) classes, they can be simply loaded by just referencing the key (i.e., there is no constructor required).

```
as.data.table(mlr_loop_functions)[, .(key, label, instance)]
```

```
                key                          label     instance
1:     bayesopt_ego Efficient Global Optimization single-crit
2:     bayesopt_emo           Multi-Objective EGO  multi-crit
3:    bayesopt_mpcl      Multipoint Constant Liar single-crit
4: bayesopt_parego                        ParEGO  multi-crit
5:  bayesopt_smsego                       SMS-EGO  multi-crit
```

You could pick and use one of the loop functions included in the dictionary above, or you can write your own for finer control over the BO process. A common choice of loop function is the Efficient Global Optimization (EGO) algorithm (Jones, Schonlau, and Welch 1998) (bayesopt_ego()). A simplified version of this code is shown at the end of this section, both to help demonstrate the EGO algorithm, and to give an example of how to write a custom BO variant yourself. In short, the code sets up the relevant components discussed above and then loops the steps above: 1) update the surrogate model, 2) update the acquisition function, 3) optimize the acquisition function to yield a new candidate, and 4) evaluate the candidate and add it to the archive. If there is an error during the loop then a fallback is used where the next candidate is proposed uniformly at random, ensuring that the process continues even in the presence of potential issues, we will return to this in Section 5.4.6.

```
my_simple_ego = function(
    instance,
    surrogate,
    acq_function,
    acq_optimizer,
    init_design_size
) {

  # setting up the building blocks
  surrogate$archive = instance$archive # archive
  acq_function$surrogate = surrogate # surrogate model
  acq_optimizer$acq_function = acq_function # acquisition function

  search_space = instance$search_space

  # initial design
  design = generate_design_sobol(
    search_space, n = init_design_size)$data
  instance$eval_batch(design)

  # MBO loop
  repeat {
    candidate = tryCatch({
      # update the surrogate model
      acq_function$surrogate$update()
      # update the acquisition function
      acq_function$update()
      # optimize the acquisition function to yield a new candidate
      acq_optimizer$optimize()
    }, mbo_error = function(mbo_error_condition) {
      generate_design_random(search_space, n = 1L)$data
    })

    # evaluate the candidate and add it to the archive
    tryCatch({
      instance$eval_batch(candidate)
    }, terminated_error = function(cond) {
      # $eval_batch() throws a terminated_error if the instance is
      # already terminated, e.g. because of timeout.
```

```
      })
      if (instance$is_terminated) break
    }

    return(instance)
  }
```

We are now ready to put everything together to automate the BO process.

5.4.3   Automating BO with OptimizerMbo

OptimizerMbo can be used to assemble the building blocks described above into a single object that can then be used for optimization. We use the `bayesopt_ego` loop function provided by `mlr_loop_functions`, which works similarly to the code shown above but takes more care to offer sensible default values for its arguments and handle edge cases correctly. You do not need to pass any of these building `opt()` blocks to each other manually as the `opt()` constructor will do this for you:

```
bayesopt_ego = mlr_loop_functions$get("bayesopt_ego")
surrogate = srlrn(lrn("regr.km", covtype = "matern5_2",
  optim.method = "BFGS", control = list(trace = FALSE)))
acq_function = acqf("ei")
acq_optimizer = acqo(opt("nloptr",
  algorithm = "NLOPT_GN_ORIG_DIRECT"),
  terminator = trm("stagnation", iters = 100, threshold = 1e-5))

optimizer = opt("mbo",
  loop_function = bayesopt_ego,
  surrogate = surrogate,
  acq_function = acq_function,
  acq_optimizer = acq_optimizer)
```

> 💡 Loop Function Arguments
>
> Additional arguments for customizing certain loop functions can be passed through with the `args` parameter of `opt()`.

In this example, we will use the same initial design that we created before and will optimize our sinusoidal function using `$optimize()`:

```
instance = OptimInstanceSingleCrit$new(objective,
  terminator = trm("evals", n_evals = 20))
design = data.table(x = c(0.1, 0.34, 0.65, 1))
instance$eval_batch(design)
optimizer$optimize(instance)
```

```
        x  x_domain       y
1: 0.7922 <list[1]> -1.577
```

Using only a few evaluations, BO comes close to the true global optimum (0.792). Figure 5.6 shows the sampling trajectory of candidates as the algorithm progressed, we can see that focus is increasingly given to more regions around the global optimum. However, even in later optimization stages, the algorithm still explores new areas, illustrating that the expected improvement acquisition function indeed balances exploration and exploitation as we required.

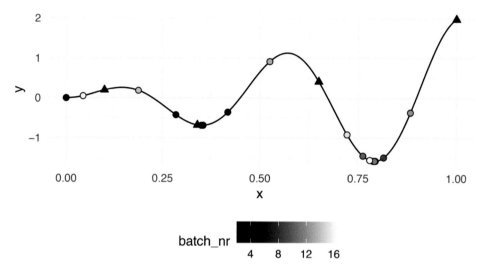

Figure 5.6: Sampling trajectory of the BO algorithm. Points of the initial design in black triangles. Sampled points are in dots with color progressing from black to white as the algorithm progresses.

If we replicate running our BO algorithm ten times (with random initial designs and varying random seeds) and compare this to a random search, we can see that BO indeed performs much better and on average reaches the global optimum after around 15 function evaluations (Figure 5.7). As expected, the performance for the initial design size is close to the performance of the random search.

5.4.4   Bayesian Optimization for HPO

`mlr3mbo` can be used for HPO by making use of `TunerMbo`, which is a wrapper   `TunerMbo`
around `OptimizerMbo` and works in the exact same way. As an example, below we will tune the `cost` and `gamma` parameters of `lrn("classif.svm")` with a radial kernel on `tsk("sonar")` with three-fold CV. We set up `tnr("mbo")` using the same objects constructed above and then run our tuning experiment as usual:

```
tuner = tnr("mbo",
  loop_function = bayesopt_ego,
  surrogate = surrogate,
  acq_function = acq_function,
  acq_optimizer = acq_optimizer)

lrn_svm = lrn("classif.svm", kernel = "radial",
  type = "C-classification",
  cost  = to_tune(1e-5, 1e5, logscale = TRUE),
```

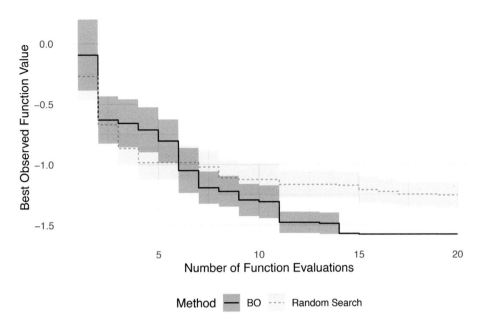

Figure 5.7: Anytime performance of BO and random search on the 1D sinusoidal function given a budget of 20 function evaluations. Solid line depicts the best observed target value averaged over 10 replications. Ribbons represent standard errors.

```
  gamma = to_tune(1e-5, 1e5, logscale = TRUE)
)

instance = tune(tuner, tsk("sonar"), lrn_svm,
  rsmp("cv", folds = 3),msr("classif.ce"), 25)

instance$result
```

```
   cost  gamma learner_param_vals  x_domain classif.ce
1: 11.51 -4.075            <list[4]> <list[2]>     0.1489
```

Multi-objective tuning is also possible with BO with algorithms using many different design choices; for example, whether they use a scalarization approach of objectives and only rely on a single surrogate model, or fit a surrogate model for each objective. More details on multi-objective BO can, for example, be found in Horn et al. (2015) or Morales-Hernández, Van Nieuwenhuyse, and Rojas Gonzalez (2022).

Below we will illustrate multi-objective tuning using the ParEGO (Knowles 2006) loop function. ParEGO (`bayesopt_parego()`) tackles multi-objective BO via a randomized scalarization approach and models a single scalarized objective function via a single surrogate model and then proceeds to find the next candidate for evaluation making use of a standard single-objective acquisition function such as the expected improvement. Other compatible loop functions can be found by looking at the `"instance"` column of `mlr_loop_functions`. We will tune three parameters of a decision tree with respect to the true positive (maximize) and false positive (minimize) rates, the Pareto front is visualized in Figure 5.8.

```
tuner = tnr("mbo",
  loop_function = bayesopt_parego,
  surrogate = surrogate,
  acq_function = acq_function,
  acq_optimizer = acq_optimizer)

lrn_rpart = lrn("classif.rpart",
  cp = to_tune(1e-04, 1e-1),
  minsplit = to_tune(2, 64),
  maxdepth = to_tune(1, 30)
)

instance = tune(tuner, tsk("sonar"), lrn_svm,
  rsmp("cv", folds = 3),
  msrs(c("classif.tpr", "classif.fpr")), 25)
```

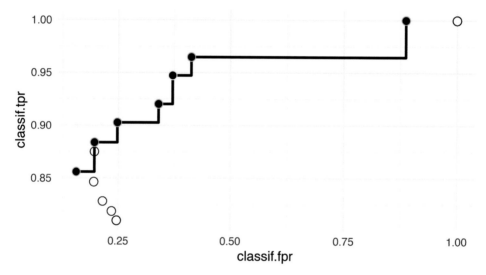

Figure 5.8: Pareto front of TPR and FPR obtained via ParEGO. White dots represent tested configurations, each black dot individually represents a Pareto-optimal configuration and all black dots together represent the Pareto front.

5.4.5   Noisy Bayesian Optimization

So far, we implicitly assumed that the black box function we are trying to optimize is deterministic, i.e., repeatedly evaluating the same point will always return the same objective function value. However, real-world black box functions are often noisy, which means that repeatedly evaluating the same point will return different objective function values due to background noise on top of the black box function. For example, if you were modeling a machine in a factory to estimate the rate of production, even if all parameters of the machine were controlled, we would still expect different performance at different times due to uncontrollable background factors such as environmental conditions.

In `bbotk`, you can mark an `Objective` object as noisy by passing the `"noisy"` tag to the `properties` parameter, which allows us to use methods that can treat such objectives differently.

```
sinus_1D_noisy = function(xs) {
  y = 2 * xs$x * sin(14 * xs$x) + rnorm(1, mean = 0, sd = 0.1)
  y
}
domain = ps(x = p_dbl(lower = 0, upper = 1))
codomain = ps(y = p_dbl(tags = "minimize"))
objective_noisy = ObjectiveRFun$new(sinus_1D_noisy,
  domain = domain, codomain = codomain, properties = "noisy")
```

Noisy objectives can be treated in different ways:

1. A surrogate model can be used to incorporate the noise.
2. An acquisition function can be used that respects noisiness.
3. The final best point(s) after optimization (i.e., the `$result` field of the instance) can be chosen in a way to reflect noisiness.

In the first case, instead of using an interpolating Gaussian process, we could instead use Gaussian process regression that estimates the measurement error by setting `nugget.estim = TRUE`:

```
srlrn(lrn("regr.km", nugget.estim = TRUE))
```

This will result in the Gaussian process not perfectly interpolating training data and the standard deviation prediction associated with the training data will be non-zero, reflecting the uncertainty in the observed function values due to the measurement error. A more in-depth discussion of noise-free vs. noisy observations in the context of Gaussian processes can be found in Chapter 2 of Williams and Rasmussen (2006).

For the second option, one example of an acquisition function that respects noisiness is the augmented expected improvement (D. Huang et al. 2012) (`acqf("aei")`), which essentially rescales the expected improvement, taking measurement error into account.

Result
Assigner
Finally, `mlr3mbo` allows for explicitly specifying how the final result after optimization is assigned to the instance (i.e., what will be saved in `instance$result`) with a result assigner, which can be specified during the construction of an `OptimizerMbo` or `TunerMbo`. `ResultAssignerSurrogate` uses a surrogate model to predict the mean of all evaluated points and proceeds to choose the point with the best mean prediction as the final optimization result. In contrast, the default method, `ResultAssignerArchive`, just picks the best point according to the evaluations logged in `archive`. Result assigners are stored in the `mlr_result_assigners` dictionary and can be constructed with `ras()`.

```
opt("mbo",
  loop_function = bayesopt_ego,
  surrogate = surrogate,
```

```
    acq_function = acq_function,
    acq_optimizer = acq_optimizer,
    result_assigner = ras("surrogate")
)
```

5.4.6   Practical Considerations in Bayesian Optimization

`mlr3mbo` tries to use reasonable defaults regarding the choice of surrogate model, acquisition function, acquisition function optimizer and even the loop function. For example, in the case of a purely numeric search space, `mlr3mbo` will by default use a Gaussian process as the surrogate model and a random forest as the fallback learner and additionally encapsulates the learner (Section 5.1.1). In the case of a mixed or hierarchical search space, `mlr3mbo` will use a random forest as the surrogate model. Therefore, users can perform BO without specifying any deviation from the defaults and still expect decent optimization performance. To see an up-to-date overview of these defaults, take a look at the help page of `mbo_defaults`. We will finish this section with some practical considerations to think about when using BO.

Error Handling

In the context of BO, there is plenty of room for potential failure of building blocks which could break the whole process. For example, if two points in the training data are too close to each other, fitting the Gaussian process surrogate model can fail.

`mlr3mbo` has several built-in safety nets to catch errors. Surrogate includes the `catch_errors` configuration control parameter, which, if set to `TRUE`, catches all errors that occur during training or updating of the surrogate model. AcqOptimizer also has the `catch_errors` configuration control parameter, which can be used to catch all errors that occur during the acquisition function optimization, either due to the surrogate model failing to predict or the acquisition function optimizer erroring. If errors are caught in any of these steps, the standard behavior of any `loop_function` is to trigger a fallback, which proposes the next candidate uniformly at random. Note, when setting `catch_errors` = `TRUE` for the `AcqOptimizer`, it is usually not necessary to also explicitly set `catch_errors` = `TRUE` for the `Surrogate`, though this may be useful when debugging.

In the worst-case scenario, if all iterations errored, the BO algorithm will simply perform a random search. Ideally, fallback learners (Section 5.1.1) should also be used, which will be employed before proposing the next candidate randomly. The value of the acquisition function is also always logged in the archive of the optimization instance so inspecting this is a good idea to ensure the algorithm behaved as expected.

Surrogate Models

In practice, users may prefer a more robust BO variant over a potentially better-performing but unstable variant. Even if the `catch_errors` parameters are turned on and are never triggered, that does not guarantee that the BO algorithm ran as intended. For instance, Gaussian processes are sensitive to the choice of kernel and kernel parameters, typically estimated through maximum likelihood estimation, suboptimal parameter values can result in white noise models with a constant mean and standard deviation prediction. In this case, the surrogate model will not provide useful mean and standard deviation predictions resulting in poor overall performance of the BO algorithm. Another practical consideration regarding the choice of surrogate model can be overhead. Fitting a vanilla Gaussian process scales cubically in the number of data points and therefore the overhead of the BO algorithm grows with the number of iterations. Furthermore, vanilla Gaussian processes natively cannot handle categorical input variables or dependencies in the search space (recall that in HPO we often deal with mixed hierarchical spaces). In contrast, a random forest – popularly used as a surrogate model in *SMAC* (Lindauer et al. 2022) – is cheap to train, quite robust in the sense that it is not as sensitive to its hyperparameters as a Gaussian process, and can easily handle mixed hierarchical spaces. On the downside, a random forest is not really Bayesian (i.e., there is no posterior predictive distribution) and suffers from poor uncertainty estimates and poor extrapolation.

Warmstarting

Warmstarting is a technique in optimization where previous optimization runs are used to improve the convergence rate and final solution of a new, related optimization run. In BO, warmstarting can be achieved by providing a set of likely well-performing configurations as part of the initial design (see, e.g., Feurer, Springenberg, and Hutter 2015). This approach can be particularly advantageous because it allows the surrogate model to start with prior knowledge of the optimization landscape in relevant regions. In `mlr3mbo`, warmstarting is straightforward by specifying a custom initial design. Furthermore, a convenient feature of `mlr3mbo` is the ability to continue optimization in an online fashion even after an optimization run has been terminated. Both `OptimizerMbo` and `TunerMbo` support this feature, allowing optimization to resume on a given instance even if the optimization was previously interrupted or terminated.

Termination

Common termination criteria include stopping after a fixed number of evaluations, once a given walltime budget has been reached, when performance reaches a certain level, or when performance improvement stagnates. In the context of BO, it can also be sensible to stop the optimization if the best acquisition function value falls below a certain threshold. For instance, terminating the optimization if the expected improvement of the next candidate(s) is negligible can be a reasonable approach. At the time of publishing, terminators based on acquisition functions have not been implemented but this feature will be coming soon.

Parallelization

The standard behavior of most BO algorithms is to sequentially propose a single candidate that should be evaluated next. Users may want to use parallelization to compute candidates more efficiently. If you are using BO for HPO, then the most efficient method is to parallelize the nested resampling, see Section 10.1.4. Alternatively, if the loop function supports candidates being proposed in batches (e.g., `bayesopt_parego()`) then the `q` argument to the loop function can be set to propose `q` candidates in each iteration that can be evaluated in parallel if the `Objective` is properly implemented.

5.5 Conclusion

In this chapter, we looked at advanced tuning methods. We started by thinking about the types of errors that can occur during tuning and how to handle these to ensure your HPO process does not crash. We presented multi-objective tuning, which can be used to optimize performance measures simultaneously. We then looked at multi-fidelity tuning, in which the Hyberband tuner can be used to efficiently tune algorithms by making use of lower-fidelity evaluations to approximate full-fidelity model performance. We will return to Hyperband in Section 8.4.4 where we will learn how to make use of pipelines in order to tune any algorithm with Hyperband. Finally, we took a deep dive into Bayesian optimization to look at how bbotk, mlr3mbo, and mlr3tuning can be used together to implement complex BO tuning algorithms in mlr3, allowing for highly flexible and sample-efficient algorithms. In the next chapter we will look at feature selection and see how mlr3filters and mlr3fselect use a very similar design interface to mlr3tuning.

Table 5.3: Important classes and functions covered in this chapter with underlying class (if applicable), class constructor or function, and important class methods (if applicable).

Class	Constructor/Function	Fields/Methods
Learner	lrn	$encapsulate; $fallback
TuningInstanceMultiCrit	ti()/tune()	$result; $archive
TunerHyperband	tnr("hyperband")	-
Objective	-	
OptimInstanceSingleCrit or OptimInstanceMultiCrit	bb_optimize()	$result; $archive
SurrogateLearner	srlrn()	
AcqFunction	acqf()	
AcqOptimizer	acqo()	
-	loop_function	-
OptimizerMbo	bbotk::opt("mbo")	
TunerMbo	tnr("mbo")	
Design	generate_design_random; generate_design_grid; generate_design_lhs; generate_design_sobol;	$data

5.6 Exercises

1. Tune the mtry, sample.fraction, and num.trees hyperparameters of lrn("regr.ranger") on tsk("mtcars") and evaluate this with a three-fold CV and the root mean squared error (same as in Chapter 4, Exercise

1). Use `tnr("mbo")` with 50 evaluations. Compare this with the performance progress of a random search run from Chapter 4, Exercise 1. Plot the progress of performance over iterations and visualize the spatial distribution of the evaluated hyperparameter configurations for both algorithms.

2. Minimize the 2D Rastrigin function $f : [-5.12, 5.12] \times [-5.12, 5.12] \to \mathbb{R}$, $\mathbf{x} \mapsto 10D + \sum_{i=1}^{D} \left[x_i^2 - 10\cos(2\pi x_i) \right]$, $D = 2$ via BO (standard sequential single-objective BO via `bayesopt_ego()`) using the lower confidence bound with `lambda = 1` as acquisition function and `"NLOPT_GN_ORIG_DIRECT"` via `opt("nloptr")` as acquisition function optimizer. Use a budget of 40 function evaluations. Run this with both the "default" Gaussian process surrogate model with Matérn 5/2 kernel, and the "default" random forest surrogate model. Compare their anytime performance (similarly as in Figure 5.7). You can construct the surrogate models with default settings using:

```
surrogate_gp = srlrn(default_gp())
surrogate_rf = srlrn(default_rf())
```

3. Minimize the following function: $f : [-10, 10] \to \mathbb{R}^2, x \mapsto \left(x^2, (x-2)^2 \right)$ with respect to both objectives. Use the ParEGO algorithm. Construct the objective function using the `ObjectiveRFunMany` class. Terminate the optimization after a runtime of 100 evals. Plot the resulting Pareto front and compare it to the analytical solution, $y_2 = \left(\sqrt{y_1} - 2 \right)^2$ with y_1 ranging from 0 to 4.

6

Feature Selection

Marvin N. Wright
Leibniz Institute for Prevention Research and Epidemiology – BIPS, and University of Bremen, and University of Copenhagen

Feature selection, also known as variable or descriptor selection, is the process of finding a subset of features to use with a given task and learner. Using an *optimal set* of features can have several benefits:

- improved predictive performance, since we reduce overfitting on irrelevant features,
- robust models that do not rely on noisy features,
- simpler models that are easier to interpret,
- faster model fitting, e.g. for model updates,
- faster prediction, and
- no need to collect potentially expensive features.

However, these objectives will not necessarily be optimized by the same set of features and thus feature selection can be seen as a multi-objective optimization problem. In this chapter, we mostly focus on feature selection as a means of improving predictive performance, but also briefly cover the optimization of multiple criteria (Section 6.2.5).

Reducing the number of features can improve models across many scenarios, but it can be especially helpful in datasets that have a high number of features in comparison to the number of data points. Many learners perform implicit, also called embedded, feature selection, e.g., via the choice of variables used for splitting in a decision tree. Most other feature selection methods are model agnostic, i.e., they can be used together with any learner. Of the many different approaches to identifying relevant features, we will focus on two general concepts, which are described in detail below: Filter and Wrapper methods (Guyon and Elisseeff 2003; Chandrashekar and Sahin 2014).

6.1 Filters

Filter methods are preprocessing steps that can be applied before training a model. A very simple filter approach could look like this:

1. calculate the correlation coefficient ρ between each feature and a numeric target variable, and
2. select all features with $\rho > 0.2$ for further modeling steps.

DOI: 10.1201/9781003402848-6

This approach is a *univariate* filter because it only considers the univariate relationship between each feature and the target variable. Further, it can only be applied to regression tasks with continuous features and the threshold of $\rho > 0.2$ is quite arbitrary. Thus, more advanced filter methods, e.g., *multivariate* filters based on feature importance, usually perform better (Bommert et al. 2020). On the other hand, a benefit of univariate filters is that they are usually computationally cheaper than more complex filter or wrapper methods. In the following, we describe how to calculate univariate, multivariate, and feature importance filters, how to access implicitly selected features, how to integrate filters in a machine learning pipeline, and how to optimize filter thresholds.

Filter algorithms select features by assigning numeric scores to each feature, e.g., correlation between features and target variable, use these to rank the features and select a feature subset based on the ranking. Features that are assigned lower scores are then omitted in subsequent modeling steps. All filters are implemented via the package `mlr3filters`. Below, we cover how to

- instantiate a `Filter` object,
- calculate scores for a given task, and
- use calculated scores to select or drop features.

Special cases of filters are feature importance filters (Section 6.1.2) and embedded methods (Section 6.1.3). Feature importance filters select features that are important according to the model induced by a selected `Learner`. They rely on the learner to extract information on feature importance from a trained model, for example, by inspecting a learned decision tree and returning the features that are used as split variables, or by computing model-agnostic feature importance (Chapter 12) values for each feature. Embedded methods use the feature selection that is implicitly performed by some learners and directly retrieve the internally selected features from the learner.

> 💡 Independent Learners and Filters
>
> The learner used in a feature importance or embedded filter is independent of learners used in subsequent modeling steps. For example, one might use feature importance of a random forest for feature selection and train a neural network on the reduced feature set.

Many filter methods are implemented in `mlr3filters`, including:

- Correlation, calculating Pearson or Spearman correlation between numeric features and numeric targets (`flt("correlation")`).
- Information gain, i.e., mutual information of the feature and the target or the reduction of uncertainty of the target due to a feature (`flt("information_gain")`).
- Minimal joint mutual information maximization (`flt("jmim")`).
- Permutation score, which calculates permutation feature importance (see Chapter 12) with a given learner for each feature (`flt("permutation")`).
- Area under the ROC curve calculated for each feature separately (`flt("auc")`).

Most of the filter methods have some limitations, for example, the correlation filter can only be calculated for regression tasks with numeric features. For a full list of all implemented filter methods, we refer the reader to https://mlr3filters.mlr-org.com, which also shows the supported task and features types. A benchmark of filter

methods was performed by Bommert et al. (2020), who recommend not to rely
on a single filter method but to try several ones if the available computational
resources allow. If only a single filter method is to be used, the authors recommend
to use a feature importance filter using random forest permutation importance (see
Section 6.1.2), similar to the permutation method described above, but also the
JMIM and AUC filters performed well in their comparison.

6.1.1 Calculating Filter Values

The first step is to create a new R object using the class of the desired filter method.
These are accessible from the `mlr_filters` dictionary with the sugar function
`flt()`. Each object of class `Filter` has a `$calculate()` method, which computes   `flt()`
the filter values and ranks them in a descending order. For example, we can use the   `$calculate(`
information gain filter described above:

```r
library(mlr3filters)
flt_gain = flt("information_gain")
```

Such a `Filter` object can now be used to calculate the filter on `tsk("penguins")`
and get the results:

```r
tsk_pen = tsk("penguins")
flt_gain$calculate(tsk_pen)

as.data.table(flt_gain)
```

```
          feature    score
1: flipper_length 0.581168
2:    bill_length 0.544897
3:     bill_depth 0.538719
4:         island 0.520157
5:      body_mass 0.442880
6:            sex 0.007244
7:           year 0.000000
```

This shows that the flipper and bill measurements are the most informative features
for predicting the species of a penguin in this dataset, whereas sex and year are
the least informative. Some filters have hyperparameters that can be changed in
the same way as `Learner` hyperparameters. For example, to calculate `"spearman"`
instead of `"pearson"` correlation with the correlation filter:

```r
flt_cor = flt("correlation", method = "spearman")
flt_cor$param_set
```

```
<ParamSet>
       id    class lower upper nlevels    default    value
1:    use ParamFct    NA    NA       5 everything
2: method ParamFct    NA    NA       3    pearson spearman
```

6.1.2   Feature Importance Filters

To use feature importance filters, we can use a learner with an `$importance()` method that reports feature importance. All learners with the property "importance" have this functionality. A list of all learners with this property can be found with

```
as.data.table(mlr_learners)[
   sapply(properties, function(x) "importance" %in% x)]
```

For some learners, the desired filter method needs to be set as a hyperparameter. For example, `lrn("classif.ranger")` comes with multiple integrated methods, which can be selected during construction: To use the feature importance method `"impurity"`, select it during learner construction:

```
lrn("classif.ranger")$param_set$levels$importance
```

```
[1] "none"                    "impurity"               "impurity_corrected"
[4] "permutation"
```

```
lrn_ranger = lrn("classif.ranger", importance = "impurity")
```

We first have to remove missing data because the learner cannot handle missing data, i.e., it does not have the property "missing". Note we use the `$filter()` method presented in Section 2.1.3 here to remove rows; the "filter" name is unrelated to feature filtering, however.

```
tsk_pen = tsk("penguins")
tsk_pen$filter(tsk_pen$row_ids[complete.cases(tsk_pen$data())])
```

Now we can use `flt("importance")` to calculate importance values:

```
flt_importance = flt("importance", learner = lrn_ranger)
flt_importance$calculate(tsk_pen)
as.data.table(flt_importance)
```

```
          feature  score
1:      bill_length 76.164
2: flipper_length 50.032
3:      bill_depth 35.531
4:           island 24.880
5:        body_mass 22.422
6:              sex  1.419
7:             year  1.046
```

6.1.3   Embedded Methods

Many learners internally select a subset of the features which they find helpful for prediction, but ignore other features. For example, a decision tree might never select some features for splitting. These subsets can be used for feature selection, which we call embedded methods because the feature selection is embedded in the learner.

The selected features (and those not selected) can be queried if the learner has the
`"selected_features"` property. As above, we can find those learners with

```
as.data.table(mlr_learners)[
  sapply(properties, function(x) "selected_features" %in% x)]
```

For example, we can use `lrn("classif.rpart")`:

```
tsk_pen = tsk("penguins")
lrn_rpart = lrn("classif.rpart")
lrn_rpart$train(tsk_pen)
lrn_rpart$selected_features()
```

```
[1] "flipper_length" "bill_length"    "island"
```

The features selected by the model can be extracted by a `Filter` object, where
`$calculate()` corresponds to training the learner on the given task:

```
flt_selected = flt("selected_features", learner = lrn_rpart)
flt_selected$calculate(tsk_pen)
as.data.table(flt_selected)
```

```
         feature score
1:        island     1
2: flipper_length    1
3:   bill_length     1
4:   bill_depth      0
5:          sex      0
6:         year      0
7:    body_mass      0
```

Contrary to other filter methods, embedded methods just return values of 1 (selected
features) and 0 (dropped feature).

6.1.4   Filter-Based Feature Selection

After calculating a score for each feature, one has to select the features to be kept
or those to be dropped from further modeling steps. For the `"selected_features"`
filter described in embedded methods (Section 6.1.3), this step is straight-forward
since the methods assign either a value of 1 for a feature to be kept or 0 for a
feature to be dropped. Below, we find the names of features with a value of 1
and select those features with `task$select()`. At first glance it may appear a
bit convoluted to have a filter assign scores based on the feature names returned
by `$selected_features()`, only to turn these scores back into the names of the
features to be kept. However, this approach allows us to use the same interface for
all filter methods, which is especially useful when we want to automate the feature
selection process in pipelines, as we will see in Section 8.4.5.

```
flt_selected$calculate(tsk_pen)

# select all features used by rpart
keep = names(which(flt_selected$scores == 1))
tsk_pen$select(keep)
tsk_pen$feature_names
```

```
[1] "bill_length"   "flipper_length" "island"
```

For filter methods that assign continuous scores, there are essentially two ways to select features:

- Select the top k features; or
- Select all features with a score above a threshold τ.

The first option is equivalent to dropping the bottom $p-k$ features. For both options, one has to decide on a threshold, which is often quite arbitrary. For example, to implement the first option with the information gain filter:

```
tsk_pen = tsk("penguins")
flt_gain = flt("information_gain")
flt_gain$calculate(tsk_pen)

# select top three features from information gain filter
keep = names(head(flt_gain$scores, 3))
tsk_pen$select(keep)
tsk_pen$feature_names
```

```
[1] "bill_depth"   "bill_length"   "flipper_length"
```

Or, the second option with $\tau = 0.5$:

```
tsk_pen = tsk("penguins")
flt_gain = flt("information_gain")
flt_gain$calculate(tsk_pen)

# select all features with score > 0.5 from information gain
# filter
keep = names(which(flt_gain$scores > 0.5))
tsk_pen$select(keep)
tsk_pen$feature_names
```

```
[1] "bill_depth"   "bill_length"   "flipper_length" "island"
```

In Section 8.4.5, we will return to filter-based feature selection and how we can use pipelines and tuning to automate and optimize the feature selection process.

6.2   Wrapper Methods

Wrapper methods work by fitting models on selected feature subsets and evaluating their performance (Kohavi and John 1997). This can be done in a sequential fashion, e.g., by iteratively adding features to the model in sequential forward selection, or in a parallel fashion, e.g., by evaluating random feature subsets in a random search. Below, we describe these simple approaches in a common framework along with more advanced methods such as genetic search. We further show how to select features by optimizing multiple performance measures and how to wrap a learner with feature selection to use it in pipelines or benchmarks.

In more detail, wrapper methods iteratively evaluate subsets of features by resampling a learner restricted to this feature subset and with a chosen performance metric (with holdout or a more expensive CV), and using the resulting performance to guide the search. The specific search strategy iteration is defined by a `FSelector` object. A simple example is the sequential forward selection that starts with computing each single-feature model, selects the best one, and then iteratively always adds the feature that leads to the largest performance improvement (Figure 6.1).

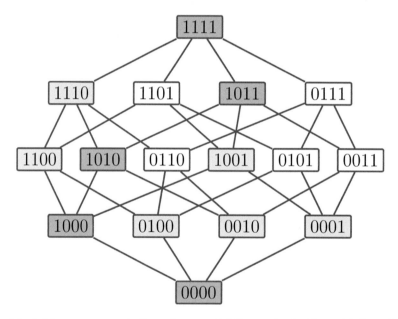

Figure 6.1: A binary representation of sequential forward selection with four features. Gray indicates feature sets that were evaluated, with dark gray indicating the best feature set in each iteration; white indicates feature sets that were not evaluated. We start at the bottom with no selected features (all are "0"). In the next iteration all features are separately tested (each is "1" separately) and the best option (darkest in row two) is selected. This continues for selecting the second, third, and fourth features.

Wrapper methods can be used with any learner, but need to train or even resample the learner potentially many times, leading to a computationally intensive method. All wrapper methods are implemented via the package `mlr3fselect`.

> 💡 Feature Selection and HPO
>
> The wrapper-based feature selection explained above is very similar to the black box optimization approach in HPO (Chapter 4), see also Figure 4.1. The major difference is that we search for well-performing feature subsets instead of hyperparameter configurations. This similarity is not only true in terms of underlying concepts and structure, but also with respect to `mlr3` classes and API. The API is in many places nearly identical, we can use the same terminators, results are logged into an archive in a similar fashion to tuning, and we can also optimize multiple performance measures to create Pareto-optimal solutions in a similar way.

6.2.1 Simple Forward Selection Example

We start with the simple example from above and do sequential forward selection with `tsk("penguins")`, similarly to how the sugar function `tune()` shown in Section 4.2 works, we can use `fselect()` to directly start the optimization and select features.

`fselect()`

```
library(mlr3fselect)

# subset features to ease visualization
tsk_pen = tsk("penguins")
tsk_pen$select(c("bill_depth", "bill_length", "body_mass",
  "flipper_length"))

instance = fselect(
  fselector = fs("sequential"),
  task =  tsk_pen,
  learner = lrn_rpart,
  resampling = rsmp("cv", folds = 3),
  measure = msr("classif.acc")
)
```

To show all analyzed feature subsets and the corresponding performance, we use `as.data.table(instance$archive)`. In this example, the `batch_nr` column represents the iteration of the sequential forward selection and we start by looking at the first iteration.

```
dt = as.data.table(instance$archive)
dt[batch_nr == 1, 1:5]
```

	bill_depth	bill_length	body_mass	flipper_length	classif.acc
1:	TRUE	FALSE	FALSE	FALSE	0.7557
2:	FALSE	TRUE	FALSE	FALSE	0.7353
3:	FALSE	FALSE	TRUE	FALSE	0.7064
4:	FALSE	FALSE	FALSE	TRUE	0.7936

We see that the feature `flipper_length` achieved the highest prediction performance in the first iteration and is thus selected. We plot the performance over the iterations:

```
autoplot(instance, type = "performance")
```

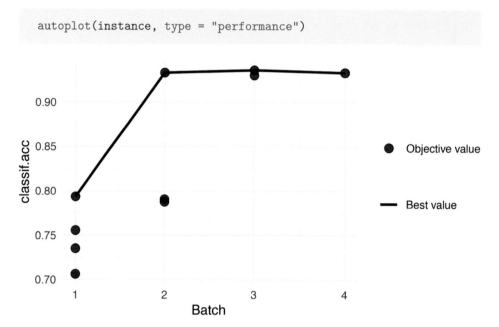

Figure 6.2: Model performance in iterations of sequential forward selection.

In the plot, we can see that adding a second feature further improves the performance to over 90%. To see which feature was added, we can go back to the archive and look at the second iteration:

```
dt[batch_nr == 2, 1:5]
```

```
   bill_depth bill_length body_mass flipper_length classif.acc
1:       TRUE       FALSE     FALSE           TRUE      0.7907
2:      FALSE        TRUE     FALSE           TRUE      0.9331
3:      FALSE       FALSE      TRUE           TRUE      0.7878
```

The improvement in batch three is small so we may even prefer to select a marginally worse model with two features to reduce data size.

To directly show the best feature set, we can use `$result_feature_set` which returns the features in alphabetical order (not order selected):

```
instance$result_feature_set
```

```
[1] "bill_depth"    "bill_length"    "flipper_length"
```

At the heart of `mlr3fselect` are the R6 classes:

- `FSelectInstanceSingleCrit`, `FSelectInstanceMultiCrit`: These two classes describe the feature selection problem and store the results.
- `FSelector`: This class is the base class for implementations of feature selection algorithms.

Internally, the `fselect()` function creates an `FSelectInstanceSingleCrit` object and executes the feature selection with an `FSelector` object, based on the selected method, in this example an `FSelectorSequential` object. This is similar to what happens in the `tune()` function and will be explained in more detail in the following section. It uses the supplied resampling and measure to evaluate all feature subsets provided by the `FSelector` on the task.

In the following two sections, these classes will be created manually, to learn more about the `mlr3fselect` package.

6.2.2 The FSelectInstance Classes

`fsi()` To create an `FSelectInstanceSingleCrit` object, we use the sugar function `fsi()`:

```
instance = fsi(
  task = tsk_pen,
  learner = lrn_rpart,
  resampling = rsmp("cv", folds = 3),
  measure = msr("classif.acc"),
  terminator = trm("evals", n_evals = 20)
)
```

Note that we have not selected a feature selection algorithm and thus did not select any features, yet. We have also supplied a `Terminator`, which is used to stop the feature selection, these are the same objects as we saw in Section 4.1.2.

To start the feature selection, we still need to select an algorithm which are defined via the `FSelector` class, described in the next section.

6.2.3 The FSelector Class

The `FSelector` class is the base class for different feature selection algorithms. The following algorithms are currently implemented in `mlr3fselect`:

- Random search, trying random feature subsets until termination (`fs("random_search")`).
- Exhaustive search, trying all possible feature subsets (`fs("exhaustive_search")`).
- Sequential search, i.e., sequential forward or backward selection (`fs("sequential")`).
- Recursive feature elimination, which uses a learner's importance scores to iteratively remove features with low feature importance (`fs("rfe")`).
- Design points, trying all user-supplied feature sets (`fs("design_points")`).
- Genetic search, implementing a genetic algorithm which treats the features as a binary sequence and tries to find the best subset with mutations (`fs("genetic_search")`).
- Shadow variable search, which adds permuted copies of all features (shadow variables), performs forward selection, and stops when a shadow variable is selected (`fs("shadow_variable_search")`).

Note that all these methods can be stopped (early) with a terminator, e.g., an exhaustive search can be stopped after a given number of evaluations. In this example, we will use a simple random search and retrieve it from the `mlr_fselectors`
`fs()` dictionary with `fs()`.

```
fselector = fs("random_search")
```

6.2.4   Starting the Feature Selection

To start the feature selection, we pass the `FSelectInstanceSingleCrit` object to the `$optimize()` method of the initialized `FSelector` object:

```
fselector$optimize(instance)
```

The algorithm proceeds as follows

1. The `FSelector` proposes at least one feature subset or may propose multiple subsets to be evaluated in parallel, which can be controlled via the setting `batch_size`.
2. For each feature subset, the given learner is fitted on the task using the provided resampling and evaluated with the given measure.
3. All evaluations are stored in the archive of the `FSelectInstanceSingleCrit` object.
4. The terminator is queried. If the termination criteria are not triggered, go to 1).
5. Determine the feature subset with the best-observed performance.
6. Store the best feature subset as the result in the instance object.

The best feature subset and the corresponding measured performance can be accessed from the instance:

```
as.data.table(instance$result)[, .(features, classif.acc)]
```

```
                                    features classif.acc
1: bill_length,body_mass,flipper_length        0.936
```

As in the forward selection example above, one can investigate all subset evaluations, which are stored in the archive of the `FSelectInstanceSingleCrit` object and can be accessed by using `as.data.table()`:

```
as.data.table(instance$archive)[1:5,
  .(bill_depth, bill_length, body_mass, flipper_length,
  classif.acc)]
```

```
   bill_depth bill_length body_mass flipper_length classif.acc
1:      FALSE        TRUE     FALSE          FALSE      0.7558
2:      FALSE        TRUE     FALSE          FALSE      0.7558
3:      FALSE       FALSE      TRUE          FALSE      0.7210
4:      FALSE        TRUE      TRUE           TRUE      0.9360
5:      FALSE        TRUE      TRUE           TRUE      0.9360
```

Now the optimized feature subset can be used to subset the task and fit the model on all observations:

```
tsk_pen = tsk("penguins")

tsk_pen$select(instance$result_feature_set)
lrn_rpart$train(tsk_pen)
```

The trained model can now be used to make a prediction on external data.

6.2.5   Optimizing Multiple Performance Measures

You might want to use multiple criteria to evaluate the performance of the feature subsets. With `mlr3fselect`, the result is the collection of all feature subsets which are not Pareto-dominated by another subset. Again, we point out the similarity with HPO and refer to multi-objective hyperparameter optimization (see Section 5.2 and Karl et al. (2022)).

In the following example, we will perform feature selection on the sonar dataset. This time, we will use `FSelectInstanceMultiCrit` to select a subset of features that has high sensitivity, i.e., TPR, and high specificity, i.e., TNR. The feature selection process with multiple criteria is similar to that with a single criterion, except that we select two measures to be optimized:

```
instance = fsi(
  task = tsk("sonar"),
  learner = lrn_rpart,
  resampling = rsmp("holdout"),
  measure = msrs(c("classif.tpr", "classif.tnr")),
  terminator = trm("evals", n_evals = 20)
)
```

The function `fsi` creates an instance of `FSelectInstanceMultiCrit` if more than one measure is selected. We now create an `FSelector` and call the `$optimize()` function of the `FSelector` with the `FSelectInstanceMultiCrit` object, to search for the subset of features with the best TPR and FPR. Note that these two measures cannot both be optimal at the same time (except for the perfect classifier) and we expect several Pareto-optimal solutions.

```
fselector = fs("random_search")
fselector$optimize(instance)
```

As above, the best feature subsets and the corresponding measured performance can be accessed from the instance.

```
as.data.table(instance$result)[, .(features, classif.tpr,
  classif.tnr)]
```

```
                   features classif.tpr classif.tnr
1: V16,V21,V31,V37,V48,V50,...      0.6410      0.8333
2:    V1,V11,V12,V18,V2,V25,...      0.8205      0.7667
3:    V1,V10,V12,V13,V14,V16,...      0.8718      0.7333
4:    V1,V13,V15,V17,V18,V19,...      0.9231      0.6333
```

We see different tradeoffs of sensitivity and specificity but no feature subset is dominated by another, i.e., has worse sensitivity *and* specificity than any other subset.

6.2.6 Nested Resampling

As in tuning, the performance estimate of the finally selected feature subset is usually optimistically biased. To obtain unbiased performance estimates, nested resampling is required and can be set up analogously to HPO (see Section 4.3). We now show this as an example on the `sonar` task. The `AutoFSelector` class wraps a learner and augments it with automatic feature selection. Because the `AutoFSelector` itself inherits from the `Learner` base class, it can be used like any other learner. In the example below, a logistic regression learner is created. This learner is then wrapped in a random search feature selector that uses holdout (inner) resampling for performance evaluation. The sugar function `auto_fselector` can be used to create an instance of `AutoFSelector`:

auto_fselec

```
afs = auto_fselector(
  fselector = fs("random_search"),
  learner = lrn("classif.log_reg"),
  resampling = rsmp("holdout"),
  measure = msr("classif.acc"),
  terminator = trm("evals", n_evals = 10)
)
afs
```

```
<AutoFSelector:classif.log_reg.fselector>
* Model: list
* Packages: mlr3, mlr3fselect, mlr3learners, stats
* Predict Type: response
* Feature Types: logical, integer, numeric, character, factor,
  ordered
* Properties: loglik, twoclass
```

The `AutoFSelector` can then be passed to `benchmark()` or `resample()` for nested resampling (Section 4.3). Below we compare our wrapped learner `afs` with a normal logistic regression `lrn("classif.log_reg")`.

```
grid = benchmark_grid(tsk("sonar"),
  list(afs, lrn("classif.log_reg")),
  rsmp("cv", folds = 3))

bmr = benchmark(grid)$aggregate(msr("classif.acc"))
as.data.table(bmr)[, .(learner_id, classif.acc)]
```

```
               learner_id classif.acc
1: classif.log_reg.fselector      0.7061
2:          classif.log_reg      0.6776
```

We can see that, in this example, the feature selection improves prediction performance.

6.3 Conclusion

In this chapter, we learned how to perform feature selection with `mlr3`. We introduced filter and wrapper methods and covered the optimization of multiple performance measures. Once you have learned about pipelines we will return to feature selection in Section 8.4.5.

If you are interested in learning more about feature selection then we recommend an overview of methods in Chandrashekar and Sahin (2014); a more formal and detailed introduction to filters and wrappers is in Guyon and Elisseeff (2003); and a benchmark of filter methods was performed by Bommert et al. (2020).

Table 6.1: Important classes and functions covered in this chapter with underlying class (if applicable), class constructor or function, and important class fields and methods (if applicable).

Class	Constructor/Function	Fields/Methods
`Filter`	`flt()`	`$calculate()`
`FSelectInstanceSingleCrit` or `FSelectInstanceMultiCrit`	`fselect()`	-
`FSelector`	`fs()`	`$optimize()`
`AutoFSelector`	`auto_fselector()`	`$train(); $predict()`

6.4 Exercises

1. Compute the correlation filter scores on `tsk("mtcars")` and use the filter to select the five features most strongly correlated with the target. Resample `lrn("regr.kknn")` on both the full dataset and the reduced one, and compare both performances based on 10-fold CV with respect to MSE. NB: Here, we have performed the feature filtering outside of CV, which is generally not a good idea as it biases the CV performance estimation. To do this properly, filtering should be embedded inside the CV via pipelines – try to come back to this exercise after you read Chapter 8 to implement this with less bias.

2. Apply backward selection to `tsk("penguins")` with `lrn("classif.rpart")` and holdout resampling by the classification accuracy measure. Compare the results with those in Section 6.2.1 by also running the forward selection from that section. Do the selected features differ? Which feature selection method reports a higher classification accuracy in its `$result`?

3. There is a problem in the performance comparison in Exercise 2 as feature selection is performed on the test-set. Change the process by applying forward feature selection with `auto_fselector()`. Compare the performance to backward feature selection from Exercise 2 using nested resampling.

4. (*) Write a feature selection algorithm that is a hybrid of a filter and a wrapper method. This search algorithm should compute filter scores for all features and then perform a forward search. But instead of tentatively adding all remaining features to the current feature set, it should only stochastically try a subset of the available features. Features with high filter scores should be added with higher probability. Start by coding a stand-alone R method for this search (based on a learner, task, resampling, performance measure and some control settings). Then, as a stretch goal, see if you can implement this as an R6 class inheriting from `FSelector`.

Part III

Pipelines and Preprocessing

7

Sequential Pipelines

Martin Binder
Ludwig-Maximilians-Universität München, and Munich Center for Machine Learning (MCML)

Florian Pfisterer
Ludwig-Maximilians-Universität München

`mlr3` aims to provide a layer of abstraction for ML practitioners, allowing users to quickly swap one algorithm for another without needing expert knowledge of the underlying implementation. A unified interface for `Task`, `Learner`, and `Measure` objects means that complex benchmark and tuning experiments can be run in just a few lines of code for any off-the-shelf model, i.e., if you just want to run an experiment using the basic implementation from the underlying algorithm, we hope we have made this easy for you to do.

`mlr3pipelines` (Binder et al. 2021) takes this modularity one step further, extending it to workflows that may also include data preprocessing (Chapter 9), building ensemble-models, or even more complicated meta-models. `mlr3pipelines` makes it possible to build individual steps within a `Learner` out of building blocks, which inherit from the `PipeOp` class. `PipeOp`s can be connected using directed edges to form a `Graph` or "pipeline", which represent the flow of data between operations. During model training, the `PipeOp`s in a `Graph` transform a given `Task` and subsequent `PipeOp`s receive the transformed `Task` as input. As well as transforming data, `PipeOp`s generate a *state*, which is used to inform the `PipeOp`s operation during prediction, similar to how learners learn and store model parameters/weights during training that go on to inform model prediction. This is visualized in Figure 7.1 using the "Scaling" `PipeOp`, which scales features during training and saves the scaling factors as a state to be used in predictions.

We refer to pipelines as either sequential or non-sequential. These terms should not be confused with "sequential" and "parallel" processing. In the context of pipelines, "sequential" refers to the movement of data through the pipeline from one `PipeOp` directly to the next from start to finish. Sequential pipelines can be visualized in a straight line – as we will see in this chapter. In contrast, non-sequential pipelines see data being processed through `PipeOp`s that may have multiple inputs and/or outputs. Non-sequential pipelines are characterized by multiple branches so data may be processed by different `PipeOp`s at different times. Visually, non-sequential pipelines will not be a straight line from start to finish, but a more complex graph. In this chapter, we will look at sequential pipelines and in the next we will focus on non-sequential pipelines.

DOI: 10.1201/9781003402848-7

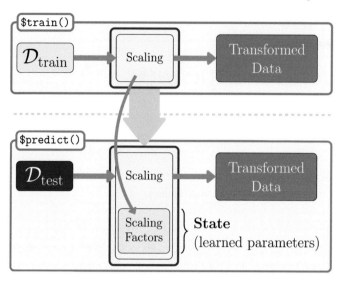

Figure 7.1: The $train() method of the "Scaling" PipeOp both transforms data (rectangles) as well as creates a state, which is the scaling factors necessary to transform data during prediction.

7.1   PipeOp: Pipeline Operators

The basic class of mlr3pipelines is the PipeOp, short for "pipeline operator". It   PipeOp
represents a transformative operation on an input (for example, a training Task),
resulting in some output. Similarly to a learner, it includes a $train() and a
$predict() method. The training phase typically generates a particular model of
the data, which is saved as the internal state. In the prediction phase, the PipeOp   State
acts on the prediction Task using information from the saved state. Therefore,
just like a learner, a PipeOp has "parameters" (i.e., the state) that are trained.
As well as "parameters", PipeOps also have hyperparameters that can be set by
the user when constructing the PipeOp or by accessing its $param_set. As with
other classes, PipeOps can be constructed with a sugar function, po(), or pos() for   po()
multiple PipeOps, and all available PipeOps are made available in the dictionary
mlr_pipeops. An up-to-date list of PipeOps contained in mlr3pipelines with links   mlr_pipeops
to their documentation can be found at https://mlr-org.com/pipeops.html, a small
subset of these are printed below. If you want to extend mlr3pipelines with a
PipeOp that has not been implemented, have a look at our vignette on extending
PipeOps by running: vignette("extending", package = "mlr3pipelines").

```
as.data.table(po())[1:6, 1:2]
```

```
              key                              label
1:         boxcox Box-Cox Transformation of Numeric Features
2:         branch                        Path Branching
3:          chunk        Chunk Input into Multiple Outputs
4: classbalancing                      Class Balancing
5:      classifavg              Majority Vote Prediction
6:    classweights    Class Weights for Sample Weighting
```

Let us now take a look at a `PipeOp` in practice using principal component analysis (PCA) as an example, which is implemented in `PipeOpPCA`. Below we construct the `PipeOp` using its ID `"pca"` and inspect it.

```
library(mlr3pipelines)

po_pca = po("pca", center = TRUE)
po_pca
```

```
PipeOp: <pca> (not trained)
values: <center=TRUE>
Input channels <name [train type, predict type]>:
  input [Task,Task]
Output channels <name [train type, predict type]>:
  output [Task,Task]
```

On printing, we can see that the `PipeOp` has not been trained and that we have changed some of the hyperparameters from their default values. The `Input channels` and `Output channels` lines provide information about the input and output types of this PipeOp. The PCA `PipeOp` takes one input (named "input") of type "Task", both during training and prediction ("`input [Task,Task]`"), and produces one called "output" that is also of type "Task" in both phases ("`output [Task,Task]`"). This highlights a key difference from the `Learner` class: `PipeOps` can return results after the training phase.

A `PipeOp` can be trained using `$train()`, which can have multiple inputs and outputs. Both inputs and outputs are passed as elements in a single `list`. The `"pca"` PipeOp takes as input the original task and after training returns the task with features replaced by their principal components.

```
tsk_small = tsk("penguins_simple")$select(c("bill_depth",
  "bill_length"))
poin = list(tsk_small$clone()$filter(1:5))
poout = po_pca$train(poin) # poin: Task in a list
poout # list with a single element 'output'
```

```
$output
<TaskClassif:penguins> (5 x 3): Simplified Palmer Penguins
* Target: species
* Properties: multiclass
* Features (2):
  - dbl (2): PC1, PC2
```

```
poout[[1]]$head()
```

```
   species     PC1       PC2
1:  Adelie  0.1561  0.005716
2:  Adelie  1.2677  0.789534
3:  Adelie  1.5336 -0.174460
4:  Adelie -2.1096  0.998977
5:  Adelie -0.8478 -1.619768
```

During training, PCA transforms incoming data by rotating it in such a way that features become uncorrelated and are ordered by their contribution to the total variance. The rotation matrix is also saved in the internal $state field during training (shown in Figure 7.1), which is then used during predictions and applied to new data.

```
po_pca$state
```

```
Standard deviations (1, .., p=2):
[1] 1.513 1.034

Rotation (n x k) = (2 x 2):
                PC1     PC2
bill_depth   -0.6116 -0.7911
bill_length   0.7911 -0.6116
```

Once trained, the $predict() function can then access the saved state to operate on the test data, which again is passed as a list:

```
tsk_onepenguin = tsk_small$clone()$filter(42)
poin = list(tsk_onepenguin)
poout = po_pca$predict(poin)
poout[[1]]$data()
```

```
   species    PC1     PC2
1:  Adelie 1.555 -1.455
```

7.2   Graph: Networks of PipeOps

PipeOps represent individual computational steps in machine learning pipelines. These pipelines themselves are defined by Graph objects. A Graph is a collection of PipeOps with "edges" that guide the flow of data.

The most convenient way of building a Graph is to connect a sequence of PipeOps using the %>>%-operator (read "double-arrow") operator. When given two PipeOps,   %>>%
this operator creates a Graph that first executes the left-hand PipeOp, followed by the right-hand one. It can also be used to connect a Graph with a PipeOp, or with another Graph. The following example uses po("mutate") to add a new feature to the task, and po("scale") to then scale and center all numeric features.

```
po_mutate = po("mutate",
  mutation = list(bill_ratio = ~bill_length / bill_depth)
)
po_scale = po("scale")
graph = po_mutate %>>% po_scale
graph
```

```
Graph with 2 PipeOps:
     ID          State sccssors prdcssors
 mutate <<UNTRAINED>>    scale
  scale <<UNTRAINED>>              mutate
```

The output provides information about the layout of the Graph. For each `PipOp` (`ID`), we can see information about the state (`State`), as well as a list of its successors (`sccssors`), which are `PipeOps` that come directly after the given `PipeOp`, and its predecessors (`prdcssors`), the `PipeOps` that are connected to its input. In this simple `Graph`, the output of the `"mutate"` `PipeOp` is passed directly to the `"scale"` `PipeOp` and neither takes any other inputs or outputs from other `PipeOps`. The `$plot()` method can be used to visualize the graph.

`$plot()`

```
    graph$plot(horizontal = TRUE)
```

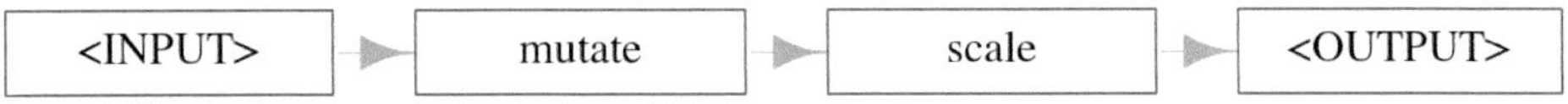

Figure 7.2: Simple sequential pipeline plot.

The plot demonstrates how a `Graph` is simply a collection of `PipeOps` that are connected by "edges". The collection of `PipeOps` inside a `Graph` can be accessed through the `$pipeops` field. The `$edges` field can be used to access edges, which returns a `data.table` listing the "source" (`src_id`, `src_channel`) and "destination" (`dst_id`, `dst_channel`) of data flowing along each edge .

`$edges/`
`$pipeops`

```
    graph$pipeops

$mutate
PipeOp: <mutate> (not trained)
values: <mutation=<list>, delete_originals=FALSE>
Input channels <name [train type, predict type]>:
  input [Task,Task]
Output channels <name [train type, predict type]>:
  output [Task,Task]

$scale
PipeOp: <scale> (not trained)
values: <robust=FALSE>
Input channels <name [train type, predict type]>:
  input [Task,Task]
Output channels <name [train type, predict type]>:
  output [Task,Task]

    graph$edges

    src_id src_channel dst_id dst_channel
1: mutate      output  scale       input
```

Instead of using `%>>%`, you can also create a `Graph` explicitly using the `$add_pipeop()` and `$add_edge()` methods to create `PipeOps` and the edges connecting them:

```
graph = Graph$new()$
  add_pipeop(po_mutate)$
  add_pipeop(po_scale)$
  add_edge("mutate", "scale")
```

> 💡 Graphs and DAGs
>
> The Graph class represents an object similar to a directed acyclic graph (DAG),
> since the input of a PipeOp cannot depend on its output and hence cycles
> are not allowed. However, the resemblance to a DAG is not perfect, since the
> Graph class allows for multiple edges between nodes. A term such as "directed
> acyclic multigraph" would be more accurate, but we use "graph" for simplicity.

Once built, a Graph can be used by calling $train() and $predict() as if it were
a Learner (though it still outputs a list during training and prediction):

```
result = graph$train(tsk_small)
result
```

```
$scale.output
<TaskClassif:penguins> (333 x 4): Simplified Palmer Penguins
* Target: species
* Properties: multiclass
* Features (3):
  - dbl (3): bill_depth, bill_length, bill_ratio
```

```
result[[1]]$data()[1:3]
```

```
   species bill_depth bill_length bill_ratio
1:  Adelie     0.7796     -0.8947    -1.0421
2:  Adelie     0.1194     -0.8216    -0.6804
3:  Adelie     0.4241     -0.6753    -0.7435
```

```
result = graph$predict(tsk_onepenguin)
result[[1]]$head()
```

```
   species bill_depth bill_length bill_ratio
1:  Adelie     0.9319      -0.529    -0.8963
```

7.3   Sequential Learner-Pipelines

Possibly the most common application for mlr3pipelines is to use it to perform
preprocessing tasks, such as missing value imputation or factor encoding, and to
then feed the resulting data into a Learner – we will see more of this in practice in
Chapter 9. A Graph representing this workflow manipulates data and fits a Learner-
model during training, ensuring that the data is processed the same way during the
prediction stage. Conceptually, the process may look as shown in Figure 7.3.

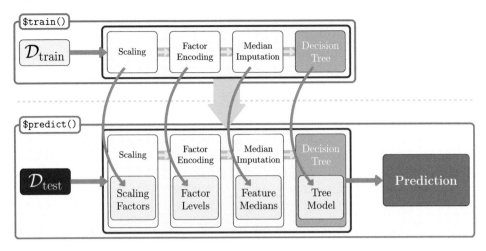

Figure 7.3: Conceptualization of training and prediction process inside a sequential learner-pipeline. During training (top row), the data is passed along the preprocessing operators, each of which modifies the data and creates a `$state`. Finally, the learner receives the data and a model is created. During prediction (bottom row), data is likewise transformed by preprocessing operators, using their respective `$state` (gray boxes) information in the process. The learner then receives data that has the same format as the data seen during training, and makes a prediction.

7.3.1 Learners as PipeOps and Graphs as Learners

In Figure 7.3, the final `PipeOp` is a `Learner`. `Learner` objects can be converted to `PipeOps` with `as_pipeop()`, however, this is only necessary if you choose to manually create a graph instead of using `%>>%`. With either method, internally `Learners` are passed to `po("learner")`. The following code creates a `Graph` that uses `po("imputesample")` to impute missing values by sampling from observed values (Section 9.3) then fits a logistic regression on the transformed task.

```
lrn_logreg = lrn("classif.log_reg")
graph = po("imputesample") %>>% lrn_logreg
graph$plot(horizontal = TRUE)
```

Figure 7.4: `"imputesample"` and `"learner"` PipeOps in a sequential pipeline.

We have seen how training and predicting `Graphs` is possible but has a slightly different design to `Learner` objects, i.e., inputs and outputs during both training and predicting are `list` objects. To use a `Graph` as a `Learner` with an identical interface, it can be wrapped in a `GraphLearner` object with `as_learner()`. The `GraphLearner` `Graph` can then be used like any other `Learner`, so now we can benchmark our pipeline to decide if we should impute by sampling or with the mode of observed values (`po("imputemode")`):

```
glrn_sample = as_learner(graph)
glrn_mode = as_learner(po("imputemode") %>>% lrn_logreg)

design = benchmark_grid(tsk("pima"), list(glrn_sample, glrn_mode),
  rsmp("cv", folds = 3))
bmr = benchmark(design)
aggr = bmr$aggregate()[, .(learner_id, classif.ce)]
aggr
```

```
                      learner_id classif.ce
1: imputesample.classif.log_reg     0.2357
2:   imputemode.classif.log_reg     0.2396
```

In this example, we can see that the sampling imputation method worked slightly better, although the difference is likely not significant.

💡 Automatic Conversion to Learner

In this book, we always use `as_learner()` to convert a `Graph` to a `Learner` explicitly for clarity. While this conversion is necessary when you want to use `Learner`-specific functions like `$predict_newdata()`, builtin `mlr3` methods like `resample()` and `benchmark_grid()` will make this conversion automatically and it is therefore not strictly needed. In the above example, it is therefore also possible to use

```
design = benchmark_grid(tsk("pima"),
  list(graph, po("imputesample") %>>% lrn_logreg),
  rsmp("cv", folds = 3))
```

7.3.2   Inspecting Graphs

You may want to inspect pipelines and the flow of data to learn more about your pipeline or to debug them. We first need to set the `$keep_results` flag to be `TRUE` so that intermediate results are retained, which is turned off by default to save memory.

```
glrn_sample$graph_model$keep_results = TRUE
glrn_sample$train(tsk("pima"))
```

The `Graph` can be accessed through the `$graph_model` field and then `PipeOps` can be accessed with `$pipeops` as before. In this example, we can see that our `Task` no longer has missing data after training the `"imputesample"` `PipeOp`. This can be used to access arbitrary intermediate results:

```
imputesample_output = glrn_sample$graph_model$pipeops$imputesample$
  .result
imputesample_output[[1]]$missings()
```

diabetes	age	pedigree	pregnant	glucose	insulin	mass	pressure
0	0	0	0	0	0	0	0

triceps
0

We could also use `$pipeops` to access our underlying Learner, note we need to use `$learner_model` to get the learner from the `PipeOpLearner`. We could use a similar method to peek at the state of any `PipeOp` in the graph:

```
pipeop_logreg = glrn_sample$graph_model$pipeops$classif.log_reg
learner_logreg = pipeop_logreg$learner_model
learner_logreg
```

```
<LearnerClassifLogReg:classif.log_reg>
* Model: glm
* Parameters: list()
* Packages: mlr3, mlr3learners, stats
* Predict Types:  [response], prob
* Feature Types: logical, integer, numeric, character, factor,
  ordered
* Properties: loglik, twoclass
```

> 💡 **`$base_learner()`**
>
> In this example we could have used **`glrn_sample$base_learner()`** to immediately access our trained learner, however, this does not generalize to more complex pipelines that may contain multiple learners.

7.3.3 Configuring Pipeline Hyperparameters

`PipeOp` hyperparameters are collected together in the `$param_set` of a graph and prefixed with the ID of the `PipeOp` to avoid parameter name clashes. Below we use the same `PipeOp` twice but set the `id` to ensure their IDs are unique.

```
graph = po("scale", center = FALSE, scale = TRUE,
  id = "scale") %>>%
  po("scale", center = TRUE, scale = FALSE, id = "center") %>>%
  lrn("classif.rpart", cp = 1)
unlist(graph$param_set$values)
```

scale.robust	scale.center	scale.scale
0	0	1
center.robust	**center.center**	**center.scale**
0	1	0
classif.rpart.xval	**classif.rpart.cp**	
0	1	

> ⚠ PipeOp IDs in Graphs
>
> If you need to change the ID of a `PipeOp` in a `Graph` then use the
> `$set_names` method from the `Graph` class, e.g., `some_graph$set_names(old`
> `= "old_name", new = "new_name")`. Do not change the ID of a `PipeOp`
> through `graph$pipeops$<old_id>$id = <new_id>`, as this will only alter
> the `PipeOp`'s record of its own ID, and not the `Graph`'s record, which will lead
> to errors.

Whether a pipeline is treated as a `Graph` or `GraphLearner`, hyperparameters are
updated and accessed in the same way.

```
graph$param_set$values$classif.rpart.maxdepth = 5
graph_learner = as_learner(graph)
graph_learner$param_set$values$classif.rpart.minsplit = 2
unlist(graph_learner$param_set$values)
```

scale.center	scale.scale	scale.robust
0	1	0
center.center	center.scale	center.robust
1	0	0
classif.rpart.cp	classif.rpart.maxdepth	classif.rpart.minsplit
1	5	2
classif.rpart.xval		
0		

7.4 Conclusion

In this chapter, we introduced `mlr3pipelines` and its building blocks: `Graph` and
`PipeOp`. We saw how to create pipelines as `Graph` objects from multiple `PipeOp`
objects and how to access `PipeOps` from a `Graph`. We also saw how to treat a
`Learner` as a `PipeOp` and how to treat a `Graph` as a `Learner`. In Chapter 8, we will
take this functionality a step further and look at pipelines where `PipeOps` are not
executed sequentially, as well as looking at how you can use `mlr3tuning` to tune
pipelines. A lot of practical examples that use sequential pipelines can be found in
Chapter 9, where we look at pipelines for data preprocessing.

Table 7.1: Important classes and functions covered in this chapter with underlying class (if applicable), class constructor or function, and important class fields and methods (if applicable).

Class	Constructor/Function	Fields/Methods
PipeOp	po()	$train(); $predict(); $state; $id; $param_set
Graph	%>>%	$add_pipeop(); $add_edge(); $pipeops; $edges;$train(); $predict()
GraphLearner	as_learner	$graph
PipeOpLearner	as_pipeop	$learner_model

7.5 Exercises

1. Create a learner containing a `Graph` that first imputes missing values using `po("imputeoor")`, standardizes the data using `po("scale")`, and then fits a logistic linear model using `lrn("classif.log_reg")`.

2. Train the learner created in the previous exercise on `tsk("pima")` and display the coefficients of the resulting model. What are two different ways to access the model?

3. Verify that the `"age"` column of the input task of `lrn("classif.log_reg")` from the previous exercise is indeed standardized. One way to do this would be to look at the `$data` field of the `lrn("classif.log_reg")` model; however, that is specific to that particular learner and does not work in general. What would be a different, more general way to do this? Hint: use the `$keep_results` flag.

8

Non-sequential Pipelines and Tuning

Martin Binder
Ludwig-Maximilians-Universität München, and Munich Center for Machine Learning (MCML)

Florian Pfisterer
Ludwig-Maximilians-Universität München

Marc Becker
Ludwig-Maximilians-Universität München, and Munich Center for Machine Learning (MCML)

Marvin N. Wright
Leibniz Institute for Prevention Research and Epidemiology – BIPS, and University of Bremen, and University of Copenhagen

In Chapter 7, we looked at simple sequential pipelines that can be built using the `Graph` class and a few `PipeOp` objects. In this chapter, we will take this further and look at non-sequential pipelines that can perform more complex operations. We will then look at tuning pipelines by combining methods in `mlr3tuning` and `mlr3pipelines` and will consider some concrete examples using multi-fidelity tuning (Section 5.3) and feature selection (Chapter 6).

We saw the power of the `%>>%`-operator in Chapter 7 to assemble graphs from combinations of multiple `PipeOps` and `Learners`. Given a single `PipeOp` or `Learner`, the `%>>%`-operator will arrange these objects into a linear `Graph` with each `PipeOp` acting in sequence. However, by using the `gunion()` function, we can instead combine multiple `PipeOps`, `Graphs`, or a mixture of both, into a parallel `Graph`.

In the following example, we create a `Graph` that centers its inputs (`po("scale")`) and then copies the centered data to two parallel streams: one replaces the data with columns that indicate whether data is missing (`po("missind")`), and the other imputes missing data using the median (`po("imputemedian")`), which we will return to in Section 9.3. The outputs of both streams are then combined into a single dataset using `po("featureunion")`.

```
library(mlr3pipelines)

graph = po("scale", center = TRUE, scale = FALSE) %>>%
  gunion(list(
    po("missind"),
    po("imputemedian")
  )) %>>%
  po("featureunion")

graph$plot(horizontal = TRUE)
```

DOI: 10.1201/9781003402848-8

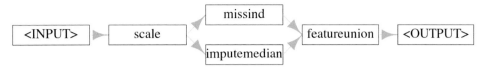

Figure 8.1: Simple parallel pipeline plot showing a common data source being scaled then the same data being passed to two `PipeOps` in parallel whose outputs are combined and returned to the user.

When applied to the first three rows of the `"pima"` task we can see how this imputes missing data and adds a column indicating where values were missing.

```
tsk_pima_head = tsk("pima")$filter(1:3)
tsk_pima_head$data(cols = c("diabetes", "insulin", "triceps"))
```

```
   diabetes insulin triceps
1:      pos      NA      35
2:      neg      NA      29
3:      pos      NA      NA
```

```
result = graph$train(tsk_pima_head)[[1]]
result$data(cols = c("diabetes", "insulin", "missing_insulin",
  "triceps","missing_triceps"))
```

```
   diabetes insulin missing_insulin triceps missing_triceps
1:      pos       0         missing       3         present
2:      neg       0         missing      -3         present
3:      pos       0         missing       0         missing
```

8.1 Selectors and Parallel Pipelines

It is common in `Graphs` for an operation to be applied to a subset of features. In `mlr3pipelines` this can be achieved in two ways (Figure 8.2): either by passing the column subset to the `affect_columns` hyperparameter of a `PipeOp` (assuming it has that hyperparameter), which controls which columns should be affected by the `PipeOp`; or, one can use the `PipeOpSelect` operator to create operations in parallel on specified feature subsets, and then unite the result using `PipeOpFeatureUnion`.

Selector

Both methods make use of `Selector`-functions. These are helper functions that indicate to a `PipeOp` which features it should apply to. `Selectors` may match column names by regular expressions (`selector_grep()`), or by column type (`selector_type()`). `Selectors` can also be used to join variables (`selector_union()`), return their set difference (`selector_setdiff()`), or select the complement of features from another `Selector` (`selector_invert()`).

For example, in Section 7.1, we applied PCA to the bill length and depth of penguins from `tsk("penguins_simple")` by first selecting these columns using the `Task` method `$select()` and then applying the `PipeOp`. We can now do this more simply with `selector_grep`, and could go on to use `selector_invert` to apply

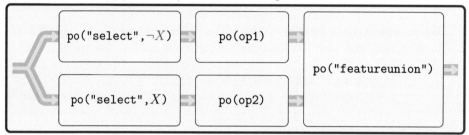

(a) The `affect_columns` hyperparameter can be used to restrict operations to a subset of features. When used, pipelines may still be run in sequence.

(b) Operating on subsets of tasks using concurrent paths by first splitting the inputs with `po("select")` and then combining outputs with `po("featureunion")`.

Figure 8.2: Two methods of setting up `PipeOps` (`po(op1)` and `po(op2)`) that operate on complementary features (X and ¬X) of an input task.

some other `PipeOp` to other features, below we use `po("scale")` and make use of the `affect_columns` hyperparameter:

```
sel_bill = selector_grep("^bill")
sel_not_bill = selector_invert(sel_bill)

graph = po("scale", affect_columns = sel_not_bill) %>>%
  po("pca", affect_columns = sel_bill)

result = graph$train(tsk("penguins_simple"))
result[[1]]$data()[1:3, 1:5]
```

```
   species     PC1     PC2 body_mass flipper_length
1:  Adelie -5.015  1.0717   -0.5676        -1.4246
2:  Adelie -4.495 -0.1853   -0.5055        -1.0679
3:  Adelie -3.755  0.4868   -1.1886        -0.4257
```

The biggest advantage of this method is that it creates a very simple, sequential `Graph`. However, one disadvantage of the `affect_columns` method is that it is relatively easy to have unexpected results if the ordering of `PipeOps` is mixed up. For example, if we had reversed the order of `po("pca")` and `po("scale")` above then we would have first created columns `"PC1"` and `"PC2"` and then erroneously scaled these, since their names do not start with "bill" and they are therefore matched by `sel_not_bill`. Creating parallel paths with `po("select")` can help mitigate such errors by selecting features given by the `Selector` and creating independent data processing streams with the given feature subset. Below we pass the parallel pipelines to `gunion()` as a `list` to ensure they receive the same input, and then combine the outputs with `po("featureunion")`.

```
po_select_bill = po("select", id = "s_bill", selector = sel_bill)
po_select_not_bill = po("select", id = "s_notbill",
  selector = sel_not_bill)

path_pca =  po_select_bill %>>% po("pca")
path_scale = po_select_not_bill %>>% po("scale")

graph = gunion(list(path_pca, path_scale)) %>>% po("featureunion")
graph$plot(horizontal = TRUE)
```

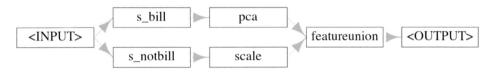

Figure 8.3: Visualization of a `Graph` where features are split into two paths, one with PCA and one with scaling, then combined and returned.

The `po("select")` method also has the significant advantage that it allows the same set of features to be used in multiple operations simultaneously, or to both transform features and keep their untransformed versions (by using `po("nop")` in one path). `PipeOpNOP` performs no operation on its inputs and is thus useful when you only want to perform a transformation on a subset of features and leave the others untouched:

```
graph = gunion(list(
  po_select_bill %>>% po("scale"),
  po_select_not_bill %>>% po("nop")
)) %>>% po("featureunion")
graph$plot(horizontal = TRUE)
```

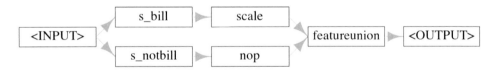

Figure 8.4: Visualization of our `Graph` where features are split into two paths, features that start with "bill" are scaled and the rest are untransformed.

```
graph$train(tsk("penguins_simple"))[[1]]$data()[1:3, 1:5]
```

```
   species bill_depth bill_length body_mass flipper_length
1:  Adelie     0.7796     -0.8947      3750            181
2:  Adelie     0.1194     -0.8216      3800            186
3:  Adelie     0.4241     -0.6753      3250            195
```

8.2   Common Patterns and ppl()

Now you have the tools to create sequential and non-sequential pipelines, you can create an infinite number of transformations on `Task`, `Learner`, and `Prediction` objects. In Section 8.3.1 and Section 8.3.2, we will work through two examples to demonstrate how you can make complex and powerful graphs using the methods and classes we have already looked at. However, many common problems in ML can be well solved by the same pipelines, and so to make your life easier we have implemented and saved these pipelines in the `mlr_graphs` dictionary; pipelines in the dictionary can be accessed with the `ppl()` sugar function.

`ppl()`

At the time of writing, this dictionary includes seven `Graph`s (required arguments included below):

- `ppl("bagging", graph)`: In `mlr3pipelines`, bagging is the process of running a `graph` multiple times on different data samples and then averaging the results. This is discussed in detail in Section 8.3.1.
- `ppl("branch", graphs)`: Uses `PipeOpBranch` to create different path branches from the given `graphs` where only one branch is evaluated. This is returned to in more detail in Section 8.4.2.
- `ppl("greplicate", graph, n)`: Create a `Graph` that replicates `graph` (which can also be a single `PipeOp`) n times. The pipeline avoids ID clashes by adding a suffix to each `PipeOp`, we will see this pipeline in use in Section 8.3.1.
- `ppl("ovr", graph)`: One-versus-rest classification for converting multiclass classification tasks into several binary classification tasks with one task for each class in the original. These tasks are then evaluated by the given `graph`, which should be a learner (or a pipeline containing a learner that emits a prediction). The predictions made on the binary tasks are combined into the multiclass prediction needed for the original task.
- `ppl("robustify")`: Performs common preprocessing steps to make any `Task` compatible with a given `Learner`. This pipeline is demonstrated in Section 9.4.
- `ppl("stacking", base_learners, super_learner)`: Stacking, returned to in detail in Section 8.3.2, is the process of using predictions from one or more models (`base_learners`) as features in a subsequent model (`super_learner`).
- `ppl("targettrafo", graph)`: Create a `Graph` that transforms the prediction target of a task and ensures that any transformations applied during training (using the function passed to the `targetmutate.trafo` hyperparameter) are inverted in the resulting predictions (using the function passed to the `targetmutate.inverter` hyperparameter); an example is given in Section 9.5.

8.3   Practical Pipelines by Example

In this section, we will put pipelines into practice by demonstrating how to turn weak learners into powerful machine learning models using bagging and stacking.

8.3.1 Bagging with "greplicate" and "subsample"

The basic idea of bagging (from **b**ootstrapp **aggregat**ing), introduced by Breiman (1996), is to aggregate multiple predictors into a single, more powerful predictor (Figure 8.5). Predictions are usually aggregated by the arithmetic mean for regression tasks or majority vote for classification. The underlying intuition behind bagging is that averaging a set of unstable and diverse (i.e., only weakly correlated) predictors can reduce the variance of the overall prediction. Each learner is trained on a different random sample of the original data.

Although we have already seen that a pre-constructed bagging pipeline is available with `ppl("bagging")`, in this section, we will build our own pipeline from scratch to showcase how to construct a complex `Graph`, which will look something like Figure 8.5.

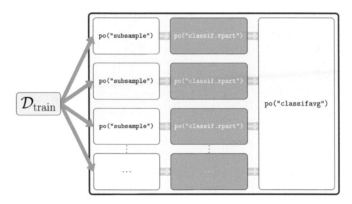

Figure 8.5: Graph that performs Bagging by independently subsampling data and fitting individual decision tree learners. The resulting predictions are aggregated by a majority vote `PipeOp`.

To begin, we use `po("subsample")` to sample a fraction of the data (here 70%), which is then passed to a classification tree (note by default `po("subsample")` samples without replacement).

```
gr_single_pred = po("subsample", frac = 0.7) %>>%
  lrn("classif.rpart")
```

Next, we use `ppl("greplicate")` to copy the graph, `gr_single_pred`, 10 times (n = 10) and finally `po("classifavg")` to take the majority vote of all predictions, note that we pass `innum = 10` to `"classifavg"` to tell the `PipeOp` to expect 10 inputs.

```
gr_pred_set = ppl("greplicate", graph = gr_single_pred, n = 10)
gr_bagging = gr_pred_set %>>% po("classifavg", innum = 10)
gr_bagging$plot()
```

Now let us see how well our bagging pipeline compares to the single decision tree and a random forest when benchmarked against `tsk("sonar")`.

Figure 8.6: Constructed bagging `Graph` with one input being sampled many times for 10 different learners.

```
# turn graph into learner
glrn_bagging = as_learner(gr_bagging)
glrn_bagging$id = "bagging"

lrn_rpart = lrn("classif.rpart")
learners = c(glrn_bagging, lrn_rpart, lrn("classif.ranger"))

bmr = benchmark(benchmark_grid(tsk("sonar"), learners,
  rsmp("cv", folds = 3)))
bmr$aggregate()[, .(learner_id, classif.ce)]
```

```
       learner_id classif.ce
1:        bagging     0.2498
2:  classif.rpart     0.2739
3: classif.ranger     0.2021
```

The bagged learner performs better than the decision tree but worse than the random forest. To automatically recreate this pipeline, you can construct `ppl("bagging")` by specifying the learner to "bag", the number of iterations, the fraction of data to sample, and the `PipeOp` to average the predictions, as shown in the code below. Note we set `collect_multiplicity = TRUE` which collects the predictions across paths, that technically use the `Multiplicity` method, which we will not discuss here but refer the reader to the documentation.

```
ppl("bagging", lrn("classif.rpart"),
  iterations = 10, frac = 0.7,
  averager = po("classifavg", collect_multiplicity = TRUE))
```

The main difference between our pipeline and a random forest is that the latter also performs feature subsampling, where only a random subset of available features is considered at each split point. While we cannot implement this directly with `mlr3pipelines`, we can use a custom `Selector` method to approximate this method. We will create this `Selector` by passing a function that takes as input the task and returns a sample of the features, we sample the square root of the number of features to mimic the implementation in `ranger`. For efficiency, we will now use `ppl("bagging")` to recreate the steps above:

```
# custom selector
selector_subsample = function(task) {
  sample(task$feature_names, sqrt(length(task$feature_names)))
}

# bagging pipeline with our selector
gr_bagging_quasi_rf = ppl("bagging",
  graph = po("select", selector = selector_subsample) %>>%
    lrn("classif.rpart", minsplit = 1),
  iterations = 100,
  averager = po("classifavg", collect_multiplicity = TRUE)
)

# bootstrap resampling
gr_bagging_quasi_rf$param_set$values$subsample.replace = TRUE

# convert to learner
glrn_quasi_rf = as_learner(gr_bagging_quasi_rf)
glrn_quasi_rf$id = "quasi.rf"

# benchmark
design = benchmark_grid(tsks("sonar"),
  c(glrn_quasi_rf, lrn("classif.ranger", num.trees = 100)),
  rsmp("cv", folds = 5)
)
bmr = benchmark(design)
bmr$aggregate()[, .(learner_id, classif.ce)]
```

```
      learner_id classif.ce
1:      quasi.rf     0.1826
2: classif.ranger     0.1779
```

In only a few lines of code, we took a weaker learner and turned it into a powerful model that we can see is comparable to the implementation in `ranger::ranger`. In the next section, we will look at a second example, which makes use of cross-validation within pipelines.

8.3.2 Stacking with po("learner_cv")

Stacking (Wolpert 1992) is another very popular ensembling technique that can significantly improve predictive performance. The basic idea behind stacking is to use predictions from multiple models (usually referred to as level 0 models) as features for a subsequent model (the level 1 model) which in turn combines these predictions (Figure 8.7). A simple combination can be a linear model (possibly regularized if you have many level 0 models), since a weighted sum of level 0 models is often plausible and good enough. Though, non-linear level 1 models can also be used, and it is also possible for the level 1 model to access the input features as well as the level 0 predictions. Stacking can be built with more than two levels (both conceptually, and in `mlr3`) but we limit ourselves to this simpler setup here, which often also performs well in practice.

As with bagging, we will demonstrate how to create a stacking pipeline manually, although a pre-constructed pipeline is available with `ppl("stacking")`.

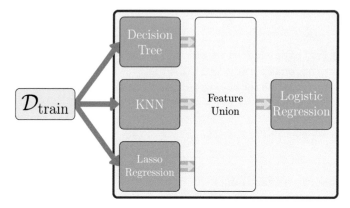

Figure 8.7: Graph that performs Stacking by fitting three models and using their outputs as features for another model after combining with `PipeOpFeatureUnion`.

Stacking pipelines depend on the level 0 learners returning predictions during the `$train()` phase. This is possible in `mlr3pipelines` with `PipeOpLearnerCV`. During training, this operator performs cross-validation and passes the out-of-sample predictions to the level 1 model. Using cross-validated predictions is recommended to reduce the risk of overfitting.

We first create the level 0 learners to produce the predictions that will be used as features. In this example, we use a classification tree, k-nearest neighbors (KNN), and a regularized GLM. Each learner is wrapped in `po("learner_cv")` which performs cross-validation on the input data and then outputs the predictions from the `Learner` in a new `Task` object.

```
lrn_rpart = lrn("classif.rpart", predict_type = "prob")
po_rpart_cv = po("learner_cv", learner = lrn_rpart,
  resampling.folds = 2, id = "rpart_cv"
)


lrn_knn = lrn("classif.kknn", predict_type = "prob")
po_knn_cv = po("learner_cv",
  learner = lrn_knn,
  resampling.folds = 2, id = "knn_cv"
)


lrn_glmnet = lrn("classif.glmnet", predict_type = "prob")
po_glmnet_cv = po("learner_cv",
  learner = lrn_glmnet,
  resampling.folds = 2, id = "glmnet_cv"
)
```

These learners are combined using `gunion()`, and `po("featureunion")` is used to merge their predictions. This is demonstrated in the output of `$train()`:

```
gr_level_0 = gunion(list(po_rpart_cv, po_knn_cv, po_glmnet_cv))
gr_combined = gr_level_0 %>>% po("featureunion")

gr_combined$train(tsk("sonar"))[[1]]$head()
```

	Class	rpart_cv.prob.M	rpart_cv.prob.R	knn_cv.prob.M	knn_cv.prob.R
1:	R	0.57895	0.4211	0.3857	0.6143
2:	R	0.88636	0.1136	0.3170	0.6830
3:	R	0.04348	0.9565	0.4396	0.5604
4:	R	0.03030	0.9697	0.4762	0.5238
5:	R	0.04348	0.9565	0.4753	0.5247
6:	R	0.23077	0.7692	0.4020	0.5980

```
2 variables not shown: [glmnet_cv.prob.M, glmnet_cv.prob.R]
```

> 💡 Retaining Features
>
> In this example, the original features were removed as each `PipeOp` only
> returns the predictions made by the respective learners. To retain the original
> features, include `po("nop")` in the list passed to `gunion()`.

The resulting task contains the predicted probabilities for both classes made from
each of the level 0 learners. However, as the probabilities always add up to 1, we
only need the predictions for one of the classes (as this is a binary classification
task), so we can use `po("select")` to only keep predictions for one class (we choose
`"M"` in this example).

```
gr_stack = gr_combined %>>%
  po("select", selector = selector_grep("\\.M$"))
```

Finally, we can combine our pipeline with the final model that will take these
predictions as its input. Below we use logistic regression, which combines the level 0
predictions in a weighted linear sum.

```
gr_stack = gr_stack %>>% po("learner", lrn("classif.log_reg"))
gr_stack$plot(horizontal = TRUE)
```

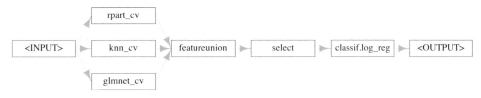

Figure 8.8: Constructed stacking Graph with one input being passed to three weak
learners whose predictions are passed to the logistic regression.

As our final model was an interpretable logistic regression, we can inspect the weights
of the level 0 learners by looking at the final trained model:

```
glrn_stack = as_learner(gr_stack)
glrn_stack$train(tsk("sonar"))
glrn_stack$base_learner()$model
```

```
Call:  stats::glm(formula = task$formula(), family = "binomial",
    data = data, model = FALSE)

Coefficients:
    (Intercept)    rpart_cv.prob.M    knn_cv.prob.M  glmnet_cv.prob.M
        -3.120             -0.134            4.040             1.804

Degrees of Freedom: 207 Total (i.e. Null);   204 Residual
Null Deviance:       287
Residual Deviance: 176   AIC: 184
```

The model weights suggest that knn influences the predictions the most with the
largest coefficient. To confirm this we can benchmark the individual models alongside
the stacking pipeline.

```
glrn_stack$id = "stacking"
design = benchmark_grid(tsk("sonar"),
    list(lrn_rpart, lrn_knn, lrn_glmnet, glrn_stack),
    rsmp("repeated_cv"))
bmr = benchmark(design)
bmr$aggregate()[, .(learner_id, classif.ce)]
```

```
        learner_id classif.ce
1:   classif.rpart      0.2876
2:    classif.kknn      0.1505
3: classif.glmnet      0.2559
4:        stacking      0.1438
```

This experiment confirms that of the individual models, the KNN learner performs
the best, however, our stacking pipeline outperforms them all. Now that we have
seen the inner workings of this pipeline, next time you might want to more efficiently
create it using ppl("stacking"), to copy the example above you would run:

```
ppl("stacking",
    base_learners = lrns(c("classif.rpart", "classif.kknn",
        "classif.glmnet")),
    super_learner = lrn("classif.log_reg")
)
```

Having covered the building blocks of mlr3pipelines and seen these in practice,
we will now turn to more advanced functionality, combining pipelines with tuning.

8.4 Tuning Graphs

By wrapping a pipeline inside a GraphLearner, we can tune it at two levels of complexity using mlr3tuning:

1. Tuning of a fixed, usually sequential pipeline, where preprocessing is combined with a given learner. This simply means the joint tuning of any subset of selected hyperparameters of operations in the pipeline. Conceptually and also technically in mlr3, this is not much different from tuning a learner that is not part of a pipeline.

2. Tuning not only the hyperparameters of a pipeline, whose structure is not completely fixed in terms of its included operations, but also which concrete PipeOps should be applied to data. This allows us to select these operations (e.g., which learner to use, which preprocessing to perform) in a data-driven manner known as "Combined Algorithm Selection and Hyperparameter optimization" (Thornton et al. 2013). As we will soon see, we can do this in mlr3pipelines by using the powerful branching (Section 8.4.2) and proxy (Section 8.4.3) meta operators. Through this, we can conveniently create our own "mini AutoML systems" (Hutter, Kotthoff, and Vanschoren 2019) in mlr3, which can even be geared for specific tasks.

8.4.1 Tuning Graph Hyperparameters

Let us consider a simple, sequential pipeline using po("pca") followed by lrn("classif.kknn"):

```
graph_learner = as_learner(po("pca") %>>% lrn("classif.kknn"))
```

The optimal setting of the **rank.** hyperparameter of our PCA PipeOp may realistically depend on the value of the k hyperparameter of the KNN model so jointly tuning them is reasonable. For this, we can simply use the syntax for tuning **Learners**, which was introduced in Chapter 4.

```
lrn_knn = lrn("classif.kknn", k = to_tune(1, 32))
po_pca = po("pca", rank. = to_tune(2, 20))
graph_learner = as_learner(po_pca %>>% lrn_knn)
graph_learner$param_set$values
```

```
$pca.rank.
Tuning over:
range [2, 20]


$classif.kknn.k
Tuning over:
range [1, 32]
```

We can see how the pipeline's `$param_set` includes the tune tokens for all selected hyperparameters, creating a joint search space. We can compare the tuned and untuned pipeline in a benchmark experiment with nested resampling by using an `AutoTuner`:

```
glrn_tuned = auto_tuner(tnr("random_search"), graph_learner,
  rsmp("holdout"), term_evals = 10)
glrn_untuned = po("pca") %>>% lrn("classif.kknn")
design = benchmark_grid(tsk("sonar"), c(glrn_tuned, glrn_untuned),
  rsmp("cv", folds = 5))
benchmark(design)$aggregate()[, .(learner_id, classif.ce)]
```

```
              learner_id classif.ce
1: pca.classif.kknn.tuned     0.2063
2:        pca.classif.kknn     0.2553
```

Tuning pipelines will usually take longer than tuning individual learners as training steps are often more complex and the search space will be larger. Therefore, parallelization is often appropriate (Section 10.1) and/or more efficient tuning methods for searching large tuning spaces such as Bayesian optimization (Section 5.4).

8.4.2   Tuning Alternative Paths with po("branch")

In the previous section, we tuned the KKNN and decision tree in the stacking pipeline, as well as tuning the rank of the PCA. However, we tuned the PCA without first considering if it was even beneficial at all, in this section we will answer that question by making use of `PipeOpBranch` and `PipeOpUnbranch`, which make it possible to specify multiple alternative paths in a pipeline. `po("branch")` creates multiple paths such that data can only flow through *one* of these as determined by the `selection` hyperparameter (Figure 8.13). This concept makes it possible to use tuning to decide which `PipeOps` and `Learners` to include in the pipeline, while also allowing all options in every path to be tuned.

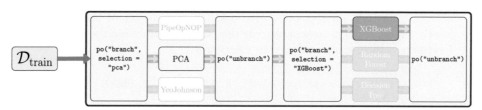

Figure 8.9: Figure demonstrates the `po("branch")` and `po("unbranch")` operators where three separate branches are created and data only flows through the PCA, which is specified with the argument to `selection`.

To demonstrate alternative paths we will make use of the MNIST (LeCun et al. 1998) data, which is useful for demonstrating preprocessing. The data is loaded from OpenML, which is described in Section 11.1, we subset the data to make the example run faster.

```
library(mlr3oml)
otsk_mnist = otsk(id = 3573)
tsk_mnist = as_task(otsk_mnist)$
  filter(sample(70000, 1000))$
  select(otsk_mnist$feature_names[sample(700, 100)])
```

po("branch") is initialized either with the number of branches or with a character-vector indicating the names of the branches, the latter makes the selection hyperparameter (discussed below) more readable. Below we create three branches: do nothing (po("nop")), apply PCA (po("pca")), remove constant features (po("removeconstants")) then apply the Yeo-Johnson transform (po("yeojohnson")). It is important to use po("unbranch") (with the same arguments as "branch") to ensure that the outputs are merged into one result object.

```
paths = c("nop", "pca", "yeojohnson")

graph = po("branch", paths, id = "brnchPO") %>>%
  gunion(list(
    po("nop"),
    po("pca"),
    po("removeconstants", id = "rm_const") %>>%
      po("yeojohnson", id = "YJ")
  )) %>>% po("unbranch", paths, id = "unbrnchPO")

graph$plot(horizontal = TRUE)
```

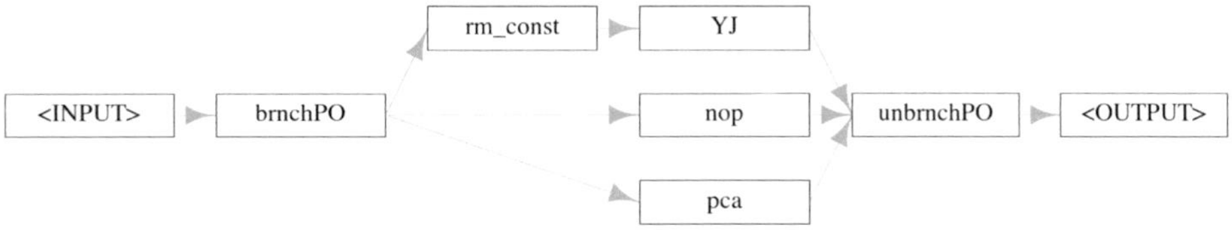

Figure 8.10: Graph with branching to three different paths that are split with po("branch") and combined with po("unbranch").

We can see how the output of this Graph depends on the setting of the branch.selection hyperparameter:

```
# use the "PCA" path
graph$param_set$values$brnchPO.selection = "pca"
# new PCA columns
head(graph$train(tsk_mnist)[[1]]$feature_names)
```

```
[1] "PC1" "PC2" "PC3" "PC4" "PC5" "PC6"
```

```
# use the "No-Op" path
graph$param_set$values$brnchPO.selection = "nop"
# same features
head(graph$train(tsk_mnist)[[1]]$feature_names)
```

```
[1] "pixel4"  "pixel10" "pixel11" "pixel14" "pixel34" "pixel39"
```

`ppl("branch")` simplifies the above by allowing you to just pass the different paths to the `graphs` argument (omitting "`rm_const`" for simplicity here):

```
ppl("branch", graphs = pos(c("nop", "pca", "yeojohnson")))
```

Branching can even be used to tune which of several learners is most appropriate for a given dataset. We extend our example further and add the choice between a decision tree and KKNN:

```
graph_learner = graph %>>%
  ppl("branch", lrns(c("classif.rpart", "classif.kknn")))
graph_learner$plot(horizontal = TRUE)
```

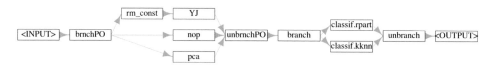

Figure 8.11: Graph with branching to three different paths that are split with `po("branch")` and combined with `po("unbranch")` then branch and recombine again.

Tuning the `selection` hyperparameters can help determine which of the possible options work best in combination. We additionally tune the `k` hyperparameter of the KNN learner, as it may depend on the type of preprocessing performed. As this hyperparameter is only active when the "`classif.kknn`" path is chosen, we will set a dependency (Section 4.4.4):

```
graph_learner = as_learner(graph_learner)

graph_learner$param_set$set_values(
  brnchPO.selection = to_tune(paths),
  branch.selection = to_tune(c("classif.rpart", "classif.kknn")),
  classif.kknn.k = to_tune(p_int(1, 32,
    depends = branch.selection == "classif.kknn"))
)

instance = tune(tnr("grid_search"), tsk_mnist, graph_learner,
  rsmp("repeated_cv", folds = 3, repeats = 3), msr("classif.ce"))

instance$archive$data[order(classif.ce)[1:5],
  .(brnchPO.selection, classif.kknn.k, branch.selection,
  classif.ce)]
```

	brnchPO.selection	classif.kknn.k	branch.selection	classif.ce
1:	yeojohnson	11	classif.kknn	0.2293
2:	yeojohnson	15	classif.kknn	0.2370
3:	yeojohnson	18	classif.kknn	0.2400
4:	yeojohnson	8	classif.kknn	0.2400
5:	yeojohnson	22	classif.kknn	0.2467

```
autoplot(instance)
```

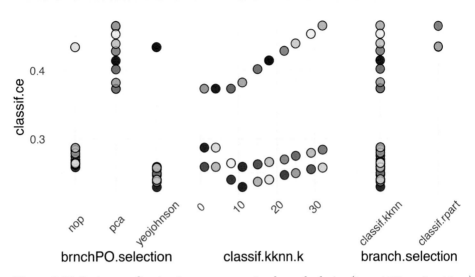

Figure 8.12: Instance after tuning preprocessing branch choice (`brnchPO.selection`), KNN k parameter (`classif.kknn.k`), and learning branch choice (`branch.selection`). Dots are different hyperparameter configurations that were tested during tuning, colors separate hyperparameter configurations.

As we can see in the results and Figure 8.12, the KNN-learner with k set to 11 was selected, which performs best in combination with the Yeo-Johnson transform.

8.4.3 Tuning with po("proxy")

> **i** **This section covers advanced ML or technical details.**

po("proxy") is a meta-operator that performs the operation that is stored in its content hyperparameter, which could be another PipeOp or Graph. It can therefore be used to tune over and select different PipeOps or Graphs that could be passed to this hyperparameter (Figure 8.13).

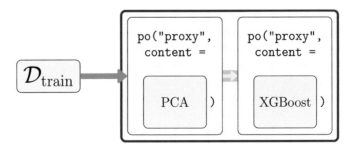

Figure 8.13: Figure demonstrates the po("proxy") operator with a PipeOp as its argument.

To recreate the example above with `po("proxy")`, the first step is to create place-holder `PipeOpProxy` operators to stand in for the operations (i.e., different paths) that should be tuned.

```
graph_learner = po("proxy", id = "preproc") %>>%
  po("proxy", id = "learner")
graph_learner = as_learner(graph_learner)
```

The tuning space for the `content` hyperparameters should be a discrete set of possibilities to be evaluated, passed as a `p_fct` (Section 4.4.2). For the `"preproc"` proxy operator this would simply be the different `PipeOps` that we want to consider:

```
# define content for the preprocessing proxy operator
preproc.content = p_fct(list(
  nop = po("nop"),
  pca = po("pca"),
  yeojohnson = po("removeconstants") %>>% po("yeojohnson")
))
```

For the `"learner"` proxy, this is more complicated as the selection of the learner depends on more than one search space component: The choice of the learner itself (`lrn("classif.rpart")` or `lrn("classif.kknn")`) and the tuned k hyperparameter of the KNN learner. To enable this we pass a transformation to `.extra_trafo` (Section 4.4.3). Note that inside this transformation we clone `learner.content`, otherwise, we would end up modifying the original `Learner` object inside the search space by reference (Section 1.5.1).

```
# define content for the learner proxy operator
learner.content = p_fct(list(
    classif.rpart = lrn("classif.rpart"),
    classif.kknn = lrn("classif.kknn")
))

# define transformation to set the content values
trafo = function(x, param_set) {
    if (!is.null(x$classif.kknn.k)) {
        x$learner.content = x$learner.content$clone(deep = TRUE)
        x$learner.content$param_set$values$k = x$classif.kknn.k
        x$classif.kknn.k = NULL
    }
    x
}
```

We can now put this all together, add the KNN tuning, and run the experiment.

```
search_space = ps(
  preproc.content = preproc.content,
  learner.content = learner.content,
  # tune KKNN parameter as normal
  classif.kknn.k = p_int(1, 32,
```

```
      depends = learner.content == "classif.kknn"),
   .extra_trafo = trafo
)

instance = tune(tnr("grid_search"), tsk_mnist, graph_learner,
   rsmp("repeated_cv", folds = 3, repeats = 3), msr("classif.ce"),
   search_space = search_space)

as.data.table(instance$result)[,
   .(preproc.content,
     classif.kknn.k = x_domain[[1]]$learner.content$param_set$values$k,
     learner.content, classif.ce)
]
```

```
   preproc.content classif.kknn.k learner.content classif.ce
1:      yeojohnson              11    classif.kknn       0.23
```

Once again, the best configuration is a KNN learner with the Yeo-Johnson transform. In practice `po("proxy")` offers complete flexibility and may be more useful for more complicated use cases, whereas `ppl("branch")` is more efficient in more straightforward scenarios.

8.4.4 Hyperband with Subsampling

> **i** **This section covers advanced ML or technical details.**

In Section 5.3, we learned about the Hyperband tuner and how it can make use of fidelity parameters to efficiently tune learners. Now that you have learned about pipelines and how to tune them, in this short section we will briefly return to Hyperband to showcase how we can put together everything we have learned in this chapter to allow Hyperband to be used with any `Learner`.

We previously saw how some learners have hyperparameters that can act naturally as fidelity parameters, such as the number of trees in a random forest. However, using pipelines, we can now create a fidelity parameter for any model using `po("subsample")`. The `frac` parameter of `po("subsample")` controls the amount of data fed into the subsequent `Learner`. In general, feeding less data to a `Learner` results in quicker model training but poorer quality predictions compared to when more training data is supplied. Resampling with less data will still give us some information about the relative performance of different model configurations, thus making the fraction of data to subsample the perfect candidate for a fidelity parameter.

In this example, we will optimize the SVM hyperparameters, `cost` and `gamma`, on `tsk("sonar")`:

```
library(mlr3tuning)

learner = lrn("classif.svm", id = "svm", type = "C-classification",
   kernel = "radial", cost  = to_tune(1e-5, 1e5, logscale = TRUE),
   gamma = to_tune(1e-5, 1e5, logscale = TRUE))
```

We then construct `po("subsample")` and specify that we want to use the `frac` parameter between $[3^{-3}, 1]$ as our fidelity parameter and set the `"budget"` tag to pass this information to Hyperband. We add this to our SVM and create a GraphLearner.

```
graph_learner = as_learner(
  po("subsample", frac = to_tune(p_dbl(3^-3, 1, tags = "budget"))) %>>%
  learner
)
```

As good practice, we encapsulate our learner and add a fallback to prevent fatal errors (Section 5.1).

```
graph_learner$encapsulate = c(train = "evaluate", predict = "evaluate")
graph_learner$timeout = c(train = 30, predict = 30)
graph_learner$fallback = lrn("classif.featureless")
```

Now we can tune our SVM by tuning our `GraphLearner` as normal, below we set `eta = 3` for Hyperband.

```
instance = tune(tnr("hyperband", eta = 3), tsk("sonar"), graph_learner,
  rsmp("cv", folds = 3), msr("classif.ce"))

instance$result_x_domain
```

```
$subsample.frac
[1] 1

$svm.cost
[1] 5126

$svm.gamma
[1] 0.03179
```

8.4.5 Feature Selection with Filter Pipelines

> **i** **This section covers advanced ML or technical details.**

In Section 6.1.4, we learnt about filter-based feature selection and how we can manually run a filter and then extract the selected features, often using an arbitrary choice of thresholds that were not tuned. Now that we have covered pipelines and tuning, we will briefly return to feature selection to demonstrate how to automate filter-based feature selection by making use of `po("filter")`. `po("filter")` includes the `filter` construction argument, which takes a `Filter` object to be used as the filter method as well as a choice of parameters for different methods of selecting features:

- `filter.nfeat` – Number of features to select
- `filter.frac` – Fraction of features to select
- `filter.cutoff` – Minimum value of filter such that features with filter values greater than or equal to the cutoff are kept
- `filter.permuted` – Random permutation of features added to task before applying the filter and all features before the **permuted**-th permuted features are kept

Below we use the information gain filter and select the top three features:

```
library(mlr3filters)
library(mlr3fselect)

task_pen = tsk("penguins")

# combine filter (keep top 3 features) with learner
po_flt = po("filter", filter = flt("information_gain"), filter.nfeat = 3)
graph = po_flt %>>% po("learner", lrn("classif.rpart"))

po("filter", filter = flt("information_gain"), filter.nfeat = 3)$
  train(list(task_pen))[[1]]$feature_names
```

```
[1] "bill_depth"     "bill_length"     "flipper_length"
```

Choosing 3 as the cutoff was fairly arbitrary but by tuning a graph we can optimize this cutoff:

```
# tune between 1 and total number of features
po_filter = po("filter", filter = flt("information_gain"),
  filter.nfeat = to_tune(1, task_pen$ncol))

graph = as_learner(po_filter %>>% po("learner", lrn("classif.rpart")))

instance = tune(tnr("random_search"), task_pen, graph,
  rsmp("cv", folds = 3), term_evals = 10)
instance$result
```

```
   information_gain.filter.nfeat learner_param_vals  x_domain classif.ce
1:                             5          <list[2]> <list[1]>     0.06972
```

In this example, 5 is the optimal number of features. It can be especially useful in feature selection to visualize the tuning results as there may be cases where the optimal result is only marginally better than a result with less features (which would lead to a model that is quicker to train and possibly easier to interpret).

```
autoplot(instance)
```

Now we can see that four variables may be equally as good in this case so we could consider going forward by selecting four features and not six as suggested by `instance$result`.

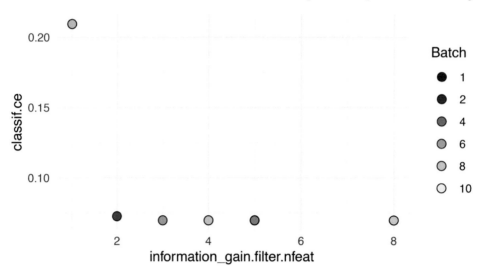

Figure 8.14: Model performance with different numbers of features, selected by an information gain filter.

8.5   Conclusion

In this chapter, we built on what we learned in Chapter 7 to develop complex non-sequential Graphs. We saw how to build our own graphs, as well as how to make use of ppl() to load Graphs that are available in mlr3pipelines. We then looked at different ways to tune pipelines, including joint tuning of hyperparameters and tuning the selection of PipeOps in a Graph, enabling the construction of simple, custom AutoML systems. In Chapter 9, we will study in more detail how to use pipelines for data preprocessing.

Table 8.1: Important classes and functions covered in this chapter with underlying class (if applicable), class constructor or function, and important class fields and methods (if applicable).

Class	Constructor/Function	Fields/Methods
Graph	ppl()	$train(); $predict()
Selector	selector_grep(); selector_type(); selector_invert()	-
PipeOpBranch; PipeOpUnbranch	po("branch"); po("unbranch")	-
PipeOpProxy	po("proxy")	-

8.6 Exercises

1. Create a graph that replaces all numeric columns that do not contain missing values with their PCA transform. Solve this in two ways, using `affect_columns` in a sequential graph, and using `po("select")` in a non-sequential graph. Train the graph on `tsk("pima")` to check your result. Hint: You may find `selector_missing()` useful.

2. The `po("select")` in Section 8.3.2 is necessary to remove redundant predictions (recall this is a binary classification task so we do not require predictions of both classes). However, if this was a multiclass classification task, then using `selector_grep()` would need to be called with a pattern for *all* prediction columns that should be *kept*, which would be inefficient. Instead it would be more appropriate to provide a pattern for the single class to remove. How would you do this using the `Selector` functions provided by `mlr3pipelines`? Implement this and train the modified stacking pipeline on `tsk("wine")`, using `lrn("classif.multinom")` as the level 1 learner.

3. How would you solve the previous exercise without explicitly naming the class you want to exclude, so that your graph works for any classification task? Hint: look at the `selector_subsample` in Section 8.3.1.

4. (*) Create your own "minimal AutoML system" by combining pipelines, branching and tuning. It should allow automatic preprocessing and the automatic selection of a well-performing learning algorithm. Both your `PipeOps` and models should be tuned. Your system should feature options for two preprocessing steps (imputation and factor encoding) and at least three learning algorithms to choose from. You can optimize this via random search, or try to use a more advanced tuning algorithm. Test it on at least three different data sets and compare its performance against an untuned random forest via nested resampling.

9

Preprocessing

Janek Thomas

Ludwig-Maximilians-Universität München, and Munich Center for Machine Learning (MCML), and Essential Data Science Training GmbH

Chapters 7 and 8 provided a technical introduction to `mlr3pipelines`, this chapter will now demonstrate how to use those pipelines to tackle common problems when preprocessing data for ML, including factor encoding, imputation of missing values, feature and target transformations, and functional feature extraction. Feature selection, an important preprocessing method, is covered in Chapter 6.

In this book, preprocessing refers to everything that happens with *data* before it is used to fit a model, while postprocessing encompasses everything that occurs with *predictions* after the model is fitted.

Data cleaning is an important part of preprocessing that involves the removal of errors, noise, and redundancy in the data; we only consider data cleaning very briefly as it is usually performed outside of `mlr3` on the raw dataset.

Data Cleaning

Another aspect of preprocessing is feature engineering, which covers all other transformations of data before it is fed to the machine learning model, including the creation of features from possibly unstructured data, such as written text, sequences, or images. The goal of feature engineering is to enable the data to be handled by a given learner, and/or to further improve predictive performance. It is important to note that feature engineering helps mostly for simpler algorithms, while highly complex models usually gain less from it and require little data preparation to be trained. Common difficulties in data that can be solved with feature engineering include features with skewed distributions, high-cardinality categorical features, missing observations, high dimensionality and imbalanced classes in classification tasks. Deep learning has shown promising results in automating feature engineering; however, its effectiveness depends on the complexity and nature of the data being processed, as well as the specific problem being addressed. Typically, it can work well with natural language processing and computer vision problems, while for standard tabular data, tree-based ensembles such as a random forest or gradient boosting are often still superior (and easier to handle). However, tabular deep learning approaches are currently catching up quickly. Hence, manual feature engineering is still often required, but with `mlr3pipelines`, which can simplify the process as much as possible.

Feature Engineering

As we work through this chapter, we will use an adapted version of the Ames housing data (De Cock 2011). We changed the data slightly and introduced some additional (artificial) problems to showcase as many aspects of preprocessing as possible on a single dataset. The modified version is shipped with `mlr3data` and the code to recreate this version of the data from the original raw data can be found at https://github.com/mlr-org/mlr3data/ in the directory **data-raw**. This original dataset was collected as an alternative to the Boston Housing data and is commonly

DOI: 10.1201/9781003402848-9

used to demonstrate feature engineering in ML. Raw and processed versions of the data can be directly loaded from the `AmesHousing` package. The dataset includes 2,930 residential properties (rows) situated in Ames, Iowa, sold between 2006 and 2010. It contains 81 features about various aspects of the property, the size and shape of the lot, and information about its condition and quality. The prediction target is the sale price in USD; hence, it is a regression task.

```
ames = mlr3data::ames_housing
```

9.1 Data Cleaning

As a first step, we explore the data and look for simple problems such as constant or duplicated features. This can be done quite efficiently with a package like `DataExplorer` or `skimr` which can be used to create a large number of informative plots.

Below we summarize the most important findings for data cleaning, but we only consider this aspect in a cursory manner:

```
# 1. `Misc_Feature_2` is a factor with only a single level `Othr`.
summary(ames$Misc_Feature_2)
```

```
Othr
2930
```

```
# 2. `Condition_2` and `Condition_3` are identical.
identical(ames$Condition_2, ames$Condition_3)
```

```
[1] TRUE
```

```
# 3. `Lot_Area` and `Lot_Area_m2` are same data on different scales
cor(ames$Lot_Area, ames$Lot_Area_m2)
```

```
[1] 1
```

For all three problems, simply removing the problematic features (or feature in a pair) might be the best course of action.

```
to_remove = c("Lot_Area_m2", "Condition_3", "Misc_Feature_2")
```

Other typical problems that should be checked are:

1. ID columns, i.e., columns that are unique for every observation should be removed or tagged.
2. NAs not correctly encoded, e.g., as `"NA"` or `""`
3. Semantic errors in the data, e.g., negative `Lot_Area`
4. Numeric features encoded as categorical for learners that can not handle such features.

Before we continue with feature engineering we will create a task, measure, and resampling strategy to use throughout the chapter.

```
tsk_ames = as_task_regr(ames, target = "Sale_Price", id = "ames")
# remove problematic features
tsk_ames$select(setdiff(tsk_ames$feature_names, to_remove))

msr_mae = msr("regr.mae")
rsmp_cv3 = rsmp("cv", folds = 3)
rsmp_cv3$instantiate(tsk_ames)
```

Lastly, we run a very simple experiment to verify our setup works as expected with a simple featureless baseline, note below we set `robust = TRUE` to always predict the *median* sale price as opposed to the *mean*.

```
lrn_baseline = lrn("regr.featureless", robust = TRUE)
lrn_baseline$id = "Baseline"
rr_baseline = resample(tsk_ames, lrn_baseline, rsmp_cv3)
rr_baseline$aggregate(msr_mae)
```

```
regr.mae
   56056
```

9.2   Factor Encoding

Many machine learning algorithm implementations, such as XGBoost (Chen and Guestrin 2016), cannot handle categorical data and so categorical features must be encoded into numerical variables.

```
lrn_xgb = lrn("regr.xgboost", nrounds = 100)
lrn_xgb$train(tsk_ames)
```

```
Error: <TaskRegr:ames> has the following unsupported feature types: factor
```

Categorical features can be grouped by their cardinality, which refers to the number of levels they contain: binary features (two levels), low-cardinality features, and high-cardinality features; there is no universal threshold for when a feature should be considered high-cardinality, and this threshold can even be tuned. For now, we will consider high-cardinality to be features with more than 10 levels:

```
names(which(lengths(tsk_ames$levels()) > 10))
```

```
[1] "Exterior_1st" "Exterior_2nd" "MS_SubClass"  "Neighborhood"
```

Binary features can be trivially encoded by setting one of the feature levels to 1 and the other to 0.

```
names(which(lengths(tsk_ames$levels()) == 2))
```

```
[1] "Alley"        "Central_Air" "Street"
```

One-hot Encoding

Low-cardinality features can be handled by one-hot encoding. One-hot encoding is a process of converting categorical features into a binary representation, where each possible category is represented as a separate binary feature. Theoretically, it is sufficient to create one less binary feature than levels, as setting all binary features to zero is also a valid representation. This is typically called dummy or treatment encoding and is required if the learner is a generalized linear model (GLM) or additive model (GAM).

Impact Encoding

Some learners support handling categorical features but may still crash for high-cardinality features if they internally apply encodings that are only suitable for low-cardinality features, such as one-hot encoding. Impact encoding (Micci-Barreca 2001) is a good approach for handling high-cardinality features. Impact encoding converts categorical features into numeric values. The idea behind impact encoding is to use the target feature to create a mapping between the categorical feature and a numerical value that reflects its importance in predicting the target feature. Impact encoding involves the following steps:

1. Group the target variable by the categorical feature.
2. Compute the mean of the target variable for each group.
3. Compute the global mean of the target variable.
4. Compute the impact score for each group as the difference between the mean of the target variable for the group and the global mean of the target variable.
5. Replace the categorical feature with the impact scores.

Impact encoding preserves the information of the categorical feature while also creating a numerical representation that reflects its importance in predicting the target. Compared to one-hot encoding, the main advantage is that only a single numeric feature is created regardless of the number of levels of the categorical features, hence it is especially useful for high-cardinality features. As information from the target is used to compute the impact scores, the encoding process must be embedded in cross-validation to avoid leakage between training and testing data (Chapter 3).

As well as encoding features, other basic preprocessing steps for categorical features include removing constant features (which only have one level and may have been removed as part of data cleaning), and collapsing levels that occur very rarely. These types of problems can occur as artifacts of resampling as the dataset size is further reduced. Stratification on such features would be an alternative way to mitigate this (Section 3.2.5).

In the code below we use `po("removeconstants")` to remove features with only one level, `po("collapsefactors")` to collapse levels that occur less than 1% of the time in the data, `po("encodeimpact")` to impact-encode high-cardinality features, `po("encode", method = "one-hot")` to one-hot encode low-cardinality features, and finally `po("encode", method = "treatment")` to treatment encode binary features.

```
factor_pipeline =
    po("removeconstants") %>>%
    po("collapsefactors", no_collapse_above_prevalence = 0.01) %>>%
    po("encodeimpact",
        affect_columns = selector_cardinality_greater_than(10),
        id = "high_card_enc") %>>%
    po("encode", method = "one-hot",
        affect_columns = selector_cardinality_greater_than(2),
        id = "low_card_enc") %>>%
    po("encode", method = "treatment",
        affect_columns = selector_type("factor"), id = "binary_enc")
```

Now we can apply this pipeline to our xgboost model to use it in a benchmark experiment; we also compare a simpler pipeline that only uses one-hot encoding to demonstrate performance differences resulting from different strategies.

```
glrn_xgb_impact = as_learner(factor_pipeline %>>% lrn_xgb)
glrn_xgb_impact$id = "XGB_enc_impact"

glrn_xgb_one_hot = as_learner(po("encode") %>>% lrn_xgb)
glrn_xgb_one_hot$id = "XGB_enc_onehot"

bmr = benchmark(benchmark_grid(tsk_ames,
    c(lrn_baseline, glrn_xgb_impact, glrn_xgb_one_hot), rsmp_cv3))
bmr$aggregate(measure = msr_mae)[, .(learner_id, regr.mae)]
```

```
        learner_id regr.mae
1:        Baseline    56056
2: XGB_enc_impact    16068
3: XGB_enc_onehot    16098
```

In this small experiment, we see that the difference between the extended factor encoding pipeline and the simpler one-hot encoding strategy pipeline is only very small. If you are interested in learning more about different encoding strategies, including a benchmark study comparing them, we recommend Pargent et al. (2022).

9.3 Missing Values

A common problem in real-world data is missing values in features. In the Ames dataset, several variables have at least one missing data point:

```
# print first five with missing data
names(which(tsk_ames$missings() > 0))[1:5]
```

```
[1] "Alley"         "BsmtFin_SF_1"   "BsmtFin_SF_2"   "BsmtFin_Type_1"
[5] "BsmtFin_Type_2"
```

Many learners cannot handle missing values automatically (e.g., `lrn("regr.ranger")` and `lrn("regr.lm")`) and others may be able to handle missing values but may use simple methods that are not ideal (e.g., just omitting rows with missing data).

Data
Imputation

The simplest data imputation method is to replace missing values by the feature's mean (`po("imputemean")`) (Figure 9.1), median (`po("imputemedian")`), or mode (`po("imputemode")`). Alternatively, one can impute by sampling from the empirical distribution of the feature, for example, a histogram (`po("imputehist")`). Instead of guessing at what a missing feature might be, missing values could instead be replaced by a new level, for example, called `.MISSING` (`po("imputeoor")`). For numeric features, Ding and Simonoff (2010) show that for binary classification and tree-based models, encoding missing values out-of-range (OOR), e.g., a constant value above the largest observed value, is a reasonable approach.

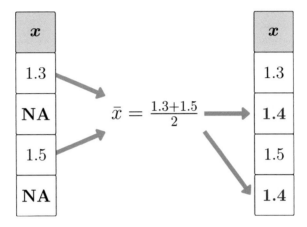

Figure 9.1: Mean imputation of missing values using observed values.

It is often important for predictive tasks that you keep track of missing data as it is common for missing data to be informative in itself. To preserve the information about which data was missing, imputation should be tracked by adding binary indicator features (one for each imputed feature) that are 1 if the feature was missing for an observation and 0 if it was present (`po("missind")`). It is important to note that recording this information will not prevent problems in model interpretation on its own. As a real-world example, medical data are typically collected more extensively for White communities than for racially minoritized communities. Imputing data from minoritized communities would at best mask this data bias, and at worst would make the data bias even worse by making vastly inaccurate assumptions (see Chapter 14 for data bias and algorithmic fairness).

In the code below we create a pipeline from the `PipeOps` listed above as well as making use of `po("featureunion")` to combine multiple `PipeOps` acting on the `"integer"` columns.

```
impute_hist = list(
    po("missind", type = "integer",
        affect_columns = selector_type("integer")
    ),
    po("imputehist", affect_columns = selector_type("integer"))
) %>>%
```

```
    po("featureunion") %>>%
    po("imputeoor", affect_columns = selector_type("factor"))

impute_hist$plot(horizontal = TRUE)
```

Figure 9.2: Pipeline to impute missing values of numeric features by histogram with binary indicators and missings in categoricals out-of-range with a new level.

Using this pipeline, we can now run experiments with `lrn("regr.ranger")`, which cannot handle missing data; we also compare a simpler pipeline that only uses OOR imputation to demonstrate performance differences resulting from different strategies.

```
glrn_rf_impute_hist = as_learner(impute_hist %>>% lrn("regr.ranger"))
glrn_rf_impute_hist$id = "RF_imp_Hist"

glrn_rf_impute_oor = as_learner(po("imputeoor") %>>% lrn("regr.ranger"))
glrn_rf_impute_oor$id = "RF_imp_OOR"

design = benchmark_grid(tsk_ames,
  c(glrn_rf_impute_hist, glrn_rf_impute_oor), rsmp_cv3)
bmr_new = benchmark(design)
bmr$combine(bmr_new)
bmr$aggregate(measure = msr_mae)[, .(learner_id, regr.mae)]
```

```
        learner_id regr.mae
1:        Baseline    56056
2: XGB_enc_impact    16068
3: XGB_enc_onehot    16098
4:     RF_imp_Hist    16377
5:     RF_imp_OOR    16393
```

Similarly to encoding, we see limited differences in performance between the different imputation strategies. This is expected here and confirms the findings of Ding and Simonoff (2010) – out-of-range imputation is a simple yet effective imputation for tree-based methods.

Many more advanced imputation strategies exist, including model-based imputation where machine learning models are used to predict missing values, and multiple imputation where data is repeatedly resampled and imputed in each sample (e.g., by mean imputation) to attain more robust estimates. However, these more advanced techniques rarely improve the models predictive performance substantially and the simple imputation techniques introduced above are usually sufficient (Poulos and Valle 2018). Nevertheless, these methods are still important, as finding imputations that fit well to the distribution of the observed values allows a model to be fitted that can be interpreted and analyzed in a second step.

9.4 Pipeline Robustify

`mlr3pipelines` offers a simple and reusable pipeline for (among other things) imputation and factor encoding called `ppl("robustify")`, which includes sensible ppl("robustify") defaults that can be used most of the time when encoding or imputing data. The pipeline includes the following `PipeOps` (some are applied multiple times and most use selectors):

1. `po("removeconstants")` – Constant features are removed.
2. `po("colapply")` – Character and ordinal features are encoded as categorical, and date/time features are encoded as numeric.
3. `po("imputehist")` – Numeric features are imputed by histogram sampling.
4. `po("imputesample")` – Logical features are imputed by sampling from the empirical distribution – this only affects the `$predict()`-step.
5. `po("missind")` – Missing data indicators are added for imputed numeric and logical variables.
6. `po("imputeoor")` – Missing values of categorical features are encoded with a new level.
7. `po("fixfactors")` – Fixes levels of categorical features such that the same levels are present during prediction and training (which may involve dropping empty factor levels).
8. `po("imputesample")` – Missing values in categorical features introduced from dropping levels in the previous step are imputed by sampling from the empirical distributions.
9. `po("collapsefactors")` – Categorical features levels are collapsed (starting from the rarest factors in the training data) until there are less than a certan number of levels, controlled by the `max_cardinality` argument (with a conservative default of 1000).
10. `po("encode")` – Categorical features are one-hot encoded.
11. `po("removeconstants")` – Constant features that might have been created in the previous steps are removed.

`ppl("robustify")` has optional arguments `task` and `learner`. If these are provided, then the resulting pipeline will be set up to handle the given task and learner specifically, for example, it will not impute missing values if the learner has the `"missings"` property, or if there are no missing values in the task to begin with. By default, when `task` and `learner` are not provided, the graph is set up to be defensive: it imputes all missing values and converts all feature types to numerics.

Linear regression is a simple model that cannot handle most problems that we may face when processing data, but with the `ppl("robustify")` we can now include it in our experiment:

```
glrn_lm_robust = as_learner(ppl("robustify") %>>% lrn("regr.lm"))
glrn_lm_robust$id = "lm_robust"

bmr_new = benchmark(benchmark_grid(tsk_ames, glrn_lm_robust,  rsmp_cv3))
bmr$combine(bmr_new)
bmr$aggregate(measure = msr_mae)[, .(learner_id, regr.mae)]
```

```
        learner_id regr.mae
1:        Baseline     56056
2: XGB_enc_impact       16068
3: XGB_enc_onehot       16098
4:     RF_imp_Hist      16377
5:     RF_imp_OOR       16393
6:      lm_robust       16298
```

Robustifying the linear regression results in a model that vastly outperforms the featureless baseline and is competitive when compared to more complex machine learning models.

9.5   Transforming Features and Targets

Simple transformations of features and the target can be beneficial (and sometimes essential) for certain learners. In particular, log transformation of the target can help in making the distribution more symmetrical and can help reduce the impact of outliers. Similarly, log transformation of skewed features can help to reduce the influence of outliers. In Figure 9.3, we plot the distribution of the target in the ames dataset and then the log-transformed target, we can see how simply taking the log of the variable results in a distribution that is much more symmetrical and with fewer outliers.

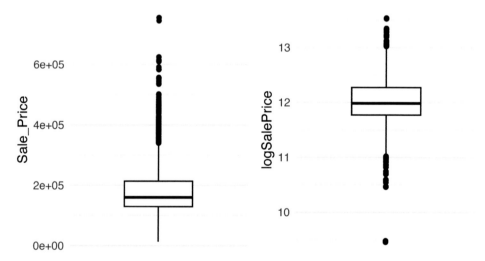

Figure 9.3: Distribution of house sales prices (in USD) in the ames dataset before (left) and after (right) log transformation. Before transformation there is a skewed distribution of prices towards cheaper properties with a few outliers of very expensive properties. After transformation the distribution is much more symmetrical with the majority of points evenly spread around the same range.

```
library(patchwork)

# copy ames data
log_ames = copy(ames)
# log transform target
log_ames[, logSalePrice := log(Sale_Price)]
# plot
autoplot(as_task_regr(log_ames, target = "Sale_Price")) +
  autoplot(as_task_regr(log_ames, target = "logSalePrice"))
```

Normalization of features may also be necessary to ensure features with a larger scale do not have a higher impact, which is especially important for distance-based methods such as k-nearest neighbors models or regularized parametric models such as Lasso or Elastic net. Many models internally scale the data if required by the algorithm, so most of the time we do not need to manually do this in preprocessing, though if this is required then `po("scale")` can be used to center and scale numeric features.

Any transformations applied to the target during training must be inverted during model prediction to ensure predictions are made on the correct scale. By example, say we are interested in log transforming the target, then we would take the following steps:

```
df = data.table(x = runif(5), y = runif(5, 10, 20))
df
```

```
        x       y
1:  0.48004 10.25
2:  0.14466 10.75
3:  0.05795 18.30
4:  0.65004 17.34
5:  0.37355 10.48
```

```
# 1. log transform the target
df[, y := log(y)]
df$y
```

```
[1] 2.327 2.375 2.907 2.853 2.350
```

```
# 2. make linear regression predictions
#    predictions on the log-transformed scale
yhat = predict(lm(y ~ x, df), df)
yhat
```

```
    1     2     3     4     5
2.556 2.571 2.575 2.548 2.561
```

```
# 3. transform to correct scale with inverse of log function
#    predictions on the original scale
exp(yhat)
```

```
   1     2     3     4     5
12.88 13.08 13.13 12.79 12.95
```

In this simple experiment, we could manually transform and invert the target, however, this is much more complex when dealing with resampling and benchmarking experiments and so the pipeline `ppl("targettrafo")` will do this heavy lifting for you. The pipeline includes a parameter `targetmutate.trafo` for the transformation to be applied during training to the target, as well as `targetmutate.inverter` for the transformation to be applied to invert the original transformation during prediction. So now let us consider the log transformation by adding this pipeline to our robust linear regression model:

```
glrn_log_lm_robust = as_learner(ppl("targettrafo",
  graph = glrn_lm_robust,
  targetmutate.trafo = function(x) log(x),
  targetmutate.inverter = function(x) list(response = exp(x$response))))
glrn_log_lm_robust$id = "lm_robust_logtrafo"

bmr_new = benchmark(benchmark_grid(tsk_ames, glrn_log_lm_robust,
  rsmp_cv3))
bmr$combine(bmr_new)
bmr$aggregate(measure = msr_mae)[, .(learner_id, regr.mae)]
```

```
             learner_id regr.mae
1:             Baseline    56056
2:        XGB_enc_impact    16068
3:        XGB_enc_onehot    16098
4:          RF_imp_Hist    16377
5:           RF_imp_OOR    16393
6:            lm_robust    16298
7: lm_robust_logtrafo    15557
```

With the target transformation and the `ppl("robustify")`, the simple linear regression now appears to be the best-performing model.

9.6 Functional Feature Extraction

As a final step of data preprocessing, we will look at feature extraction from functional features. In Chapter 6, we look at automated feature selection and how automated approaches with filters and wrappers can be used to reduce a dataset to an optimized set of features. Functional feature extraction differs from this process as we are now interested in features that are dependent on one another and together may provide useful information but not individually. Figure 9.4 visualizes the difference between regular and functional features.

As a concrete example, consider the power consumption of kitchen appliances in houses in the Ames dataset.

```
energy_data = mlr3data::energy_usage
```

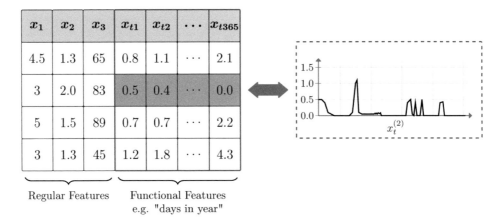

x_1	x_2	x_3	x_{t1}	x_{t2}	$\cdots$	x_{t365}
4.5	1.3	65	0.8	1.1	$\cdots$	2.1
3	2.0	83	0.5	0.4	$\cdots$	0.0
5	1.5	89	0.7	0.7	$\cdots$	2.2
3	1.3	45	1.2	1.8	$\cdots$	4.3

Regular Features   Functional Features
e.g. "days in year"

Figure 9.4: Variables x1, x2, x3 are regular features, variables xt1,..., xt365 are functional features that could be plotted to identify important properties.

In this dataset, each row represents one house and each feature is the total power consumption from kitchen appliances at a given time (Bagnall et al. 2017). The consumption is measured in two-minute intervals, resulting in 720 features.

```
library(ggplot2)
ggplot(data.frame(y = as.numeric(energy_data[1, ])),
    aes(y = y, x = 1:720)) +
  geom_line() + theme_minimal() +
  labs(x = "2-Minute Interval", y = "Power Consumption")
```

Adding these 720 features to our full dataset is a bad idea as each individual feature does not provide meaningful information, similarly, we cannot automate selection of the best feature subset for the same reason. Instead, we can *extract* information about the curves to gain insights into the kitchen's overall energy usage. For example, we could extract the maximum used wattage, overall used wattage, number of peaks, and other similar features.

To extract features we will write our own `PipeOp` that inherits from `PipeOpTaskPreprocSimple`. To do this we add a private method called `.transform_dt` that hardcodes the operations in our task. In this example, we select the functional features (which all start with "att"), extract the mean, minimum, maximum, and variance of the power consumption, and then remove the functional features. To read more about building custom `PipeOps`, open the corresponding vignette by running `vignette("extending", package = "mlr3pipelines")` in R.

```
PipeOpFuncExtract = R6::R6Class("PipeOpFuncExtract",
  inherit = mlr3pipelines::PipeOpTaskPreprocSimple,
  private = list(
    .transform_dt = function(dt, levels) {
      ffeat_names = paste0("att", 1:720)
      ffeats = dt[, ..ffeat_names]
```

```
            dt[, energy_means := apply(ffeats, 1, mean)]
            dt[, energy_mins := apply(ffeats, 1, min)]
            dt[, energy_maxs := apply(ffeats, 1, max)]
            dt[, energy_vars := apply(ffeats, 1, var)]
            dt[, (ffeat_names) := NULL]
            dt
        }
    )
)
```

Before using this in an experiment, we first test that the `PipeOp` works as expected.

```
tsk_ames_ext = cbind(ames, energy_data)
tsk_ames_ext = as_task_regr(tsk_ames_ext, "Sale_Price", "ames_ext")
# remove the redundant variables identified at the start of this chapter
tsk_ames_ext$select(setdiff(tsk_ames_ext$feature_names, to_remove))

func_extractor = PipeOpFuncExtract$new("energy_extract")
tsk_ames_ext = func_extractor$train(list(tsk_ames_ext))[[1]]
tsk_ames_ext$data(1,
  c("energy_means", "energy_mins", "energy_maxs", "energy_vars"))
```

```
   energy_means energy_mins energy_maxs energy_vars
1:        1.062     0.01427       21.98       3.708
```

These outputs look sensible compared to Figure 9.5 so we can now run our final benchmark experiment using feature extraction. We do not need to add the `PipeOp` to each learner as we can apply it once (as above) before any model training by applying it to all available data.

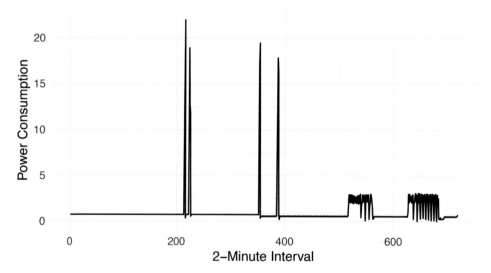

Figure 9.5: Energy consumption of one example house in a day, recorded in two-minute intervals.

```
learners = list(lrn_baseline, lrn("regr.rpart"), glrn_xgb_impact,
    glrn_rf_impute_oor, glrn_lm_robust, glrn_log_lm_robust)

bmr_final = benchmark(benchmark_grid(c(tsk_ames_ext, tsk_ames), learners,
    rsmp_cv3))

perf = bmr_final$aggregate(measure = msr_mae)
perf[order(learner_id, task_id), .(task_id, learner_id, regr.mae)]
```

```
        task_id        learner_id regr.mae
 1:        ames          Baseline    56056
 2:    ames_ext          Baseline    56056
 3:        ames         RF_imp_OOR    16433
 4:    ames_ext         RF_imp_OOR    14317
 5:        ames     XGB_enc_impact    16068
 6:    ames_ext     XGB_enc_impact    14400
 7:        ames          lm_robust    16291
 8:    ames_ext          lm_robust    15093
 9:        ames lm_robust_logtrafo    15555
10:    ames_ext lm_robust_logtrafo    13905
11:        ames          regr.rpart    27371
12:    ames_ext          regr.rpart    27111
```

The final results indicate that adding these extracted features improved the performance of all models (except the featureless baseline).

In this example, we could have just applied the transformations to the dataset directly and not used a `PipeOp`. However, the advantage of using the `PipeOp` is that we could have chained it to a subset of learners to prevent a blow-up of experiments in the benchmark experiment.

9.7 Conclusion

In this chapter, we built on everything learned in Chapters 7 and 8 to look at concrete usage of pipelines for data preprocessing. We focused primarily on feature engineering, which can make use of `mlr3pipelines` to automate preprocessing as much as possible while still ensuring user control. We looked at factor encoding for categorical variables, imputing missing data, transforming variables, and feature extraction. Preprocessing is almost always required in machine learning experiments, and applying the `ppl("robustify")` will help in many cases to simplify this process by applying the most common preprocessing steps, we will see this in use in Chapter 11.

We have not introduced any new classes in this chapter, so instead Table 9.1 lists the `PipeOps` and `Graphs` we discussed.

Table 9.1: `PipeOps` and `Graphs` discussed in this chapter.

PipeOp/Graph	Description
PipeOpRemoveConstants	Remove variables consisting of one value
PipeOpCollapseFactors	Combine rare factor levels
PipeOpEncodeImpact	Impact encoding
PipeOpEncode	Other factor encoding methods
PipeOpMissInd	Add an indicator column to track missing data
PipeOpImputeHist	Impute missing data by sampling from a histogram
PipeOpImputeOOR	Impute missing data with out-of-range values
pipeline_robustify	Graph with common imputation and encoding methods
pipeline_targettrafo	Graph to transform target during training and invert transformation during prediction

9.8 Exercises

We will consider a prediction problem similar to the one from this chapter, but using the King County Housing regression data instead (available with `tsk("kc_housing")`). To evaluate the models, we again use 10-fold CV, mean absolute error and `lrn("regr.glmnet")`. For now we will ignore the `date` column and simply remove it:

```
library("mlr3data")
kc_housing = tsk("kc_housing")
kc_housing$select(setdiff(kc_housing$feature_names, "date"))
```

1. Have a look at the features, are there any features which might be problematic? If so, change or remove them. Check the dataset and learner properties to understand which preprocessing steps you need to do.
2. Build a suitable pipeline that allows `glmnet` to be trained on the dataset. Construct a new `glmnet` model with `ppl("robustify")`. Compare the two pipelines in a benchmark experiment.
3. Now consider the `date` feature: How can you extract information from this feature in a way that `glmnet` can use? Does this improve the performance of your pipeline? Finally, consider the spatial nature of the dataset. Can you extract an additional feature from the lat / long coordinates? (Hint: Downtown Seattle has lat/long coordinates `47.605/122.334`).

Part IV

Advanced Topics

10

Advanced Technical Aspects of mlr3

Michel Lang
Research Center Trustworthy Data Science and Security, and TU Dortmund University

Sebastian Fischer
Ludwig-Maximilians-Universität München, and Munich Center for Machine Learning (MCML)

Raphael Sonabend
Imperial College London

In the previous chapters, we demonstrated how to turn machine learning concepts and methods into code. In this chapter, we will turn to those technical details that can be important for more advanced uses of `mlr3`, including:

- Parallelization with the `future` framework (Section 10.1);
- Error handling and debugging (Section 10.2);
- Adjusting the logger to your needs (Section 10.3);
- Working with out-of-memory data, e.g., data stored in databases (Section 10.4); and
- Adding new classes to `mlr3` (Section 10.5).

10.1 Parallelization

The term parallelization refers to running multiple algorithms in parallel, i.e., executing them simultaneously on multiple CPU cores, CPUs, or computational nodes. Not all algorithms can be parallelized, but when they can, parallelization allows significant savings in computation time.

In general, there are many possibilities to parallelize, depending on the hardware to run the computations. If you only have a single CPU with multiple cores, then *threads* or *processes* are ways to utilize all cores on a local machine. If you have multiple machines on the other hand, they can communicate and exchange information via protocols such as *network sockets* or the *Message Passing Interface*. Larger computational sites rely on scheduling systems to orchestrate the computation for multiple users and usually offer a shared network file system all machines can access. Interacting with scheduling systems on compute clusters is covered in Section 11.2 using the R package `batchtools`.

There are a few pieces of terminology associated with parallelization that we will use in this section:

- The parallelization backend is the hardware to parallelize with a respective interface provided by an R package. Many parallelization backends have different APIs, so we use the `future` package as a unified, abstraction layer for many parallelization backends. From a user perspective, `mlr3` interfaces with `future` directly so all you will need to do is configure the backend before starting any computations. *[Parallelization Backend]*
- The Main process is the R session or process that orchestrates the computational work, called jobs.
- Workers are the R sessions, processes, or machines that receive the jobs, perform calculations, and then send the results back to Main.

An important step in parallel programming involves the identification of sections of the program flow that are both time-consuming ("bottlenecks") and can run independently of a different section, i.e., section A's operations are not dependent on the results of section B's operations, and vice versa. Fortunately, these sections are usually relatively easy to spot for machine learning experiments:

1. Training of a learning algorithm (or other computationally intensive parts of a machine learning pipeline) *may* contain independent sections which can run in parallel, e.g.,
   - A single decision tree iterates over all features to find the best split point, for each feature independently.
   - A random forest usually fits hundreds of trees independently.

   The key principle that makes parallelization possible for these examples (and in general in many fields of statistics and ML) is called data parallelism, which means the same operation is performed concurrently on different elements of the input data. Parallelization of learning algorithms is covered in Section 10.1.1. *[Data Parallelism]*

2. Resampling consists of independent repetitions of train-test-splits and benchmarking consists of multiple independent resamplings (Section 10.1.2).

3. Tuning (Chapter 4) often is iterated benchmarking, embedded in a sequential procedure that determines the hyperparameter configurations to try next. While many tuning algorithms are inherently sequential to some degree, there are some (e.g., random search) that can propose multiple configurations in parallel to be evaluated independently, providing another level for parallelization (Section 10.1.4).

4. Predictions of a single learner for multiple observations can be computed independently (Section 10.1.5).

These examples are referred to as "embarrassingly parallel" as they are so easy to parallelize. If we can formulate the problem as a function that can be passed to map-like functions such as `lapply()`, then you have an embarrassingly parallel problem. However, just because a problem *can* be parallelized, it does not follow that every operation in a problem *should* be parallelized. Starting and terminating workers as well as possible communication between workers comes at a price in the form of additionally required runtime which is called parallelization overhead. This overhead strongly varies between parallelization backends and must be carefully weighed against the runtime of the sequential execution to determine if parallelization is worth the effort. If the sequential execution is comparably fast, enabling parallelization may introduce additional complexity with little runtime savings, or could even slow down the execution. It is possible to control the granularity of the parallelization to *[Embarrassingly Parallel]* *[Parallelization Overhead]* *[Granularity]* reduce the parallelization overhead. For example, we could reduce the overhead of parallelizing a `for`-loop with 1000 iterations on four CPU cores by chunking the work of the 1000 jobs into four computational jobs performing 250 iterations each, resulting in four big jobs and not 1000 small ones.

This effect is illustrated in the following code chunk using a socket cluster with the `parallel` package, which has a `chunk.size` option so we do not need to manually create chunks:

```
# set up a socket cluster with 4 workers on the local machine
library(parallel)
cores = 4
cl = makeCluster(cores)

# vector to operate on
x = 1:10000

# fast function to parallelize
f = function(y) sqrt(y + 1)

# unchunked approach: 1000 jobs
system.time({parSapply(cl, x, f, chunk.size = 1)})
```

```
 user  system elapsed
0.504   0.120   0.780
```

```
# chunked approach: 4 jobs
system.time({parSapply(cl, x, f, chunk.size = 2500)})
```

```
 user  system elapsed
0.003   0.001   0.013
```

Synchronization Overhead

Whenever you have the option to control the granularity by setting the chunk size, you should aim for at least as many jobs as workers. However, if there are too few job chunks with strongly dissimilar runtimes, the system may end up waiting for the last chunk to finish, while other resources are idle. This is referred to as synchronization overhead. You should therefore aim for chunks with a runtime of at least several seconds, so that the parallelization overhead remains reasonable, while still having enough chunks to ensure that you can fully utilize the system. If you have heterogeneous runtimes, you can consider grouping jobs so that the runtimes of the chunks are more homogeneous. If runtimes can be estimated, then both `batchtools::binpack()` and `batchtools::lpt()` (documented together with the `chunk()` function) are useful for chunking jobs. If runtimes cannot be estimated, then it can be useful to randomize the order of jobs. Otherwise jobs could be accidentally ordered by runtime, for example, because they are sorted by a hyperparameter that has a strong influence on training time. Naively chunking jobs could then lead to some chunks containing much more expensive jobs than others, resulting in avoidable underutilization of resources. `mlr3misc` ships with the functions `chunk()` and `chunk_vector()` that conveniently chunk jobs and also shuffle them by default. There are also options to control the chunk size for parallelization in `mlr3`, which are discussed in Section 10.1.2.

> 💡 Reproducibility
>
> Reproducibility is often a concern during parallelization because special Pseudorandom number generators (PRNGs) may be required (Bengtsson 2020). However, `future` ensures that all workers will receive the same PRNG streams, independent of the number of workers (Bengtsson 2020). Therefore, `mlr3` experiments will be reproducible as long as you use `set.seed` at the start of your scripts (with the PRNG of your choice).

10.1.1 Parallelization of Learners

At the lowest level, external code can be parallelized if available in underlying implementations. For example, while fitting a single decision tree, each split that divides the data into two disjoint partitions requires a search for the best cut point on all p features. Instead of iterating over all features sequentially, the search can be broken down into p threads, each searching for the best cut point on a single feature. These threads can then be scheduled depending on available CPU cores, as there is no need for communication between the threads. After all the threads have finished, the results are collected and merged before terminating the threads. The p best-cut points per feature are collected and aggregated to the single best-cut point across all features by iterating over the p results sequentially.

> 💡 GPU Computation
>
> Parallelization on GPUs is not covered in this book. `mlr3` only distributes the fitting of multiple learners, e.g., during resampling, benchmarking, or tuning. On this rather abstract level, GPU parallelization does not work efficiently. However, some learning procedures can be compiled against CUDA/OpenCL to utilize the GPU while fitting a single model. We refer to the respective documentation of the learner's implementation, e.g., https://xgboost.readth edocs.io/en/stable/gpu/ for XGBoost.

Threading is implemented in the compiled code of the package (e.g., in C or C++), which means that the R interpreter calls the external code and waits for the results to be returned, without noticing that the computations are executed in parallel. Therefore, threading can conflict with certain parallel backends, leading the system to be overutilized in the best-case scenario, or causing hangs or segfaults in the worst case. For this reason, we introduced the convention that threading parallelization is turned off by default. Hyperparameters that control the number of threads are tagged with the label `"threads"`:

```
lrn_ranger = lrn("classif.ranger")

# show all hyperparameters tagged with "threads"
lrn_ranger$param_set$ids(tags = "threads")
```

```
[1] "num.threads"
```

```
# The number of threads is initialized to 1
lrn_ranger$param_set$values$num.threads
```

[1] 1

To enable the parallelization for this learner, `mlr3` provides the helper function `set_threads()`, which automatically adjusts the hyperparameters associated with builtin learner parallelization:

```
# use four CPUs
set_threads(lrn_ranger, n = 4)
```

```
<LearnerClassifRanger:classif.ranger>
* Model: -
* Parameters: num.threads=4
* Packages: mlr3, mlr3learners, ranger
* Predict Types:  [response], prob
* Feature Types: logical, integer, numeric, character, factor,
  ordered
* Properties: hotstart_backward, importance, multiclass,
  oob_error, twoclass, weights
```

If we did not specify an argument for the **n** parameter then the default is a heuristic to detect the correct number using `availableCores()`. This heuristic is not always ideal (interested readers might want to look up "Amdahl's Law") and utilizing all available cores is occasionally counterproductive and can slow down overall runtime (Bengtsson 2022), moreover using all cores is not ideal if:

- You want to simultaneously use your system for other purposes.
- You are on a multi-user system and want to spare some resources for other users.
- You have linked R to a threaded BLAS implementation like OpenBLAS and your learners make heavy use of linear algebra.

```
# auto-detect cores on the local machine
set_threads(lrn_ranger)
```

```
<LearnerClassifRanger:classif.ranger>
* Model: -
* Parameters: num.threads=8
* Packages: mlr3, mlr3learners, ranger
* Predict Types:  [response], prob
* Feature Types: logical, integer, numeric, character, factor,
  ordered
* Properties: hotstart_backward, importance, multiclass,
  oob_error, twoclass, weights
```

To control how many cores are set, we recommend manually setting the number of CPUs in your system's `.Rprofile` file:

```
options(mc.cores = 4)
```

There are also other approaches for parallelization of learners, e.g., by directly supporting one specific parallelization backend or a parallelization framework like foreach. If this is supported, parallelization must be explicitly activated, e.g., by setting a hyperparameter. If you need to parallelize on the learner level because a single model fit takes too much time, and you only fit a few of these models, consult the documentation of the respective learner. In many scenarios, it makes more sense to parallelize on a different level like resampling or benchmarking which is covered in the following subsections.

10.1.2 Parallelization of Resamplings and Benchmarks

In addition to parallel learners, most machine learning experiments can be easily parallelized during resampling. By definition, resampling is performed by aggregating over independent repetitions of multiple train-test splits.

mlr3 makes use of future to enable parallelization over resampling iterations using the parallel backend, which can be configured by the user via the plan() function.

By example, we will look at parallelizing three-fold CV for a decision tree on the sonar task (Figure 10.1). We use the multisession plan (which internally uses socket clusters from the **parallel** package) that should work on all operating systems.

```
library(future)

# select the multisession backend to use
future::plan("multisession")

# run our experiment
tsk_sonar = tsk("sonar")
lrn_rpart = lrn("classif.rpart")
rsmp_cv3 = rsmp("cv", folds = 3)
system.time({resample(tsk_sonar, lrn_rpart, rsmp_cv3)})
```

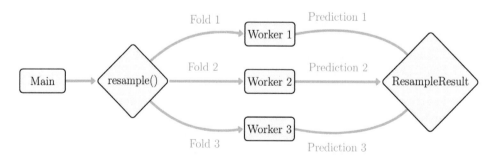

Figure 10.1: Parallelization of a resampling using three-fold CV. The main process calls the resample() function, which starts the parallelization process and the computational task is split into three parts for three-fold CV. The folds are passed to three workers, each fitting a model on the respective subset of the task and predicting on the left-out observations. The predictions (and trained models) are communicated back to the main process which combines them into a ResampleResult.

```
  user   system elapsed
 0.068   0.003   0.481
```

By default, all CPUs of your machine are used unless you specify the argument `workers` in `future::plan()` (see the previous section for issues that this might cause). In contrast to threads, the technical overhead for starting workers, communicating objects, sending back results, and shutting down the workers is quite large for the `"multisession"` backend.

The `multicore` backend comes with more overhead than threading, but considerably less overhead than `"multisession"`, as the `"multicore"` backend only copies R objects when modified ("copy-on-write"), whereas objects are always copied to the respective session before any computation for `"multisession"`. The `"multicore"` backend has the major disadvantage that it is not supported on Windows systems – for this reason, we will stick with the `"multisession"` backend for all examples here.

In general, it is advised to only consider parallelization for resamplings where each iteration runs at least a few seconds. There are two `mlr3` options to control the execution and granularity:

- If `mlr3.exec_random` is set to `TRUE` (default), the order of jobs is randomized in resamplings and benchmarks. This can help if you run a benchmark or tuning with heterogeneous runtimes.
- Option `mlr3.exec_chunk_size` can be used to control how many jobs are mapped to a single `future` and defaults to 1. The value of this option is passed to `future_mapply()` and `future.scheduling` is constantly set to `TRUE`.

Tuning the chunk size can help in some rare cases to mitigate the parallelization overhead but is unlikely to be useful in larger problems or longer runtimes.

Benchmarks can be seen as a collection of multiple independent resamplings where a combination of a task, a learner, and a resampling strategy defines one resampling to perform. In pseudo-code, the calculation can be written as

```
foreach combination of (task, learner, resampling strategy) {
    foreach resampling iteration {
        execute(resampling, j)
    }
}
```

Therefore, we could either:

1. Parallelize over all resamplings and execute each resampling sequentially (parallelize outer loop); or
2. Iterate over all resamplings and execute each resampling in parallel (parallelize inner loop).

`mlr3` simplifies this decision for you by flattening all experiments to the same level, i.e., `benchmark()` iterates over the elements of the Cartesian product of the iterations of the outer and inner loops. Therefore, there is no need to decide whether you want to parallelize the tuning *or* the resampling, you always parallelize both. This approach makes the computation fine-grained and allows the `future` backend to group the jobs into chunks of suitable size (depending on the number of workers), it also makes the procedure identical to parallelizing resampling:

```
# simple benchmark design
design = benchmark_grid(tsks(c("sonar", "penguins")),
  lrns(c("classif.featureless", "classif.rpart")), rsmp_cv3)

# enable parallelization
future::plan("multisession")

# run benchmark in parallel
bmr = benchmark(design)
```

See Section 11.2 for larger benchmark experiments that may have a cumulative runtime of weeks, months or even years.

10.1.3   Parallelization of Tuning

Tuning is usually an iterative procedure, consisting of steps that are themselves embarrassingly parallel. In each iteration, a tuner proposes a batch of hyperparameter configurations (which could be of size 1), which can then be evaluated in parallel. After each iteration, most tuners adapt themselves in some way based on the obtained performance values. Random and grid search are exceptions as they do not choose configurations based on past results, instead, for these tuners, all evaluations are independent and can, in principle, be fully parallelized.

Tuning is implemented in `mlr3` as iterative benchmarks. The `Tuner` proposes a batch of learners, each with a different configuration in its `$param_set$values`, where the size of the batch can usually be controlled with the `batch_size` configuration parameter. This batch is passed to `benchmark()` with the resampling strategy of the tuning instance.

Since each call to `benchmark()` depends on previous results, it is generally not possible to parallelize tuning at a higher "level" than individual benchmarks. Instead, the individual `benchmark()` evaluations are parallelized by `mlr3` as if they were experiments without tuning. This means that the individual resampling iterations of each evaluated configuration are all parallelized at the same time. To ensure full parallelization, make sure that the `batch_size` multiplied by the number of resampling iterations is at least equal to the number of available workers. If you expect homogeneous runtimes, i.e., you are tuning over a single learner or pipeline without any hyperparameters with a large influence on the runtime, aim for a multiple of the number of workers. In general, larger batches allow for more parallelization, while smaller batches imply a more frequent evaluation of the termination criteria. Independently of whether you use parallelization, the termination criteria are only checked between evaluations of batches.

The following code shows a parallelized execution of random search with the termination criterion set to 20 iterations and a moderate batch size, where 36 resampling splits – 12 configurations of three splits each – are evaluated in parallel on four workers. The batch size, set to a multiple of the number of workers, ensures that available resources are used efficiently. However, note that the tuning only terminates after a multiple of the given batch size, in this case after 24 evaluations.

```
future::plan("multisession", workers = 4)

instance = tune(
  tnr("random_search", batch_size = 12),
  tsk("penguins"),
  lrn("classif.rpart", minsplit = to_tune(2, 128)),
  rsmp("cv", folds = 3),
  term_evals = 20
)


instance$archive$n_evals
```

```
[1] 24
```

In this example, we could have increased the batch size to 20 to make use of available resources in the most efficient way while stopping exactly at the number of evaluations, however this does not generalize to other termination criteria where we do not know the number of evaluations in advance. For example, if we used `trm("perf_reached")` with a batch size of 12, then if the first configuration of the batch yielded better performance than the given threshold, the remaining 11 configurations would still be unnecessarily evaluated.

10.1.4 Nested Resampling Parallelization

Nested resampling can conceptually be parallelized at three different levels, each corresponding to jobs of different granularity:

1. The parallelization of the outer resampling. A job is then the tuning of a learner on the respective training set of the outer resampling splits.
2. The parallel evaluation of the batch of hyperparameter configurations proposed in one tuning iteration. A job is then, for example, the cross-validation of such a configuration.
3. The parallelization of the inner resampling in tuning. A job is then a train-predict-score step of a single configuration.

This is demonstrated in the pseudocode below, which is a simplified form of Algorithm 3 from Bischl et al. (2023):

```
# outer resampling, level 1:
for (i in seq_len(n_outer_splits)) {
  # tuning instance, in this example mainly represents the archive
  tuning_inst = ti(...)
  inner_task = get_training_task(task, outer_splits[[i]])
  # tuning loop, the details of which depend on the tuner being used
  # This does not correspond to a level:
  while (!tuning_inst$is_terminated) {
    proposed_points = propose_points(tuning_inst$archive, batch_size)
    # Evaluation of configurations, level 2:
    for (hp_configuration in proposed_points) {
      split_performances = numeric()
      # Inner resampling, level 3:
```

```
    for (j in seq_len(n_inner_splits)) {
      split_performances[j] = evaluate_performance(
        learner, hp_configuration, inner_task, inner_splits[[j]]
      )
    }
    performance = aggregate(split_performances)
    update_archive(tuning_inst$archive, configuration, performance)
  }
}
evaluate_performance(
  learner, tuning_inst$result, task, outer_splits[[i]]
)
}
```

This algorithm is implemented in `mlr3` in a slightly more efficient manner. At the second level (the evaluation of hyperparameter configurations), it exploits the functionality of `benchmark()`: a `Learner` object is created for each proposed hyperparameter configuration and all learners are resampled in a benchmark experiment in the innermost for-loop, effectively executing the second level along with the third level on a finer granularity (number of proposed points times number of inner resampling iterations). Hence, when parallelizing nested resampling in `mlr3`, the user only has to choose between two options: parallelizing the outer resampling or the inner benchmarking.

By example, let us tune the `minsplit` argument of a classification tree using an `AutoTuner` (Section 4.2) and random search with only two iterations. Note that this is a didactic example to illustrate the interplay of the different parallelization levels and not a realistic setup. We use holdout for inner resampling and set the `batch_size` to 2, which yields two independent iterations in the inner benchmark experiment. A five-fold CV is used for our outer resampling. For the sake of simplicity, we will also ignore the final model fit the `AutoTuner` performs after tuning. Below, we run the example sequentially without parallelization:

```
library(mlr3tuning)
# reset to default sequential plan
future::plan("sequential")

lrn_rpart = lrn("classif.rpart",
  minsplit = to_tune(2, 128))

lrn_rpart_tuned = auto_tuner(tnr("random_search", batch_size = 2),
  lrn_rpart, rsmp("holdout"), msr("classif.ce"), 2)

rr = resample(tsk("penguins"), lrn_rpart_tuned, rsmp("cv", folds = 5))
```

We can now either opt to parallelize the outer CV or the inner benchmarking. Let us assume we have a single CPU with four cores (C1 - C4) available and each inner holdout evaluation during tuning takes four seconds. If we parallelize the outer five-fold CV (Figure 10.2), each of the four cores would run one outer resampling first, the computation of the fifth iteration has to wait as there are no more available cores.

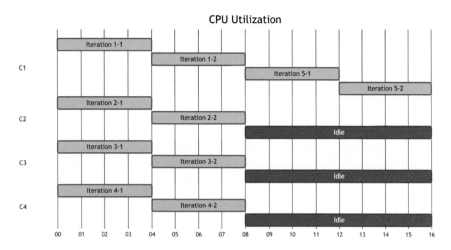

Figure 10.2: CPU utilization for four CPUs while parallelizing the outer five-fold CV with a sequential two-fold CV inside. Jobs are labeled as [iteration outer]-[iteration inner].

```
# Parallelize outer loop
future::plan(list("multisession", "sequential"))

# Alternative: skip specification of 2nd level, since future
# sets all levels after the first to "sequential" by default
future::plan("multisession")
```

This approach is illustrated in Figure 10.2. Each of the four workers starts with the computation of a different inner benchmark, each of which runs sequentially and therefore takes eight seconds on one worker. As there are more jobs than workers, the remaining fifth iteration of the outer resampling is queued on C1 **after** the first four iterations are finished after eight seconds. During the computation of the fifth outer resampling iteration, only C1 is busy, the other three cores are idle.

In contrast, if we parallelize the inner benchmark (Figure 10.3) then the outer resampling runs sequentially: the five inner benchmarks are scheduled one after the other, each of which runs its two holdout evaluations in parallel on two cores; meanwhile, C3 and C4 are idle.

```
# Parallelize inner loop
future::plan(list("sequential", "multisession"))
```

In this example, both possibilities for parallelization are not exploiting the full potential of the four cores. With parallelization of the outer loop, all results are computed after 16 seconds, if we parallelize the inner loop we obtain them after 20 seconds, and in both cases some CPU cores remain idle for at least some of the time.

`mlr3` and `future` make it possible to enable parallelization for both loops for nested parallelization, even on different parallelization backends, which can be useful in

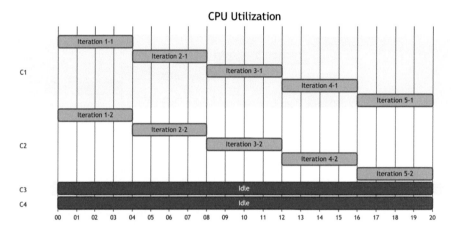

Figure 10.3: CPU utilization for four cores while parallelizing the inner benchmarking (consisting of two holdout evaluations) with a sequential five-fold CV outside. Jobs are labeled as [iteration outer]-[iteration inner].

some distributed computing setups. Note that the detection of available cores does not work for such a nested parallelization and the number of workers must be manually set instead:

```
# Runs both loops in parallel
future::plan(list(
  tweak("multisession", workers = 2),
  tweak("multisession", workers = 2)
))
```

This example would run on up to four cores on the local machine: first, two new sessions would be spawned for the outer loop. Both new sessions then spawn two additional sessions each to evaluate the inner benchmark. Although two cores are still idle when the fifth outer resampling iteration runs, this approach reduces the total runtime to 12 seconds, which is optimal in this example.

10.1.5 Parallelization of Predictions

Finally, predictions from a single learner can be parallelized as the predictions of multiple observations are independent. For most learners, training is the bottleneck and parallelizing the prediction is not a worthwhile endeavor, but there can be exceptions, e.g., if your test dataset is very large.

To predict in parallel, the test data is first split into multiple groups and the predict method of the learner is applied to each group in parallel using an active backend configured via `plan()`. The resulting predictions are then combined internally in a second step. To avoid predicting in parallel accidentally, parallel predictions must be enabled in the learner via the `parallel_predict` field:

```
# train random forest on sonar task
tsk_sonar = tsk("sonar")
lrn_rpart = lrn("classif.rpart")
lrn_rpart$train(tsk_sonar)

# set up parallel predict on four workers
future::plan("multisession", workers = 4)
lrn_rpart$parallel_predict = TRUE

# predict
prediction = lrn_rpart$predict(tsk_sonar)
```

10.2 Error Handling

In large experiments, it is not uncommon that a model fit or prediction fails with an error. This is because the algorithms have to process arbitrary data, and not all eventualities can always be handled. While we try to identify obvious problems before execution, such as when missing values occur for a learner that cannot handle them, other problems are far more complex to detect. Examples include numerical problems that may cause issues in training (e.g., due to lack of convergence), or new levels of categorical variables appearing in the prediction step. Different learners behave quite differently when encountering such problems: some models signal a warning during the training step that they failed to fit but return a baseline model, while other models stop the execution. During prediction, some learners error and refuse to predict the response for observations they cannot handle, while others may predict NA. In this section, we will discuss how to prevent these errors from causing the program to stop when we do not want it to (e.g., during a benchmark experiment).

For illustration (and internal testing) of error handling, mlr3 ships with lrn("classif.debug") and lrn("regr.debug"):

```
tsk_penguins = tsk("penguins")
lrn_debug = lrn("classif.debug")
lrn_debug
```

```
<LearnerClassifDebug:classif.debug>: Debug Learner for Classification
* Model: -
* Parameters: list()
* Packages: mlr3
* Predict Types:  [response], prob
* Feature Types: logical, integer, numeric, character, factor,
  ordered
* Properties: hotstart_forward, missings, multiclass, twoclass
```

This learner lets us simulate problems that are frequently encountered in ML. It can be configured to stochastically trigger warnings, errors, and even segfaults, during training or prediction.

With the learner's default settings, the learner will remember a random label and constantly predict this label without signaling any conditions. In the following code we tell the learner to signal an error during the training step:

```
# set probability to signal an error to `1`
lrn_debug$param_set$values$error_train = 1
lrn_debug$train(tsk_penguins)
```

```
Error in .__LearnerClassifDebug__.train(self = self, private = private, :
Error from classif.debug->train()
```

Now we can look at how to deal with errors during `mlr3` experiments.

10.2.1   Encapsulation

Encapsulation ensures that signaled conditions (e.g., messages, warnings and errors) are intercepted and that all conditions raised during the training or prediction step are logged into the learner without interrupting the program flow. This means that models can be used for fitting and predicting and any conditions can be analyzed post hoc. However, the result of the experiment will be a missing model and/or predictions, depending on where the error occurs. In Section 10.2.2, we will discuss fallback learners to replace missing models and/or predictions.

Each `Learner` contains the field `$encapsulate` to control how the train or predict $encapsulate
steps are wrapped. The first way to encapsulate the execution is provided by the package `evaluate`, which evaluates R expressions and captures and tracks conditions (outputs, messages, warnings or errors) without letting them stop the process (see documentation of `encapsulate()` for full details):

```
# trigger warning and error in training
lrn_debug = lrn("classif.debug", warning_train = 1, error_train = 1)

# enable encapsulation for train() and predict()
lrn_debug$encapsulate = c(train = "evaluate", predict = "evaluate")
lrn_debug$train(tsk_penguins)
```

Note how we passed `"evaluate"` to `train` and `predict` to enable encapsulation in both training and predicting. However, we could have only set encapsulation for one of these stages by instead passing `c(train = "evaluate", predict = "none")` or `c(train = "none", predict = "evaluate")`.

Note that encapsulation captures all output written to the standard output (stdout) and standard error (stderr) streams and stores them in the learner's log. However, in some computational setups, the calling process needs to operate on the log output, such as the `batchtools` package in Chapter 11. In this case, use the encapsulation method `"try"` instead, which catches signaled conditions but does not suppress the output.

After training the learner, one can access the log via the fields `log`, `warnings` and `errors`:

```
lrn_debug$log
```

```
       stage    class                                          msg
1:  train  warning  Warning from classif.debug->train()
2:  train    error    Error from classif.debug->train()
```

```
  lrn_debug$warnings
```

```
[1] "Warning from classif.debug->train()"
```

```
  lrn_debug$errors
```

```
[1] "Error from classif.debug->train()"
```

Another encapsulation method is implemented in the `callr` package. In contrast to `evaluate`, the computation is handled in a separate R process. This guards the calling session against segmentation faults which otherwise would tear down the complete main R session (if we demonstrate that here we would break our book). On the downside, starting new processes comes with comparably more computational overhead.

```
  lrn_debug$encapsulate = c(train = "callr", predict = "callr")
  # set segfault_train and remove warning_train and error_train
  lrn_debug$param_set$values = list(segfault_train = 1)
  lrn_debug$train(task = tsk_penguins)$errors
```

```
[1] "callr process exited with status -11"
```

As well as catching errors, we can also set a timeout, in seconds, so that learners do not run for an indefinite time (e.g., due to failing to converge) but are terminated after a specified time. This works most reliably when using `callr` encapsulation, since the `evaluate` method is sometimes not able to interrupt a learner if it gets stuck in external compiled code. If learners are interrupted, then this is logged as an error by the encapsulation process. Again, the timeout can be set separately for training and prediction:

```
  # near instant timeout for training, no timeout for predict
  lrn_debug$timeout = c(train = 1e-5, predict = Inf)
  lrn_debug$train(task = tsk_penguins)$errors
```

```
[1] "reached elapsed time limit"
```

With these methods, we can now catch all conditions and post hoc analyze messages, warnings and errors.

Unfortunately, catching errors and ensuring an upper time limit is only half the battle. If there are errors during training then we will not have a trained model to query, or if there are errors during predicting, then we will not have predictions to analyze:

```
  # no saved model as there was an error during training
  lrn("classif.debug", error_train = 1)$train(tsk_penguins)$model
```

```
Error in .__LearnerClassifDebug__.train(self = self, private = private, :
Error from classif.debug->train()
```

```
# saved model
lrn_debug = lrn("classif.debug", error_predict = 1)$train(tsk_penguins)
lrn_debug$model
```

```
$response
[1] "Adelie"

$pid
[1] 13523

$iter
NULL

$id
[1] "09512759-941c-4359-a701-9305513a021b"

attr(,"class")
[1] "classif.debug_model"
```

```
#  but no predictions due to an error during predicting
lrn_debug$predict(tsk_penguins)
```

```
Error in .__LearnerClassifDebug__.predict(self = self, private = private, :
Error from classif.debug->predict()
```

Missing learners and/or predictions are particularly problematic during automated processes such as resampling, benchmarking, or tuning (Section 5.1.1), as results cannot be aggregated properly across iterations. In the next section, we will look at fallback learners that impute missing models and predictions.

10.2.2 Fallback Learners

Say an error has occurred when training a model in one or more iterations during resampling, then there are three methods to proceed with our experiment:

1. Ignore iterations with failures – This might be the most frequent approach in practice, however, it is **not** statistically sound. Say we are trying to evaluate the performance of a model. This model might error if in some resampling splits, there are factor levels during predicting that were not seen during training, thus leading to the model being unable to handle these and erroring. If we discarded failed iterations, our model would appear to perform well despite it failing to make predictions for an entire class of features.
2. Penalize failing learners – Instead of ignoring failed iterations, we could impute the worst possible score (as defined by a given Measure) and thereby heavily penalize the learner for failing. However, this will often be too harsh for many problems, and for some measures, there is no reasonable value to impute.

Fallback
Learner

3. Train and predict with a fallback learner – Instead of imputing with the worst possible score, we could train a baseline learner and make predictions from this model.

We strongly recommend the final option, which is statistically sound and can be easily used in any practical experiment. `mlr3` includes two baseline learners: `lrn("classif.featureless")`, which, in its default configuration, always predicts the majority class, and `lrn("regr.featureless")`, which predicts the average response by default.

$fallback

To make this procedure convenient during resampling and benchmarking, we support fitting a baseline (though in theory you could use any **Learner**) as a fallback learner by passing a **Learner** to `$fallback`. In the next example, we add a classification baseline to our debug learner, so that when the debug learner errors, `mlr3` falls back to the predictions of the featureless learner internally. Note that while encapsulation is not enabled explicitly, it is automatically enabled and set to `"evaluate"` if a fallback learner is added.

```
lrn_debug = lrn("classif.debug", error_train = 1)
lrn_debug$fallback = lrn("classif.featureless")

lrn_debug$train(tsk_penguins)
lrn_debug
```

```
<LearnerClassifDebug:classif.debug>: Debug Learner for Classification
* Model: -
* Parameters: error_train=1
* Packages: mlr3
* Predict Types:  [response], prob
* Feature Types: logical, integer, numeric, character, factor,
  ordered
* Properties: hotstart_forward, missings, multiclass, twoclass
* Errors: Error from classif.debug->train()
```

The learner's log contains the captured error, and although no model is stored as the error was in training, we can still obtain predictions from our fallback:

```
lrn_debug$log
```

```
  stage class                                msg
1: train error Error from classif.debug->train()
```

```
lrn_debug$model
```

```
NULL
```

```
prediction = lrn_debug$predict(tsk_penguins)
prediction$score()
```

```
classif.ce
   0.5581
```

In the following snippet, we compare the debug learner with a simple classification tree. We re-parametrize the debug learner to fail in roughly 50% of the resampling iterations during the training step:

```
lrn_debug = lrn("classif.debug", error_train = 0.5)
lrn_debug$fallback = lrn("classif.featureless")

aggr = benchmark(benchmark_grid(
  tsk_penguins,
  list(lrn_debug, lrn("classif.rpart")),
  rsmp("cv", folds = 20)))$aggregate(conditions = TRUE)
aggr[, .(learner_id, warnings, errors, classif.ce)]
```

```
      learner_id warnings errors classif.ce
1: classif.debug        0     12    0.61944
2: classif.rpart        0      0    0.05523
```

Even though the debug learner occasionally failed to provide predictions, we still obtained a statistically sound aggregated performance value which we can compare to the aggregated performance of the classification tree. It is also possible to split the benchmark up into separate `ResampleResult` objects which sometimes helps to get more context. For example, if we only want to have a closer look into the debug learner, we can extract the errors from the corresponding resample results:

```
rr = aggr[learner_id == "classif.debug"]$resample_result[[1L]]
rr$errors[1:2]
```

```
   iteration                              msg
1:         2 Error from classif.debug->train()
2:         4 Error from classif.debug->train()
```

In summary, combining encapsulation and fallback learners makes it possible to benchmark and tune unreliable or unstable learning algorithms in a convenient and statistically sound fashion.

10.3 Logging

`mlr3` uses the `lgr` package to control the verbosity of the output, i.e., to decide how much output is shown when `mlr3` operations are run, from suppression of all non-critical messages to detailed messaging for debugging. In this section, we will cover how to change logging levels, redirect output, and finally change the timing of logging feedback.

`mlr3` uses the following verbosity levels from `lgr`:

- `"warn"` – Only non-breaking warnings are logged
- `"info"` – Information such as model runtimes are logged, as well as warnings
- `"debug"` – Detailed messaging for debugging, as well as information and warnings

The default log level in `mlr3` is `"info"`, this means that messages are only displayed for messages that are informative or worse, i.e., `"info"` and `"warn"`.

To change the logging threshold you need to retrieve the R6 logger object from `lgr`, and then call `$set_threshold()`, for example, to lower the logging threshold to enable debugging messaging we would change the threshold to `"debug"`:

```
lgr::get_logger("mlr3")$set_threshold("debug")
```

Or to suppress all messaging except warnings:

```
lgr::get_logger("mlr3")$set_threshold("warn")
```

`lgr` comes with a global option called `"lgr.default_threshold"` which can be set via `options()` to make your choice permanent across sessions (note this will affect all packages using `lgr`), e.g., `options(lgr.default_threshold = "info")`.

The packages in `mlr3` that make use of optimization, i.e., `mlr3tuning` or `mlr3fselect`, use the logger of their base package `bbotk`. This means you could disable "info"-logging from the `mlr3` logger, but keep the output from `mlr3tuning`:

```
lgr::get_logger("mlr3")$set_threshold("warn")
lgr::get_logger("bbotk")$set_threshold("info")
```

By default, output from `lgr` is printed in the console, however, you could choose to redirect this to a file in various formats, for example to a JSON file:

```
tf = tempfile("mlr3log_", fileext = ".json")

# get the logger as R6 object
logger = lgr::get_logger("mlr")

# add Json appender
logger$add_appender(lgr::AppenderJson$new(tf), name = "json")

# signal a warning
logger$warn("this is a warning from mlr3")
```

```
WARN  [15:32:51.496] this is a warning from mlr3
```

```
# print the contents of the file (splitting over two lines)
x = readLines(tf)
cat(paste0(substr(x, 1, 71), "\n", substr(x, 72, nchar(x))))
```

```
{"level":300,"timestamp":"2023-07-04 15:32:51","logger":"mlr","caller":
"eval","msg":"this is a warning from mlr3"}
```

```
# remove the appender again
logger$remove_appender("json")
```

See the vignettes in the `lgr` for more comprehensive examples.

When using parallelization and/or encapsulation, logs may be delayed, out of order, or, in case of some errors, not present at all. When it is necessary to have immediate access to log messages, e.g., when debugging, one may choose to disable `future` and encapsulation. To enable "debug mode", set `options(mlr3.debug = TRUE)` and ensure the `$encapsulate` slot of learners is set to `"none"` (default) or `"evaluate"`. Debug mode should only be enabled during debugging and not in production use as it disables parallelization and leads to unexpected RNG behavior that prevents reproducibility.

10.4 Data Backends

`Task` objects store their data in an abstract data object, the `DataBackend`. A data backend provides a unified API to retrieve subsets of the data or query information about it, regardless of how the data is stored on the system. The default backend uses `data.table` via the `DataBackendDataTable` class as a very fast and efficient in-memory database.

While storing the task's data in memory is most efficient for accessing it for model fitting, there are two major disadvantages:

1. Even if only a small proportion of the data is required, for example, when doing subsampling, the complete dataset sits in, and consumes, memory. This is especially a problem if you work with large tasks or many tasks simultaneously, e.g., for benchmarking.
2. During parallelization (Section 10.1), the complete data needs to be transferred to the workers which can increase the overhead.

To avoid these drawbacks, especially for larger data, it can be necessary to interface out-of-memory data to reduce the memory requirements. This way, only the part of the data which is currently required by the learners will be placed in the main memory to operate on. There are multiple options to handle this:

1. `DataBackendDplyr`, which interfaces the R package dbplyr, extending dplyr to work on many popular SQL databases like *MariaDB*, *PostgresSQL*, or *SQLite*.
2. `DataBackendDuckDB` for the *DuckDB* database connected via duckdb, which is a fast, zero-configuration alternative to SQLite.
3. `DataBackendDuckDB` for Parquet files. This means the data does not need to be converted to DuckDB's native storage format and instead you can work directly on directories containing one or multiple files stored in the popular Parquet format.

In the following, we will show how to work with each of these choices using `mlr3db`.

10.4.1 Databases with DataBackendDplyr

To demonstrate `DataBackendDplyr` we use the (pretty big) NYC flights dataset from the `nycflights13` package and move it into a SQLite database. Although as_sqlite_backend() provides a convenient function to perform this step, we construct the database manually here.

```
# load data
requireNamespace("DBI")
requireNamespace("RSQLite")
requireNamespace("nycflights13")
data("flights", package = "nycflights13")
dim(flights)
```

[1] 336776   19

```
# add column of unique row ids
flights$row_id = seq(nrow(flights))

# create sqlite database in temporary file
path = tempfile("flights", fileext = ".sqlite")
con = DBI::dbConnect(RSQLite::SQLite(), path)
tbl = DBI::dbWriteTable(con, "flights", as.data.frame(flights))
DBI::dbDisconnect(con)

# remove in-memory data
rm(flights)
```

With the SQLite database stored in file `path`, we now re-establish a connection and switch to dplyr/dbplyr for some essential preprocessing.

```
# establish connection
con = DBI::dbConnect(RSQLite::SQLite(), path)

# select the "flights" table
library(dplyr)
library(dbplyr)
tbl = tbl(con, "flights")
```

As databases are intended to store large volumes of data, a natural first step is to subset and filter the data to suitable dimensions. Therefore, we build up an SQL query in a step-wise fashion using `dplyr` verbs and:

1. Select a subset of columns to work on;
2. Remove observations where the arrival delay (`arr_delay`) has a missing value;
3. Filter the data to only use every second row (to reduce example runtime); and
4. Merge factor levels of the feature `carrier` so infrequent carriers are replaced by level "other".

```
# 1. subset columns
keep = c("row_id", "year", "month", "day", "hour", "minute", "dep_time",
  "arr_time", "carrier", "flight", "air_time", "distance", "arr_delay")
tbl = select(tbl, all_of(keep))

# 2. filter by missing
tbl = filter(tbl, !is.na(arr_delay))

# 3. select every other row
tbl = filter(tbl, row_id %% 2 == 0)

# 4. merge infrequent carriers
infrequent = c("OO", "HA", "YV", "F9", "AS", "FL", "VX", "WN")
tbl = mutate(tbl, carrier = case_when(
  carrier %in% infrequent ~ "other",
  TRUE ~ carrier))
```

Having prepared our data, we can now create a DataBackendDplyr and can then query basic information from our new DataBackend:

```
library(mlr3db)
backend_flights = as_data_backend(tbl, primary_key = "row_id")
c(nrow = backend_flights$nrow, ncol = backend_flights$ncol)
```

```
  nrow   ncol
163707     13
```

```
backend_flights$head()
```

```
   row_id year month day hour minute dep_time arr_time carrier flight
1:      2 2013     1   1    5     29      533      850      UA   1714
2:      4 2013     1   1    5     45      544     1004      B6    725
3:      6 2013     1   1    5     58      554      740      UA   1696
4:      8 2013     1   1    6      0      557      709      EV   5708
5:     10 2013     1   1    6      0      558      753      AA    301
6:     12 2013     1   1    6      0      558      853      B6     71
3 variables not shown: [air_time, distance, arr_delay]
```

Note that the DataBackendDplyr can only operate on the data we provided, so does not "know" about the rows and columns we already filtered out (this is in contrast to using $filter and $subset as in Section 2.1.3, which only remove row or column roles and not the rows/columns themselves).

With a backend constructed, we can now use the standard mlr3 API:

```
tsk_flights = as_task_regr(backend_flights, id = "flights_sqlite",
  target = "arr_delay")
rsmp_sub002 = rsmp("subsampling", ratio = 0.02, repeats = 3)
```

Above we created a regression task by passing a backend as the first argument and then created a resampling strategy where we will subsample 2% of the observations three times. In each resampling iteration, only the required subset of the data is queried from the SQLite database and passed to our learner:

```
rr = resample(tsk_flights, lrn("regr.rpart"), rsmp_sub002)
measures = msrs(c("regr.rmse", "time_train", "time_predict"))
rr$aggregate(measures)
```

```
 regr.rmse   time_train time_predict
   35.9536       0.6997       7.3883
```

As we have finished our experiment we can now close our connection, which we can
do by removing the `tbl` object referencing the connection and then closing it.

```
rm(tbl)
DBI::dbDisconnect(con)
```

10.4.2 Parquet Files with DataBackendDuckDB

DuckDB databases provide a modern alternative to SQLite, tailored to the needs of
ML. Parquet is a popular column-oriented data storage format supporting efficient
compression, making it far superior to other popular data exchange formats such as
CSV.

Converting a `data.frame` to DuckDB is possible by passing the `data.frame` to
convert and the `path` to store the data to as_duckdb_backend(). By example,
below we first query the location of an example dataset in a Parquet file shipped
with `mlr3db` and then convert the resulting `DataBackendDuckDB` object into a
classification task, all without loading the dataset into memory:

```
path = system.file(file.path("extdata", "spam.parquet"),
  package = "mlr3db")
backend = as_duckdb_backend(path)
as_task_classif(backend, target = "type")
```

```
<TaskClassif:backend> (4601 x 58)
* Target: type
* Properties: twoclass
* Features (57):
  - dbl (57): address, addresses, all, business, capitalAve,
    capitalLong, capitalTotal, charDollar, charExclamation,
    charHash, charRoundbracket, charSemicolon,
    charSquarebracket, conference, credit, cs, data, direct,
    edu, email, font, free, george, hp, hpl, internet, lab,
    labs, mail, make, meeting, money, num000, num1999, num3d,
    num415, num650, num85, num857, order, original, our, over,
    parts, people, pm, project, re, receive, remove, report,
    table, technology, telnet, will, you, your
```

Accessing the data internally triggers a query and the required subsets of data
are fetched to be stored in an in-memory `data.frame`. After the retrieved data is
processed, the garbage collector can release the occupied memory. The backend can
also operate on a folder with multiple parquet files.

10.5   Extending mlr3 and Defining a New `Measure`

After getting this far in the book you are well on your way to being an `mlr3` expert and may even want to add more classes to our universe. While many classes could be extended, all have a similar design interface and so, we will only demonstrate how to create a custom `Measure`. If you are interested in implementing new learners, `PipeOps`, or tuners, then check out the vignettes in the respective packages: `mlr3extralearners`, `mlr3pipelines`, or `mlr3tuning`. If you are considering creating a package that adds an entirely new task type then feel free to contact us for some support via GitHub, email, or Mattermost. This section assumes good knowledge of `R6`, see Section 1.5.1 for a brief introduction and references to further resources.

As an example, let us consider a regression measure that scores a prediction as 1 if the difference between the true and predicted values is less than one standard deviation of the truth, or scores the prediction as 0 otherwise. In maths this would be defined as $f(y, \hat{y}) = \frac{1}{n} \sum_{i=1}^{n} \mathbb{I}(|y_i - \hat{y}_i| < \sigma_y)$, where σ_y is the standard deviation of the truth and $\mathbb{I}$ is the indicator function. In code, this measure may be written as:

```
threshold_acc = function(truth, response) {
  mean(ifelse(abs(truth - response) < sd(truth), 1, 0))
}


threshold_acc(c(100, 0, 1), c(1, 11, 6))
```

```
[1] 0.6667
```

By definition of this measure, its values are bounded in $[0, 1]$ where a perfect score of 1 would mean all predictions are within a standard deviation of the truth, hence for this measure larger scores are better.

To use this measure in `mlr3`, we need to create a new `R6Class`, which will inherit from `Measure` and in this case specifically from `MeasureRegr`. The code for this new measure is in the snippet below, with an explanation following it. This code chunk can be used as a template for the majority of performance measures.

```
MeasureRegrThresholdAcc = R6::R6Class("MeasureRegrThresholdAcc",
  inherit = mlr3::MeasureRegr, # regression measure
  public = list(
    initialize = function() { # initialize class
      super$initialize(
        id = "thresh_acc", # unique ID
        packages = character(), # no package dependencies
        properties = character(), # no special properties
        predict_type = "response", # measures response prediction
        range = c(0, 1), # results in values between (0, 1)
        minimize = FALSE # larger values are better
      )
    }
  ),
```

```
  private = list(
    # define score as private method
    .score = function(prediction, ...) {
      # define loss
      threshold_acc = function(truth, response) {
        mean(ifelse(abs(truth - response) < sd(truth), 1, 0))
      }
      # call loss function
      threshold_acc(prediction$truth, prediction$response)
    }
  )
)
```

1. In the first two lines we name the class, here `MeasureRegrThresholdAcc`, and then state this is a regression measure that inherits from `MeasureRegr`.
2. We initialize the class by stating its unique ID is `"thresh_acc"`, that it does not require any external packages (`packages = character()`) and that it has no special properties (`properties = character()`).
3. We then pass specific details of the loss function which are: it measures the quality of a `"response"` type prediction, its values range between (0, 1), and that the loss is optimized as its maximum (`minimize = FALSE`).
4. Finally, we define the score itself as a private method called `.score` where we pass the predictions to the function we defined just above.

Sometimes measures require data from the training set, the task, or the learner. These are usually complex edge-cases examples, so we will not go into detail here, for working examples we suggest looking at the code for `MeasureSurvSongAUC` and `MeasureSurvAUC`. You can also consult the manual page of the `Measure` for an overview of other properties and meta-data that can be specified.

Once you have defined your measure you can load it with the R6 constructor (`$new()`), or make it available to be constructed with the `msr()` sugar function by adding it to the `mlr_measures` dictionary:

```
tsk_mtcars = tsk("mtcars")
split = partition(tsk_mtcars)
lrn_featureless = lrn("regr.featureless")$train(tsk_mtcars, split$train)
prediction = lrn_featureless$predict(tsk_mtcars, split$test)
prediction$score(MeasureRegrThresholdAcc$new())
```

```
thresh_acc
   0.7273
```

```
# or add to dictionary by passing a unique key to the first argument
#  and the class to the second
mlr3::mlr_measures$add("regr.thresh_acc", MeasureRegrThresholdAcc)
prediction$score(msr("regr.thresh_acc"))
```

```
thresh_acc
   0.7273
```

While we only covered how to create a simple regression measure, the process of adding other classes to our universe is in essence the same:

1. Find the right class to inherit from
2. Add methods that:
   a) Initialize the object with the correct properties (`$initialize()`).
   b) Implement the public and private methods that do the actual computation. In the above example, this was the private `$.score()` method.

We are always happy to chat and welcome new contributors, please get in touch if you need assistance in extending `mlr3`.

10.6   Conclusion

This chapter covered several advanced topics including parallelization, error handling, logging, working with databases, and extending the `mlr3` universe. For simple use cases, you will probably not need to know each of these topics in detail, however, we do recommend being familiar at least with error handling and fallback learners, as these are essential to preventing even simple experiments being interrupted. If you are working with large experiments or datasets, then understanding parallelization, logging, and databases will also be essential.

We have not covered any of these topics extensively and therefore recommended the following resources should you want to read more about these areas. If you are interested to learn more about parallelization in R, we recommend Schmidberger et al. (2009) and Eddelbuettel (2020). To find out more about logging, have a read of the vignettes in `lgr`, which cover everything from logging to JSON files to retrieving logged objects for debugging. For an overview of available DBMS in R, see the CRAN task view on databases at https://cran.r-project.org/view=Databases, and in particular the vignettes of the `dbplyr` package for DBMS readily available in `mlr3`.

Table 10.1: Important classes and functions covered in this chapter with underlying class (if applicable), class constructor or function, and important class fields and methods (if applicable).

Class	Constructor/Function	Fields/Methods
-	`plan()`	-
-	`set_threads()`	-
-	`tweak()`	-
Learner	`lrn()`	`$encapsulate;` `$fallback; $timeout;` `$parallel_predict;` `$log`
Logger	`get_logger`	`$set_threshold()`
DataBackendDplyr	`as_data_backend`	-
DataBackendDuckDB	`as_duckdb_backend`	-

10.7   Exercises

1. Consider the following example where you resample a learner (debug
   learner, sleeps for three seconds during train) on four workers using the
   multisession backend:

```
tsk_penguins = tsk("penguins")
lrn_debug = lrn("classif.debug", sleep_train = function() 3)
rsmp_cv6 = rsmp("cv", folds = 6)

future::plan("multisession", workers = 4)
resample(tsk_penguins, lrn_debug, rsmp_cv6)
```

   (a) Assuming you were running this experiment on a computer with four
       CPUs, and that the learner would actually calculate something and not
       just sleep: Would all CPUs be busy for the entire time of this calculation?
   (b) Prove your point by measuring the elapsed time, e.g., using
       `system.time()`.
   (c) What would you change in the setup and why?

2. Create a new custom binary classification measure which scores ("prob"-
   type) predictions. This measure should compute the absolute difference
   between the predicted probability for the positive class and a 0-1 encoding
   of the ground truth and then average these values across the test set. Test
   this with `classif.log_reg` on `tsk("sonar")`.

3. "Tune" the `error_train` hyperparameter of the `classif.debug` learner
   on a continuous interval from 0 to 1, using a simple classification tree as
   the fallback learner and the penguins task. Tune for 50 iterations using
   random search and 10-fold cross-validation. Inspect the resulting archive
   and find out which evaluations resulted in an error, and which did not.
   Now do the same in the interval 0.3 to 0.7. Are your results surprising?

11

Large-Scale Benchmarking

Sebastian Fischer
Ludwig-Maximilians-Universität München, and Munich Center for Machine Learning (MCML)

Michel Lang
Research Center Trustworthy Data Science and Security, and TU Dortmund University

Marc Becker
Ludwig-Maximilians-Universität München, and Munich Center for Machine Learning (MCML)

In machine learning, it is often difficult to evaluate methods using mathematical analysis alone. Even when formal analyses can be successfully applied, it is often an open question whether real-world datasets satisfy the necessary assumptions for the theorems to hold. Empirical benchmark experiments evaluate the performance of different algorithms on a wide range of datasets. These empirical investigations are essential for understanding the capabilities and limitations of existing methods and for developing new and improved approaches. Trustworthy benchmark experiments are often "large-scale", which means they may make use of many datasets, measures, and learners. Moreover, datasets must span a wide range of domains and problem types as conclusions can only be drawn about the kind of datasets on which the benchmark study was conducted.

Large-scale benchmark experiments consist of three primary steps: sourcing the data for the experiment, executing the experiment, and analyzing the results; we will discuss each of these in turn. In Section 11.1, we will begin by discussing `mlr3oml`, which provides an interface between `mlr3` and OpenML (Vanschoren et al. 2013), a popular tool for uploading and downloading datasets. Increasing the number of datasets leads to "large-scale" experiments that may require significant computational resources, so in Section 11.2, we will introduce `mlr3batchmark`, which connects `mlr3` with `batchtools` (Lang, Bischl, and Surmann 2017), which provides methods for managing and executing experiments on high-performance computing (HPC) clusters. Finally, in Section 11.3, we will demonstrate how to make use of `mlr3benchmark` to formally analyze the results from large-scale benchmark experiments.

Throughout this chapter, we will use the running example of benchmarking a random forest model against a logistic regression as in Couronné, Probst, and Boulesteix (2018). We will also assume that you have read Chapters 7 and 10. We make use of `ppl("robustify")` (Section 9.4) for automating common preprocessing steps. We also set a featureless baseline as a fallback learner (Section 10.2.2) and set `"try"` as our encapsulation method (Section 10.2.1), which logs errors/warnings to an external file that can be read by `batchtools` (we will return to this in Section 11.2.3).

DOI: 10.1201/9781003402848-11

```
# featureless baseline
lrn_baseline = lrn("classif.featureless", id = "featureless")

# logistic regression pipeline
lrn_lr = lrn("classif.log_reg")
lrn_lr = as_learner(ppl("robustify", learner = lrn_lr) %>>% lrn_lr)
lrn_lr$id = "logreg"
lrn_lr$fallback = lrn_baseline
lrn_lr$encapsulate = c(train = "try", predict = "try")

# random forest pipeline
lrn_rf = lrn("classif.ranger")
lrn_rf = as_learner(ppl("robustify", learner = lrn_rf) %>>% lrn_rf)
lrn_rf$id = "ranger"
lrn_rf$fallback = lrn_baseline
lrn_rf$encapsulate = c(train = "try", predict = "try")

learners = list(lrn_lr, lrn_rf, lrn_baseline)
```

As a starting example, we will compare our learners across three classification tasks using accuracy and three-fold CV.

```
design = benchmark_grid(tsks(c("german_credit", "sonar", "pima")),
  learners, rsmp("cv", folds = 10))
bmr = benchmark(design)
bmr$aggregate(msr("classif.acc"))[, .(task_id, learner_id, classif.acc)]
```

```
        task_id  learner_id classif.acc
1: german_credit      logreg      0.7460
2: german_credit      ranger      0.7610
3: german_credit featureless      0.7000
4:         sonar      logreg      0.7162
5:         sonar      ranger      0.8317
6:         sonar featureless      0.5329
7:          pima      logreg      0.7747
8:          pima      ranger      0.7683
9:          pima featureless      0.6511
```

In this small experiment, random forests appears to outperform the other learners on all three datasets. However, this analysis is not conclusive as we only considered three tasks, and the performance differences might not be statistically significant. In the following, we will introduce some techniques to improve the study.

11.1   Getting Data with OpenML

OpenML   To draw meaningful conclusions from benchmark experiments, a good choice of datasets and tasks is essential. OpenML is an open-source platform that facilitates the sharing and dissemination of machine learning research data, algorithms, and experimental results, in a standardized format enabling consistent cross-study comparison. OpenML's design ensures that all data on the platform is "FAIR" (**F**indability, **A**ccessibility, **I**nteroperability, and **R**eusability), which ensures the data is easily discoverable and reusable. All entities on the platform have unique identifiers and standardized (meta)data that can be accessed via a REST API or the web interface.

In this section, we will cover some of the main features of OpenML and how to use them via the `mlr3oml` interface package. In particular, we will discuss OpenML datasets, tasks, and task collections, but will not cover algorithms or experiment results here.

11.1.1   Datasets

Finding data from OpenML is possible via the website or its REST API that `mlr3oml` interfaces. `list_oml_data()` can be used to filter datasets for specific properties; for example, by number of features, rows, or number of classes in a classification problem:

```
library(mlr3oml)

odatasets = list_oml_data(
  number_features = c(10, 20),
  number_instances = c(45000, 50000),
  number_classes = 2
)
```

```
odatasets[NumberOfFeatures < 16,
  c("data_id", "name", "NumberOfFeatures", "NumberOfInstances")]
```

	data_id	name	NumberOfFeatures	NumberOfInstances
1:	179	adult	15	48842
2:	1590	adult	15	48842
3:	43898	adult	15	48790
4:	45051	adult-test	15	48842
5:	45068	adult	15	48842

Note that `list_oml_data()` returns a `data.table` with many more meta-features than shown here; this table can itself be used to filter further.

We can see that some datasets have duplicated names, which is why each dataset also has a unique ID. By example, let us consider the "adult" dataset with ID 1590. Metadata for the dataset is loaded with `odt()`, which returns an object of class `OMLData`.   `odt()`

```
odata = odt(id = 1590)
odata
```

```
<OMLData:1590:adult> (48842x15)
 * Default target: class
```

The `OMLData` object contains metadata about the dataset but importantly does not (yet) contain the data. This means that information about the dataset can be queried without having to load the entire data into memory; for example, the license and dimension of the data:

```
odata$license
```

```
[1] "Public"
```

```
c(nrow = odata$nrow, ncol = odata$ncol)
```

```
 nrow  ncol
48842    15
```

If we want to work with the actual data, then accessing the `$data` field will download the data, import it into R, and then store the `data.frame` in the `OMLData` object:

```
# first 5 rows and columns
odata$data[1:5, 1:5]
```

```
   age workclass fnlwgt     education education.num
1:  25   Private 226802          11th             7
2:  38   Private  89814       HS-grad             9
3:  28 Local-gov 336951    Assoc-acdm            12
4:  44   Private 160323  Some-college            10
5:  18      <NA> 103497  Some-college            10
```

> 💡 mlr3oml Cache
>
> After `$data` has been called the first time, all subsequent calls to `$data` will be transparently redirected to the in-memory `data.frame`. Additionally, many objects can be permanently cached on the local file system by setting the option `mlr3oml.cache` to either `TRUE` or to a specific path to be used as the cache folder.

Data can then be converted into `mlr3` backends (see Section 10.4) with the `as_data_backend()` function and then into tasks:

```
backend = as_data_backend(odata)
tsk_adult = as_task_classif(backend, target = "class")
tsk_adult
```

```
<TaskClassif:backend> (48842 x 15)
* Target: class
* Properties: twoclass
* Features (14):
  - fct (8): education, marital.status, native.country,
    occupation, race, relationship, sex, workclass
  - int (6): age, capital.gain, capital.loss, education.num,
    fnlwgt, hours.per.week
```

Some datasets on OpenML contain columns that should neither be used as a feature nor a target. The column names that are usually included as features are accessible through the field `$feature_names`, and we assign them to the `mlr3` task accordingly. Note that for the dataset at hand, this would not have been necessary, as all non-target columns are to be treated as predictors, but we include it for clarity.

```
tsk_adult$col_roles$feature = odata$feature_names
tsk_adult
```

```
<TaskClassif:backend> (48842 x 15)
* Target: class
* Properties: twoclass
* Features (14):
  - fct (8): education, marital.status, native.country,
    occupation, race, relationship, sex, workclass
  - int (6): age, capital.gain, capital.loss, education.num,
    fnlwgt, hours.per.week
```

11.1.2   Task

OpenML tasks are built on top of OpenML datasets and additionally specify the target variable, the train-test splits to use for resampling, and more. Note that this differs from `mlr3 Task` objects, which do not contain information about the resampling procedure. Similarly to `mlr3`, OpenML has different types of tasks, such as regression and classification. Analogously to filtering datasets, tasks can be filtered with `list_oml_tasks()`. To find a task that makes use of the data we have been using, we would pass the data ID to the `data_id` argument:

```
# tasks making use of the adult data
adult_tasks = list_oml_tasks(data_id = 1590)
```

```
adult_tasks[task_type == "Supervised Classification", task_id]
```

```
[1]    7592   14947 126025 146154 146598 168878 233099 359983 361515
```

From these tasks, we randomly select the task with ID 359983. We can load the object using `otsk()`, which returns an `OMLTask` object.   otsk()

```
otask = otsk(id = 359983)
otask
```

```
<OMLTask:359983>
 * Type: Supervised Classification
 * Data: adult (id: 1590; dim: 48842x15)
 * Target: class
 * Estimation: crossvalidation (id: 1; repeats: 1, folds: 10)
```

The `OMLData` object associated with the underlying dataset can be accessed through the `$data` field.

```
otask$data
```

```
<OMLData:1590:adult> (48842x15)
 * Default target: class
```

The data splits associated with the estimation procedure are accessible through the field `$task_splits`. In mlr3 terms, these are the instantiation of a Resampling on a specific Task.

```
    otask$task_splits
```

```
            type rowid repeat.  fold
     1: TRAIN 32427        0     0
     2: TRAIN 13077        0     0
     3: TRAIN 15902        0     0
     4: TRAIN 17703        0     0
     5: TRAIN 35511        0     0
     ---
488416:  TEST  8048        0     9
488417:  TEST 12667        0     9
488418:  TEST 43944        0     9
488419:  TEST 25263        0     9
488420:  TEST 43381        0     9
```

The OpenML task can be converted to both an `mlr3::Task` and `ResamplingCustom` instantiated on the task using `as_task()` and `as_resampling()`, respectively:

```
    tsk_adult = as_task(otask)
    tsk_adult
```

```
<TaskClassif:adult> (48842 x 15)
* Target: class
* Properties: twoclass
* Features (14):
  - fct (8): education, marital.status, native.country,
    occupation, race, relationship, sex, workclass
  - int (6): age, capital.gain, capital.loss, education.num,
    fnlwgt, hours.per.week
```

```
    resampling = as_resampling(otask)
    resampling
```

```
<ResamplingCustom>: Custom Splits
* Iterations: 10
* Instantiated: TRUE
* Parameters: list()
```

mlr3oml also allows direct construction of mlr3 tasks and resamplings with the standard `tsk()` and `rsmp()` constructors, e.g.:

```
    tsk("oml", task_id = 359983)
```

```
<TaskClassif:adult> (48842 x 15)
* Target: class
* Properties: twoclass
* Features (14):
  - fct (8): education, marital.status, native.country,
    occupation, race, relationship, sex, workclass
  - int (6): age, capital.gain, capital.loss, education.num,
    fnlwgt, hours.per.week
```

11.1.3 Task Collection

The OpenML task collection is a container object bundling existing tasks. This allows for the creation of benchmark suites, which are curated collections of tasks that satisfy certain quality criteria. Examples include the OpenML CC-18 benchmark suite (Bischl et al. 2021), the AutoML benchmark (Gijsbers et al. 2022) and the benchmark for tabular deep learning (Grinsztajn, Oyallon, and Varoquaux 2022). OMLCollection objects are loaded with ocl(), by example we will look at CC-18, `ocl()` which has ID 99:

```
otask_collection = ocl(id = 99)
```

```
otask_collection
```

```
<OMLCollection: 99> OpenML-CC18 Curated Class[...]
 * data:  72
 * tasks: 72
```

The task includes 72 classification tasks on different datasets that can be accessed through $task_ids:

```
otask_collection$task_ids[1:5] # first 5 tasks in the collection
```

```
[1]  3  6 11 12 14
```

Task collections can be used to quickly define benchmark experiments in `mlr3`. To easily construct all tasks and resamplings from the benchmarking suite, you can use as_tasks() and as_resamplings() respectively:

```
tasks = as_tasks(otask_collection)
resamplings = as_resamplings(otask_collection)
```

Alternatively, if we wanted to filter the collection further, say to a binary classification experiment with six tasks, we could run list_oml_tasks() with the task IDs from the CC-18 collection as argument task_id. We can either use the list_oml_tasks() argument to request the number of classes to be 2, or we can make use of the fact that the result of list_oml_tasks() is a data.table and subset the resulting table.

```
binary_cc18 = list_oml_tasks(
  limit = 6,
  task_id = otask_collection$task_ids,
  number_classes = 2
)
```

We now define the tasks and resamplings which we will use for comparing the logistic regression with the random forest learner. Note that all resamplings in this collection consist of exactly 10 iterations.

```
# load tasks as a list
otasks = lapply(binary_cc18$task_id, otsk)

# convert to mlr3 tasks and resamplings
tasks = as_tasks(otasks)
resamplings = as_resamplings(otasks)
```

To define the design table, we use `benchmark_grid()` and set `paired` to `TRUE`, which is used in situations where each resampling is instantiated on a corresponding task (therefore, the `tasks` and `resamplings` below must have the same length) and each learner should be evaluated on every resampled task.

```
large_design = benchmark_grid(tasks, learners, resamplings,
  paired = TRUE)
large_design[1:6] # first 6 rows
```

```
        task      learner resampling
1: kr-vs-kp       logreg     custom
2: kr-vs-kp       ranger     custom
3: kr-vs-kp featureless     custom
4: breast-w       logreg     custom
5: breast-w       ranger     custom
6: breast-w featureless     custom
```

Having set up our large experiment, we can now look at how to efficiently carry it out on a cluster.

11.2   Benchmarking on HPC Clusters

High-performance Computing

As discussed in Section 10.1, parallelization of benchmark experiments is straight-forward as they are embarrassingly parallel. However, for large experiments, parallelization on a high-performance computing (HPC) cluster is often preferable. `batchtools` provides a framework to simplify running large batches of computational experiments in parallel from R on such sites. It is highly flexible, making it suitable for a wide range of computational experiments, including machine learning, optimization, simulation, and more.

> 💡 **"batchtools" backend for future**
>
> In Section 10.1.2, we touched upon different parallelization backends. The package `future` includes a `"batchtools"` plan, however, this does not allow the additional control that comes with working with `batchtools` directly.

An HPC cluster is a collection of interconnected computers or servers providing computational power beyond what a single computer can achieve. HPC clusters typically consist of multiple compute nodes, each with multiple CPU/GPU cores, memory, and local storage. These nodes are usually connected by a high-speed network and network file system which enables the nodes to communicate and work together on a given task. The most important difference between HPC clusters and a personal computer (PC), is that the nodes often cannot be accessed directly, but instead, computational jobs are queued by a scheduling system such as Slurm (Simple Linux Utility for Resource Management). A scheduling system is a software tool that orchestrates the allocation of computing resources to users or applications on the cluster. It ensures that multiple users and applications can access the resources of the cluster fairly and efficiently, and also helps to maximize the utilization of the computing resources.

Figure 11.1 contains a rough sketch of an HPC architecture. Multiple users can log into the head node (typically via SSH) and add their computational jobs to the queue by sending a command of the form "execute computation X using resources Y for Z amount of time". The scheduling system controls when these computational jobs are executed.

For the rest of this section, we will look at how to use `batchtools` and `mlr3batchmark` for submitting jobs, adapting jobs to clusters, ensuring reproducibility, querying job status, and debugging failures.

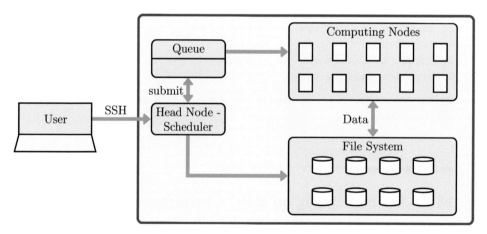

Figure 11.1: Illustration of an HPC cluster architecture.

11.2.1   Experiment Registry Setup

`batchtools` is built around experiments or "jobs". One replication of a job is defined by applying a (parameterized) algorithm to a (parameterized) problem. A benchmark experiment in `batchtools` consists of running many such experiments with different algorithms, algorithm parameters, problems, and problem parameters. Each such experiment is computationally independent of all other experiments and constitutes the basic level of computation `batchtools` can parallelize. For this section, we will define a single `batchtools` experiment as one resampling iteration of one learner on one task, in Section 11.2.4, we will look at different ways of defining an experiment.

The first step in running an experiment is to create or load an experiment registry with `makeExperimentRegistry()` or `loadRegistry()` respectively. This constructs the inter-communication object for all functions in `batchtools` and corresponds to a folder on the file system. Among other things, the experiment registry stores the algorithms, problems, and job definitions; log outputs and status of submitted, running, and finished jobs; job results; and the "cluster function" that defines the interaction with the scheduling system in a scheduling-software-agnostic way.

Below, we create a registry in a subdirectory of our working directory – on a real cluster, make sure that this folder is stored on a shared network filesystem, otherwise, the nodes cannot access it. We also set the registry's `seed` to 1 and the `packages` to `"mlr3verse"`, which will make these packages available in all our experiments.

```
library(batchtools)

# create registry
reg = makeExperimentRegistry(
  file.dir = "./experiments",
  seed = 1,
  packages = "mlr3verse"
)
```

Once the registry has been created, we need to populate it with problems and algorithms to form the jobs, this is most easily carried out with `mlr3batchmark`, although finer control is possible with `batchtools` and will be explored in Section 11.2.4. `batchmark()` converts `mlr3` tasks and resamplings to `batchtools` problems, and converts `mlr3` learners to `batchtools` algorithms; jobs are then created for all resampling iterations.

```
library(mlr3batchmark)
batchmark(large_design, reg = reg)
```

Now the registry includes six problems, one for each resampled task, and 180 jobs from 3 learners × 6 tasks × 10 resampling iterations. The single algorithm in the registry is because `mlr3batchmark` specifies a single algorithm that is parametrized with the learner IDs.

```
reg
```

```
Experiment Registry
  Backend   : Interactive
  File dir  : /your_directory/experiments
  Work dir  : /your_directory
  Jobs      : 180
  Problems  : 6
```

```
Algorithms: 1
Seed      : 1
Writeable : TRUE
```

By default, the "Interactive" cluster function (see makeClusterFunctionsInteractive()) is used – this is the abstraction for the scheduling system, and "interactive" here means to not use a real scheduler but instead to use the interactive R session for sequential computation. getJobTable() can be used to get more detailed information about the jobs. Here, we only show a few selected columns for readability and unpack the list columns algo.pars and prob.pars using unwrap().

```
job_table = getJobTable(reg = reg)
job_table = unwrap(job_table)
job_table = job_table[,
  .(job.id, learner_id, task_id, resampling_id, repl)
]

job_table
```

```
    job.id  learner_id  task_id  resampling_id  repl
1:       1      logreg  kr-vs-kp        custom     1
2:       2      logreg  kr-vs-kp        custom     2
3:       3      logreg  kr-vs-kp        custom     3
4:       4      logreg  kr-vs-kp        custom     4
5:       5      logreg  kr-vs-kp        custom     5
---
176:   176  featureless  spambase      custom     6
177:   177  featureless  spambase      custom     7
178:   178  featureless  spambase      custom     8
179:   179  featureless  spambase      custom     9
180:   180  featureless  spambase      custom    10
```

In this output, we can see how each job is now assigned a unique job.id and that each row corresponds to a single iteration (column repl) of a resample experiment.

11.2.2   Job Submission

With the experiments defined, we can now submit them to the cluster. However, it is best practice to first test each algorithm individually using testJob(). By example, we will only test the first job (id = 1) and will use an external R session (external = TRUE).   testJob()

```
result = testJob(1, external = TRUE, reg = reg)
```

Once we are confident that the jobs are defined correctly (see Section 11.2.3 for jobs with errors), we can proceed with their submission, by specifying the resource requirements for each computational job and then optionally grouping jobs.

Configuration of resources is dependent on the cluster function set in the registry. We will assume we are working with a Slurm cluster and accordingly initialize the cluster function with makeClusterFunctionsSlurm() and will make use of the `slurm-simple.tml` template file that can be found in a subdirectory of the `batchtools` package itself (the exact location can be found by running `system.file("templates", package = "batchtools")`), or the `batchtools` GitHub repository. A template file is a shell script with placeholders filled in by `batchtools` and contains the command to start the computation via `Rscript` or `R CMD batch`, as well as comments which serve as annotations for the scheduler, for example, to communicate resources or paths on the file system.

The exemplary template should work on many Slurm installations out-of-the-box, but you might have to modify it for your cluster – it can be customized to work with more advanced configurations.

```
cf = makeClusterFunctionsSlurm(template = "slurm-simple")
```

To proceed with the examples on a local machine, we recommend setting the cluster function to a Socket backend with `makeClusterFunctionsSocket()`. The chosen cluster function can be saved to the registry by passing it to the `$cluster.functions` field.

```
reg$cluster.functions = cf
saveRegistry(reg = reg)
```

With the registry setup, we can now decide if we want to run the experiments in chunks (Section 10.1) and then specify the resource requirements for the submitted jobs.

`chunk()`   For this example, we will use `chunk()` to chunk the jobs such that five iterations of one resample experiment are run sequentially in one computational job – in practice the optimal grouping will be highly dependent on your experiment (Section 10.1).

```
ids = job_table$job.id
chunks = data.table(
  job.id = ids, chunk = chunk(ids, chunk.size = 5, shuffle = FALSE)
)
chunks[1:6] # first 6 jobs
```

```
   job.id chunk
1:      1     1
2:      2     1
3:      3     1
4:      4     1
5:      5     1
6:      6     2
```

The final step is to decide the resource requirements for each job. The set of resources depends on your cluster and the corresponding template file. If you are unsure about the resource requirements, you can start a subset of jobs with liberal resource constraints, e.g., the maximum runtime allowed for your computing site. Measured runtimes and memory usage can later be queried with `getJobTable()` and used to better estimate the required resources for the remaining jobs. In this example, we will set the number of CPUs per job to 1, the walltime (time limit before jobs are stopped by the scheduler) to one hour (3600 seconds), and the RAM limit (memory limit before jobs are stopped by the scheduler) to 8000 megabytes.

```
resources = list(ncpus = 1, walltime = 3600, memory = 8000)
```

With all the elements in place, we can now submit our jobs.

```
submitJobs(ids = chunks, resources = resources, reg = reg)

# wait for all jobs to terminate
waitForJobs(reg = reg)
```

> 💡 Submitting Jobs
>
> A good approach to submit computational jobs is by using a persistent R session (e.g., with Terminal Multiplexer (TMUX)) on the head node to continue job submission (or computation, depending on the cluster functions) in the background.
>
> However, `batchtools` registries are saved to the file system and therefore persistent when the R session is terminated. This means that you can also submit jobs from an interactive R session, terminate the session, and analyze the results later in a new session.

11.2.3 Job Monitoring, Error Handling, and Result Collection

Once jobs have been submitted, they can then be queried with getStatus() to find their current status and the results (or errors) can be investigated. If you terminated your R sessions after job submission, you can load the experiment registry with loadRegistry().

getStatus()

loadRegistry

```
getStatus(reg = reg)
```

```
Status for 180 jobs at 2023-07-04 15:34:46:
  Submitted    : 180 (100.0%)
  -- Queued    :   0 (  0.0%)
  -- Started   : 180 (100.0%)
  ---- Running :   0 (  0.0%)
  ---- Done    : 180 (100.0%)
  ---- Error   :   0 (  0.0%)
  ---- Expired :   0 (  0.0%)
```

To query the ids of jobs in the respective categories, see findJobs() and, e.g., findNotSubmitted() or findDone(). In our case, we can see all experiments finished and none expired (i.e., were removed from the queue without ever starting, Expired : 0) or crashed (Error : 0). It can still be sensible to use grepLogs() to check the logs for suspicious messages and warnings before proceeding with the analysis of the results.

In any large-scale experiment many things can and will go wrong, for example, the cluster might have an outage, jobs may run into resource limits or crash, or there could be bugs in your code. In these situations, it is important to quickly determine what went wrong and to recompute only the minimal number of required jobs.

To see debugging in practice we will use the debug learner (see Section 10.2) with a 50% probability of erroring in training. When calling batchmark() again, the new experiments will be added to the registry on top of the existing jobs.

```
extra_design = benchmark_grid(tasks,
    lrn("classif.debug", error_train = 0.5), resamplings, paired = TRUE)

batchmark(extra_design, reg = reg)
```

> 💡 Registry Argument
>
> All batchtools functions that interoperate with a registry take a registry as an argument. By default, this argument is set to the last created registry, which is currently the reg object defined earlier. We pass it explicitly in this section for clarity.

Now we can get the IDs of the new jobs (which have not been submitted yet) and submit them by passing their IDs.

```
ids = findNotSubmitted(reg = reg)
submitJobs(ids, reg = reg)
```

After these jobs have terminated, we can get a summary of those that failed:

```
getStatus(reg = reg)
```

```
Status for 240 jobs at 2023-07-04 15:34:47:
  Submitted    : 240 (100.0%)
  -- Queued    :   0 (  0.0%)
  -- Started   : 240 (100.0%)
  ---- Running :   0 (  0.0%)
  ---- Done    : 213 ( 88.8%)
  ---- Error   :  27 ( 11.2%)
  ---- Expired :   0 (  0.0%)
```

```
error_ids = findErrors(reg = reg)
summarizeExperiments(error_ids, by = c("task_id", "learner_id"),
    reg = reg)
```

```
            task_id     learner_id .count
1:          kr-vs-kp classif.debug      6
2:          breast-w classif.debug      3
3: credit-approval classif.debug      5
4:          credit-g classif.debug      6
5:          diabetes classif.debug      5
6:          spambase classif.debug      2
```

In a real experiment, we would now investigate the debug learner further to understand why it errored, try to fix those bugs, and then potentially rerun those experiments only.

Assuming learners have been debugged (or we are happy to ignore them), we can then collect the results of our experiment with `reduceResultsBatchmark()`, which constructs a `BenchmarkResult` from the results. Below we filter out results from the debug learner.

```
ids = findExperiments(algo.pars = learner_id != "classif.debug",
  reg = reg)
bmr = reduceResultsBatchmark(ids, reg = reg)
bmr$aggregate()[1:5]
```

```
   nr  task_id  learner_id resampling_id iters classif.ce
1:  1 kr-vs-kp      logreg        custom    10    0.02566
2:  2 kr-vs-kp      ranger        custom    10    0.01440
3:  3 kr-vs-kp featureless        custom    10    0.47778
4:  4 breast-w      logreg        custom    10    0.03578
5:  5 breast-w      ranger        custom    10    0.03006
Hidden columns: resample_result
```

11.2.4   Custom Experiments with batchtools

> **i**   **This section covers advanced ML or technical details.**

In general, we recommend using `mlr3batchmark` for scheduling simpler `mlr3` jobs on an HPC, however, we will also briefly show you how to use `batchtools` without `mlr3batchmark` for finer control over your experiment. Again we start by creating an experiment registry.

```
reg = makeExperimentRegistry(
  file.dir = "./experiments-custom",
  seed = 1,
  packages = "mlr3verse"
)
```

"Problems" are then manually registered with `addProblem()`. In this example, we will register all task-resampling combinations of the `large_design` above using the task ids as unique names. We specify that the `data` for the problem (i.e., the static data that is trained/tested by the learner) is the task/resampling pair. Finally, we pass a function (`fun`, dynamic problem part) that takes in the static problem `data` and returns it as the problem `instance` without making changes (Figure 11.2). The `fun` shown below is the default behavior and could be omitted, we show it here for clarity. This function could be more complex and take further parameters to modify the problem instance dynamically.

```
for (i in seq_along(tasks)) {
  addProblem(
    name = tasks[[i]]$id,
    data = list(task = tasks[[i]], resampling = resamplings[[i]]),
    fun = function(data, job, ...) data,
    reg = reg
  )
}
```

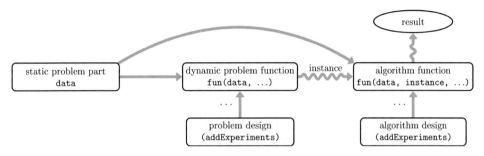

Figure 11.2: Illustration of a batchtools problem, algorithm, and experiment.

Next, we need to specify the algorithm to run with `addAlgorithm()`. Algorithms are again specified with a unique **name**, as well as a function to define the computational steps of the experiment and to return its result.

Here, we define one job to represent a complete resample experiment. In general, algorithms in `batchtools` may return arbitrary objects – those are simply stored on the file system and can be processed with a custom function while collecting the results.

```
addAlgorithm(
  "run_learner",
  fun = function(instance, learner, job, ...) {
    resample(instance$task, learner, instance$resampling)
  },
  reg = reg
)
```

Finally, we will define concrete experiments with `addExperiments()` by passing problem designs (`prob.designs`) and algorithm designs (`algo.designs`) that assign parameters to problems and algorithms, respectively (Figure 11.2).

In the code below, we add all resampling iterations for the six tasks as experiments. By leaving `prob.designs` unspecified, experiments for all existing problems are created per default. We set the **learner** parameter of our algorithm (`"run_learner"`) to be the three learners from our `large_design` object. Note that whenever an experiment is added, the current seed is assigned to the experiment and then incremented.

```
alg_des = list(run_learner = data.table(learner = learners))
addExperiments(algo.designs = alg_des, reg = reg)
summarizeExperiments()
```

Our jobs can now be submitted to the cluster; by not specifying specific job IDs, *all* experiments are submitted.

```
submitJobs(reg = reg)
```

We can retrieve the job results using `loadResult()`, which outputs the objects returned by the algorithm function, which in our case is a `ResampleResult`. To retrieve all results at once, we can use `reduceResults()` to create a single `BenchmarkResult`. For this, we use the combine function `c()` which can combine multiple objects of type `ResampleResult` or `BenchmarkResult` to a single `BenchmarkResult`.

```
rr = loadResult(1, reg = reg)
as.data.table(rr)[1:5]
```

```
                  task               learner               resampling
1: <TaskClassif[51]> <GraphLearner[38]> <ResamplingCustom[20]>
2: <TaskClassif[51]> <GraphLearner[38]> <ResamplingCustom[20]>
3: <TaskClassif[51]> <GraphLearner[38]> <ResamplingCustom[20]>
4: <TaskClassif[51]> <GraphLearner[38]> <ResamplingCustom[20]>
5: <TaskClassif[51]> <GraphLearner[38]> <ResamplingCustom[20]>
2 variables not shown: [iteration, prediction]
```

```
bmr = reduceResults(c, reg = reg)
bmr$aggregate()[1:5]
```

```
   nr   task_id  learner_id resampling_id iters classif.ce
1:  1 kr-vs-kp      logreg         custom    10    0.02566
2:  2 kr-vs-kp      ranger         custom    10    0.01377
3:  3 kr-vs-kp featureless         custom    10    0.47778
4:  4 breast-w      logreg         custom    10    0.03578
5:  5 breast-w      ranger         custom    10    0.02861
Hidden columns: resample_result
```

11.3 Statistical Analysis

The final step of a benchmarking experiment is to use statistical tests to determine which (if any) of our learners performed the best. `mlr3benchmark` provides infrastructure for applying statistical significance tests on `BenchmarkResult` objects.

Currently, Friedman tests and pairwise Friedman-Nemenyi tests (Demšar 2006) are supported to analyze benchmark experiments with at least two independent tasks and at least two learners. As a first step, we recommend performing a pairwise comparison of learners using pairwise Friedman-Nemenyi tests with `$friedman_posthoc()`. This method first performs a global comparison to see if any learner is statistically better than another. To use these methods we first convert the benchmark result to a BenchmarkAggr object using `as_benchmark_aggr()`.

`as_benchmark`
`aggr()`

```
library(mlr3benchmark)
bma = as_benchmark_aggr(bmr, measures = msr("classif.ce"))
bma$friedman_posthoc()
```

```
    Pairwise comparisons using Nemenyi-Wilcoxon-Wilcox all-pairs test for a
      two-way balanced complete block design

data: ce and learner_id and task_id

           logreg ranger
ranger     0.1932 -
featureless 0.1932 0.0015


P value adjustment method: single-step
```

These results indicate a statistically significant difference between the "featureless" learner and "ranger" (assuming $p \leq 0.05$ is significant). This table can be visualized in a critical difference plot (Figure 11.3), which typically shows the mean rank of a learning algorithm on the x-axis along with a thick horizontal line that connects learners that are pairwise not significantly different (while correcting for multiple tests).

```
autoplot(bma, type = "cd", ratio = 1/5)
```

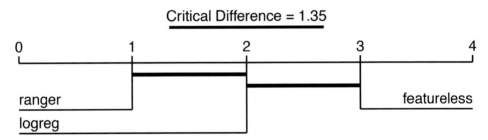

Figure 11.3: Critical difference diagram comparing the random forest, logistic regression, and featureless baseline. The critical difference of 1.35 in the title refers to the difference in mean rank required to conclude that one learner performs statistically different to another.

Using Figure 11.3, we can conclude that on average the random forest had the lowest (i.e., best) rank, followed by the logistic regression, and then the featureless baseline. While the random forest was statistically better performing than the baseline (no connecting line in Figure 11.3), it was not statistically superior to the logistic regression (connecting line in Figure 11.3). We could now further compare this with the large benchmark study conducted by Couronné, Probst, and Boulesteix (2018), where the random forest outperformed the logistic regression in 69% of 243 real-world datasets.

11.4 Conclusion

In this chapter, we have explored how to conduct large-scale machine learning experiments using `mlr3`. We have shown how to acquire diverse datasets from OpenML through the `mlr3oml` interface package, how to execute large-scale experiments with `batchtools` and `mlr3batchmark` integration, and finally how to analyze the results of these experiments with `mlr3benchmark`. For further reading about `batchtools` we recommend Lang, Bischl, and Surmann (2017) and Bischl et al. (2015).

Table 11.1: Important classes and functions covered in this chapter with underlying class (if applicable), class constructor or function, and important class fields and methods (if applicable).

Class	Constructor/Function	Fields/Methods
`OMLData`	`odt()`	`$data`; `$feature_names`
`OMLTask`	`otsk()`	`$data`; `$task_splits`
`OMLCollection`	`ocl()`	`$task_ids`
`Registry`	`makeExperimentRegistry()`	`submitJobs()`; `getStatus()`; `reduceResultsBatchmark`; `getJobTable`
	`batchmark()`	-
`BenchmarkAggr()`	`as_benchmark_aggr()`	`$friedman_posthoc()`

11.5 Exercises

In these exercises, we will conduct an empirical study analyzing whether a random forest is predictively stronger than a single decision tree. Our null hypothesis is that there is no significant performance difference.

1. Load the OpenML collection with ID 269, which contains regression tasks from the AutoML benchmark (Gijsbers et al. 2022). Peek into this suite to study the contained data sets and their characteristics. Then find all tasks with less than 4000 observations and convert them to `mlr3` tasks.
2. Create an experimental design that compares `lrn("regr.ranger")` and `lrn("regr.rpart")` on those tasks. Use the robustify pipeline for both learners and a featureless fallback learner. You can use three-fold CV instead of the OpenML resamplings to save time. Run the comparison experiments with `batchtools`. Use default hyperparameter settings and do not perform any tuning to keep the experiments simple.
3. Conduct a global Friedman test and, if appropriate, post hoc Friedman-Nemenyi tests, and interpret the results. As an evaluation measure, use the MSE.

12

Model Interpretation

Susanne Dandl
Ludwig-Maximilians-Universität München, and Munich Center for Machine Learning (MCML)

Przemysław Biecek
MI2.AI, Warsaw University of Technology, and University of Warsaw

Giuseppe Casalicchio
Ludwig-Maximilians-Universität München, and Munich Center for Machine Learning (MCML), and Essential Data Science Training GmbH

Marvin N. Wright
Leibniz Institute for Prevention Research and Epidemiology – BIPS, and University of Bremen, and University of Copenhagen

The increasing availability of data and software frameworks to create predictive models has allowed the widespread adoption of ML in many applications. However, high predictive performance of such models often comes at the cost of interpretability. Many models are called a "black box" as the decision-making process behind their predictions is often not immediately interpretable. This lack of explanation can decrease trust in ML and may create barriers to the adoption of predictive models, especially in critical applications such as medicine, engineering, and finance (Lipton 2018).

In recent years, many interpretation methods have been developed that allow developers to "peek" inside these models and produce explanations to, for example, understand how features are used by the model to make predictions (Guidotti et al. 2018). Interpretation methods can be valuable from multiple perspectives:

1. To gain global insights into a model, for example, to identify which features were the most important overall or how the features act on the predictions.
2. To improve the model if flaws are identified (in the data or model), for example, if the model depends on one feature unexpectedly.
3. To understand and control individual predictions, for example, to identify how a given prediction may change if a feature is altered.
4. To assess algorithmic fairness, for example, to inspect whether the model adversely affects certain subpopulations or individuals (see Chapter 14).

In this chapter, we will look at model-agnostic (i.e., can be applied to any model) interpretable machine learning (IML) methods that can be used to understand models post hoc (after they have been trained). We will focus on methods implemented in three R packages that nicely interface with `mlr3`: iml (Section 12.1), counterfactuals (Section 12.2), and DALEX (Section 12.3).

Inter-
pretable
Machine
Learning

iml and `DALEX` offer similar functionality but differ in design choices in that `iml` makes use of the `R6` class system whereas `DALEX` is based on the `S3` class system. `counterfactuals` also uses the `R6` class system. In contrast to `iml` and `counterfactuals`, `DALEX` focuses on comparing multiple predictive models, usually of different types. We will only provide a brief overview of the methodology discussed below, we recommend Molnar (2022) as a comprehensive introductory book about IML.

As a running example throughout this chapter, we will consider a gradient boosting machine (GBM) fit on half the features in the `"german_credit"` task. In practice, we would tune the hyperparameters of GBM as discussed in Chapter 4 and perform feature selection as discussed in Chapter 6 to select the most relevant features. However, for the sake of simplicity, we utilize an untuned GBM in these examples as it exhibited satisfactory performance even without fine-tuning.

```
library(mlr3verse)
tsk_german = tsk("german_credit")$select(
  cols = c("duration", "amount", "age", "status", "savings", "purpose",
  "credit_history", "property", "employment_duration", "other_debtors"))
split = partition(tsk_german)
lrn_gbm = lrn("classif.gbm", predict_type = "prob")
lrn_gbm$train(tsk_german, row_ids = split$train)
```

💡 Performance-based Interpretation Methods Require Test Data

Performance-based interpretation methods such as permutation feature importance (Section 12.1.1) rely on measuring the generalization performance. Hence, they should be computed on an independent test set to decrease bias in estimation (see Chapter 3).

However, the differences in interpretation between training and test data are less pronounced (Molnar et al. 2022) in prediction-based methods that do not require performance estimation such as ICE/PD (Section 12.1.2) or Shapley values (Section 12.1.4).

12.1 The iml Package

`iml` (Molnar, Bischl, and Casalicchio 2018) implements a unified interface for a variety of model-agnostic interpretation methods that facilitate the analysis and interpretation of machine learning models. `iml` supports machine learning models (for classification or regression) fitted by *any* R package, and in particular all `mlr3` models are supported by wrapping learners in an `Predictor` object, which unifies the input-output behavior of the trained models. This object contains the prediction model as well as the data used for analyzing the model and producing the desired explanation. We construct the `Predictor` object using our trained learner and holdout test data:

```
library(iml)

# features in test data
credit_x = tsk_german$data(rows = split$test,
  cols = tsk_german$feature_names)
# target in test data
credit_y = tsk_german$data(rows = split$test,
  cols = tsk_german$target_names)


predictor = Predictor$new(lrn_gbm, data = credit_x, y = credit_y)
```

With our `Predictor` setup we can now consider different model interpretation methods.

12.1.1 Feature Importance

When deploying a model in practice, it is often of interest to know which features contribute the most to the *predictive performance* of the model. This can be useful to better understand the problem at hand and the relationship between features and target. In model development, this can be used to filter features (Section 6.1) that do not contribute a lot to the model's predictive ability. In this book, we use the term "feature importance" to describe global methods that calculate a single score per feature that reflect the importance regarding a given quantity of interest, e.g., model performance, thus allowing features to be ranked.

Permutation
Feature
Importance

One of the most popular feature importance methods is the permutation feature importance (PFI), originally introduced by Breiman (2001a) for random forests and adapted by Fisher, Rudin, and Dominici (2019) as a model-agnostic feature importance measure (originally termed, "model reliance"). Feature permutation is the process of randomly shuffling observed values for a single feature in a dataset. This removes the original dependency structure of the feature with the target variable and with all other features while maintaining the marginal distribution of the feature. The PFI measures the change in the model performance before (original model performance) and after (permuted model performance) permuting a feature. If a feature is not important, then there will be little change in model performance after permuting that feature. Conversely, we would expect a clear decrease in model performance if the feature is more important. It is generally recommended to repeat the permutation process and aggregate performance changes over multiple repetitions to decrease randomness in results.

PFI is run in `iml` by constructing an object of class `FeatureImp` and specifying the performance measure, below we use classification error. By default, the permutation is repeated five times to keep computation time low (this can be changed with `n.repetitions` when calling the constructor `$new()`, below we set `n.repetitions = 100`) and in each repetition, the importance value corresponding to the change in the classification error is calculated. The `$plot()` method shows the median of the five resulting importance values (as a point) and the boundaries of the error bars in the plot refer to the 5% and 95% quantiles of the importance values (Figure 12.1).

💡 Increase the Number of Repetitions to Obtain Useful Error Bars

The default number of repetitions when constructing a `FeatureImp` object is 5. However, the number of repetitions should be increased if you want to obtain useful error bars from the resulting plot.

```
importance = FeatureImp$new(predictor, loss = "ce", n.repetitions = 100)
importance$plot()
```

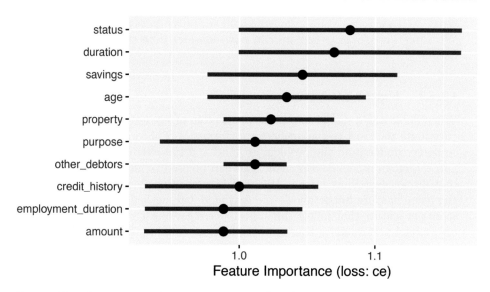

Figure 12.1: Permutation feature importance (PFI). Points indicate the median and bars the 5% and 95% quantiles of the PFI over 100 repetitions of the permutation process.

The plot automatically ranks features from most (largest median performance change) to least (smallest median performance change) important. In Figure 12.1, the feature `status` is most important, if we permute the `status` column in the data the classification error of our model increases by a factor of around 1.08. By default, `FeatureImp` calculates the *ratio* of the model performance before and after permutation as an importance value; the *difference* of the performance measures can be returned by passing `compare = "difference"` when calling `$new()`.

12.1.2   Feature Effects

Feature effect methods describe how or to what extent a feature contributes towards the *model predictions* by analyzing how the predictions change when changing a feature. These methods can be distinguished between local and global feature effect methods. Global feature effect methods refer to how a prediction changes *on average* when a feature is changed. In contrast, local feature effect methods address the question of how a *single* prediction of a given observation changes when a feature value is changed. To a certain extent, local feature effect methods can reveal interactions in the model that become visible when the local effects are heterogeneous, i.e., if changes in the local effect are different across the observations.

Partial
Dependence

Partial dependence (PD) plots (Friedman 2001) can be used to visualize global feature effects by visualizing how model predictions change on average when varying the values of a given feature of interest.

Individual
Conditional
Expectation

Individual conditional expectation (ICE) curves (Goldstein et al. 2015) (a.k.a. Ceteris Paribus Effects) are a local feature effects method that display how the prediction of a *single* observation changes when varying a feature of interest, while all other features stay constant. Goldstein et al. (2015) demonstrated that the PD plot is the average of ICE curves. ICE curves are constructed by taking a single observation and feature of interest, and then replacing the feature's value with another value and plotting the new prediction, this is then repeated for many feature values (e.g., across an equidistant grid of the feature's value range). The x-axis of an ICE curve visualizes the set of replacement feature values and the y-axis is the model prediction. Each ICE curve is a local explanation that assesses the feature effect of a single observation on the model prediction. An ICE plot contains one ICE curve (line) per observation. If the ICE curves are heterogeneous, i.e., not parallel, then the model may have estimated an interaction involving the considered feature.

> Feature Effects Can Be Non-Linear
>
> Feature effects are very similar to regression coefficients, β, in linear models which offer interpretations such as "if you increase this feature by one unit, your prediction increases on average by β if all other features stay constant". However, feature effects are not limited to linear effects and can be applied to any type of predictive model.

Let us put this into practice by considering how the feature `amount` influences the predictions in our subsetted credit classification task. Below we initialize an object of class `FeatureEffect` by passing the feature name of interest and the feature effect method, we use `"pdp+ice"` to indicate that we want to visualize ICE curves with a PD plot (average of the ICE curves). We recommend always plotting PD and ICE curves together as PD plots on their own could mask heterogeneous effects. We use `$plot()` to visualize the results (Figure 12.2).

```
effect = FeatureEffect$new(predictor, feature = "amount",
  method = "pdp+ice")
effect$plot()
```

Figure 12.2 shows that if the `amount` is smaller than roughly 10,000 then on average there is a high chance that the predicted creditworthiness will be `good`. Furthermore, the ICE curves are roughly parallel, meaning that there do not seem to be strong interactions present where `amount` is involved.

12.1.3 Surrogate Models

Interpretable models such as decision trees or linear models can be used as surrogate models to approximate or mimic an, often very complex, black box model. Inspecting the surrogate model can provide insights into the behavior of a black box model, for example by looking at the model coefficients in a linear regression or splits in a decision tree. We differentiate between local surrogate models, which approximate a model locally around a specific data point of interest, and global surrogate models

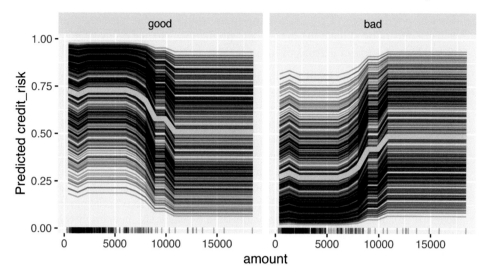

Figure 12.2: Partial dependence (PD) plot (yellow) and individual conditional expectation (ICE) curves (black) that show how the credit amount affects the predicted credit risk.

which approximate the model across the entire input space (Ribeiro, Singh, and Guestrin 2016; Molnar 2022).

The features used to train a surrogate model are usually the same features used to train the black box model or at least data with the same distribution to ensure a representative input space. However, the target used to train the surrogate model is the predictions obtained from the black box model, not the real outcome of the underlying data. Hence, conclusions drawn from the surrogate model are only valid if the surrogate model approximates the black box model very well (i.e., if the model fidelity is high). It is therefore also important to measure and report the approximation error of the surrogate model.

The data used to train the black box model may be very complex or limited, making it challenging to directly train a well-performing interpretable model on that data. Instead, we can use the black box model to generate new labeled data in specific regions of the input space with which we can augment the original data. The augmented data can then be used to train an interpretable model that captures and explains the relationships learned by the black box model (in specific regions) or to identify flaws or unexpected behavior.

12.1.3.1 Global Surrogate Model

Initializing the `TreeSurrogate` class fits a conditional inference tree (`ctree()`) surrogate model to the predictions from our trained model. This class extracts the decision rules created by the tree surrogate and the `$plot()` method visualizes the distribution of the predicted outcomes from each terminal node. Below, we pass `maxdepth = 2` to the constructor to build a tree with two binary splits, yielding four terminal nodes.

```
tree_surrogate = TreeSurrogate$new(predictor, maxdepth = 2L)
```

Before inspecting this model, we need to first check if the surrogate model approximates the prediction model accurately, which we can assess by comparing the predictions of the tree surrogate and the predictions of the black box model. For example, we could quantify the number of matching predictions and measure the accuracy of the surrogate in predicting the predictions of the black box GBM model:

```
pred_surrogate = tree_surrogate$predict(credit_x, type = "class")$.class
pred_surrogate = factor(pred_surrogate, levels = c("good", "bad"))
pred_gbm = lrn_gbm$predict_newdata(credit_x)$response
confusion = mlr3measures::confusion_matrix(pred_surrogate, pred_gbm,
  positive = "good")
confusion
```

```
        truth
response good bad
    good  269   4
    bad    38  19
acc :  0.8727; ce  :  0.1273; dor :  33.6250; f1  :  0.9276
fdr :  0.0147; fnr :  0.1238; fomr:  0.6667; fpr :  0.1739
mcc :  0.4731; npv :  0.3333; ppv :  0.9853; tnr :  0.8261
tpr :  0.8762
```

This shows an accuracy of around 87% in predictions from the surrogate compared to the black box model, which is good enough for us to use our surrogate for further interpretation, for example by plotting the splits in the terminal node:

```
tree_surrogate$plot()
```

Or we could access the trained tree surrogate via the `$tree` field of the `TreeSurrogate` object and then have access to all methods in partykit:

```
partykit::print.party(tree_surrogate$tree)
```

```
[1] root
|   [2] status in no checking account, ... < 0 DM
|   |   [3] duration <= 36: *
|   |   [4] duration > 36: *
|   [5] status in 0<= ... < 200 DM, ... >= 200 DM /
          salary for at least 1 year
|   |   [6] duration <= 42: *
|   |   [7] duration > 42: *
```

12.1.3.2 Local Surrogate Model

In general, it can be very difficult to accurately approximate the black box model with an interpretable surrogate in the entire feature space. Therefore, local surrogate models focus on a small area in the feature space surrounding a point of interest. Local surrogate models are constructed as follows:

1. Obtain predictions from the black box model for a given dataset.
2. Weight the observations in this dataset by their proximity to our point of interest.

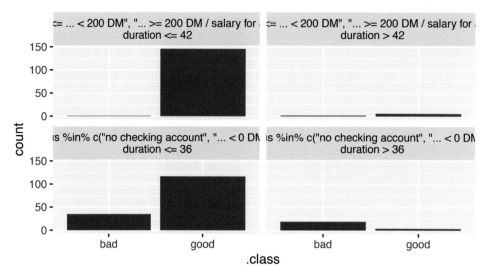

Figure 12.3: Distribution of the predicted outcomes for each terminal node identified by the tree surrogate. The top two nodes consist of applications with a positive balance in the account (`status`is either `"0 <= ... < 200 DM"`, `"... >= 200 DM"` or `"salary for at least 1 year"`) and either a duration of less or equal than 42 months (top left), or more than 42 months (top right). The bottom nodes contain applicants that either have no checking account or a negative balance (`status`) and either a duration of less than or equal to 36 months (bottom left) or more than 36 months (bottom right).

3. Fit an interpretable, surrogate model on the weighted dataset using the predictions of the black box model as the target.
4. Explain the prediction of our point of interest with the surrogate model.

To illustrate this, we will select a random data point to explain. As we are dealing with people, we will name our observation "Charlie" and first look at the black box predictions:

```
Charlie = credit_x[35, ]
gbm_predict = predictor$predict(Charlie)
gbm_predict
```

```
   good    bad
1 0.6346 0.3654
```

We can see that the model predicts the class "good" with 63.5% probability, so now we can use `LocalModel` to find out why this prediction was made. The underlying surrogate model is a locally weighted L1-penalized linear regression model such that only a pre-defined number of features per class, `k` (default is 3), will have a non-zero coefficient and as such are the `k` most influential features, below we set `k = 2`. We can also set the parameter `gower.power` which specifies the size of the neighborhood for the local model (default is `gower.power = 1`), the smaller the value, the more the model will focus on points closer to the point of interest, below we set `gower.power = 0.1`. This implementation is very closely related to Local

Interpretable Model-agnostic Explanations (LIME) (Ribeiro, Singh, and Guestrin 2016), the differences are outlined in the documentation of `iml::LocalModel`.

```
predictor$class = "good" # explain the 'good' class
local_surrogate = LocalModel$new(predictor, Charlie, gower.power = 0.1,
  k = 2)
```

If the prediction of the local model and the prediction of the black box GBM model greatly differ, then you might want to experiment with changing the k and `gower.power` parameters. These parameters can be considered as hyperparameters of the local surrogate model, which should be tuned to obtain an accurate local surrogate. First, we check if the predictions for Charlie match:

```
c(gbm = gbm_predict[[1]], local = local_surrogate$predict()[[1]])
```

```
   gbm   local
0.6346 0.6539
```

Ideally, we should assess the fidelity of the surrogate model in the local neighborhood of Charlie, i.e., how well the local surrogate model approximates the predictions of the black box GBM model for multiple data points in the vicinity of Charlie. A practical approach to assess this local model fidelity involves generating artificial data points within Charlie's local neighborhood (and potentially applying distance-based weighting) or selecting the k nearest neighbors from the original data. For illustration purposes, we now quantify the approximation error using the mean absolute error calculated from the 10 nearest neighbors (including Charlie) according to the Gower distance (Gower 1971):

```
ind_10nn = gower::gower_topn(Charlie, credit_x, n = 10)$index[, 1]
Charlie_10nn = credit_x[ind_10nn, ]

gbm_pred_10nn = predictor$predict(Charlie_10nn)[[1]]
local_pred_10nn = local_surrogate$predict(Charlie_10nn)[[1]]
mean(abs(gbm_pred_10nn - local_pred_10nn))
```

```
[1] 0.05475
```

As we see good agreement between the local and black box model (on average, the predictions of both the local surrogate and the black box model for Charlie's 10 nearest neighbors differ only by 0.055), we can move on to look at the most influential features for Charlie's predictions:

```
local_surrogate$results[, c("feature.value", "effect")]
```

```
            feature.value    effect
1              duration=12 -0.02000
2 status=no checking account -0.08544
```

In this case, "duration" and "status" were most important and both have a negative effect on the prediction of Charlie.

12.1.4 Shapley Values

Shapley values were originally developed in the context of cooperative game theory to study how the payout of a game can be fairly distributed among the players that form a team. This concept has been adapted for use in ML as a local interpretation method to explain the contributions of each input feature to the final model prediction of a single observation (Štrumbelj and Kononenko 2013). Hence, the "players" are the features, and the "payout", which should be fairly distributed among features, refers to the difference between the individual observation's prediction and the mean prediction.

Shapley values estimate how much each input feature contributed to the final prediction for a single observation (after subtracting the mean prediction). By assigning a value to each feature, we can gain insights into which features were the most important ones for the considered observation. Compared to the penalized linear model as a local surrogate model, Shapley values guarantee that the prediction is fairly distributed among the features as they also inherently consider interactions between features when calculating the contribution of each feature.

⚠ Correctly Interpreting Shapley Values

Shapley values are frequently **misinterpreted** as the difference between the predicted value after removing the feature from model training. The Shapley value of a feature is calculated by considering all possible subsets of features and computing the difference in the model prediction with and without the feature of interest included. Hence, it refers to the average marginal contribution of a feature to the difference between the actual prediction and the mean prediction, given the current set of features.

Shapley values can be calculated by passing the `Predictor` and the observation of interest to the constructor of `Shapley`. The exact computation of Shapley values is time consuming, as it involves taking into account all possible combinations of features to calculate the marginal contribution of a feature. Therefore, the estimation of Shapley values is often approximated. The `sample.size` argument (default is `sample.size = 100`) can be increased to obtain a more accurate approximation of exact Shapley values.

```
shapley = Shapley$new(predictor, x.interest = Charlie,
   sample.size = 1000)
shapley$plot()
```

In Figure 12.4, the Shapley values (`phi`) of the features show us how to fairly distribute the difference of Charlie's probability of being creditworthy to the dataset's average probability among the given features. The approximation is sufficiently good if all Shapley values (`phi`) sum up to the difference of the actual prediction and the average prediction. Here, we used `sample.size = 1000` leading to sufficiently good prediction difference of -0.079 between the actual prediction of Charlie (0.635) and the average prediction (0.706). The "purpose" variable has the most positive effect on the probability of being creditworthy, with an increase in the predicted probability of around 5%. In contrast, the "status" variable leads to a decrease in the predicted probability of over 10%.

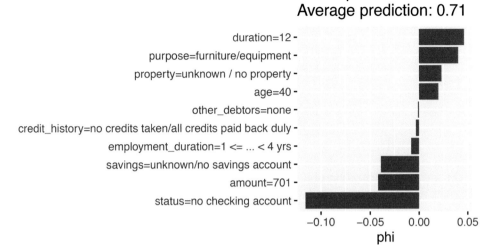

Figure 12.4: Shapley values for Charlie. The actual prediction (0.63) displays the prediction of the model for the observation we are interested in, the average prediction (0.71) displays the average prediction over the given test dataset. Each horizontal bar is the Shapley value (phi) for the given feature.

12.2   The counterfactuals Package

Counterfactual explanations try to identify the smallest possible changes to the input features of a given observation that would lead to a different prediction (Wachter, Mittelstadt, and Russell 2017). In other words, a counterfactual explanation provides an answer to the question: "What changes in the current feature values are necessary to achieve a different prediction?".

Counterfactual explanations can have many applications in different areas such as healthcare, finance, and criminal justice, where it may be important to understand how small changes in input features could affect the model's prediction. For example, a counterfactual explanation could be used to suggest lifestyle changes to a patient to reduce their risk of developing a particular disease, or to suggest actions that would increase the chance of a credit being approved. For our `tsk("german_credit")` example, we might consider what changes in features would turn a "bad" credit prediction into a "good" one (Figure 12.5).

What-If   A simple counterfactual method is the What-If approach (Wexler et al. 2019) where, for a given prediction to explain, the counterfactual is the closest data point in the dataset with the desired prediction. Usually, many possible counterfactual data points can exist. However, the approach by Wexler et al. (2019), and several other early counterfactual methods (see Guidotti (2022) for a comprehensive overview), only produce a single, somewhat arbitrary counterfactual explanation, which can be regarded as problematic when counterfactuals are used for insights or actions against the model.

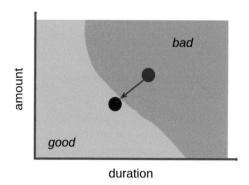

Figure 12.5: Illustration of a counterfactual explanation. The real observation (blue, right dot) is predicted to have "bad" credit. The brown (left) dot is one possible counterfactual that would result in a "good" credit prediction.

In contrast, the multi-objective counterfactuals method (MOC) (Dandl et al. 2020) generates multiple artificially-generated counterfactuals that may not be equal to observations in a given dataset. The generation of counterfactuals is based on an optimization problem that aims for counterfactuals that:

Multi-objective Counterfactuals

1) Have the desired prediction;
2) Are close to the observation of interest;
3) Only require changes in a few features; and
4) Originate from the same distribution as the observations in the given dataset.

In MOC, all four objectives are optimized simultaneously via a multi-objective optimization method. Several other counterfactual methods rely on single-objective optimization methods, where multiple objectives are combined into a single objective, e.g., using a weighted sum. However, a single-objective approach raises concerns about the appropriate weighting of objectives and is unable to account for inherent trade-offs among individual objectives. Moreover, it may restrict the solution set of the counterfactual search to a single candidate. MOC returns a set of non-dominated and, therefore equally good, counterfactuals with respect to the four objectives (similarly to the Pareto front we saw in Section 5.2).

Counterfactual explanations are available in the `counterfactuals` package, which depends on `Predictor` objects as inputs.

12.2.1   What-If Method

Continuing our previous example, we saw that the GBM model classifies Charlie as having good credit with a predicted probability of 63.5%. We can use the What-If method to understand how the features need to change for this predicted probability to increase to 75%. We initialize a `WhatIfClassif` object with our `Predictor` and state that we only want to find one counterfactual (`n_counterfactuals = 1L`), increasing `n_counterfactuals` would return the specified number of counterfactuals closest to the point of interest. The `$find_counterfactuals()` method generates a counterfactual of class `Counterfactuals`, below we set our desired predicted probability to be between 0.75 and 1 (`desired_prob = c(0.75, 1)`). The `$evaluate(show_diff = TRUE)` method tells us how features need to be changed to generate our desired class.

```
library(counterfactuals)
whatif = WhatIfClassif$new(predictor, n_counterfactuals = 1L)
cfe = whatif$find_counterfactuals(Charlie,
  desired_class = "good", desired_prob = c(0.75, 1))
data.frame(cfe$evaluate(show_diff = TRUE))

 age amount credit_history duration employment_duration other_debtors
1  -3   1417           <NA>       -3                <NA>          <NA>
 property purpose savings     status dist_x_interest no_changed
1     <NA>    <NA>    <NA> ... < 0 DM          0.1176          4
 dist_train dist_target minimality
1          0           0          1
```

Here we can see that, to achieve a predicted probability of at least 75% for good credit, Charlie would have to be three years younger, the duration of credit would have to be reduced by three months, the amount would have to be increased by 1417 DM and the status would have to be "... < 0 DM" (instead of "no checking account").

12.2.2   MOC Method

Calling the MOC method is similar to the What-If method but with a `MOCClassif()` object. We set the **epsilon** parameter to 0 to penalize counterfactuals in the optimization process with predictions outside the desired range. With MOC, we can also prohibit changes in specific features via the **fixed_features** argument, below we restrict changes in the "age" variable. For illustrative purposes, we only run the multi-objective optimizer for 30 generations.

```
moc = MOCClassif$new(predictor, epsilon = 0, n_generations = 30L,
  fixed_features = "age")
cfe_multi = moc$find_counterfactuals(Charlie,
  desired_class = "good", desired_prob = c(0.75, 1))
```

The multi-objective approach does not guarantee that all counterfactuals have the desired prediction so we use `$subset_to_valid()` to restrict counterfactuals to those we are interested in:

```
cfe_multi$subset_to_valid()
cfe_multi
```

```
6 Counterfactual(s)

Desired class: good
Desired predicted probability range: [0.75, 1]

Head:
    age amount                           credit_history duration
1:  40     701 no credits taken/all credits paid back duly        12
2:  40     701 no credits taken/all credits paid back duly        12
3:  40     701 no credits taken/all credits paid back duly        12
6 variables not shown: [employment_duration, other_debtors,
    property, purpose, savings, status]
```

This method generated 6 counterfactuals but as these are artificially generated they are not necessarily equal to actual observations in the underlying dataset. For a concise overview of the required feature changes, we can use the `plot_freq_of_feature_changes()` method, which visualizes the frequency of feature changes across all returned counterfactuals.

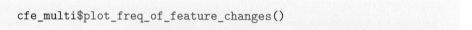

Figure 12.6: Barplots of the relative frequency of feature changes of the counterfactuals found by MOC.

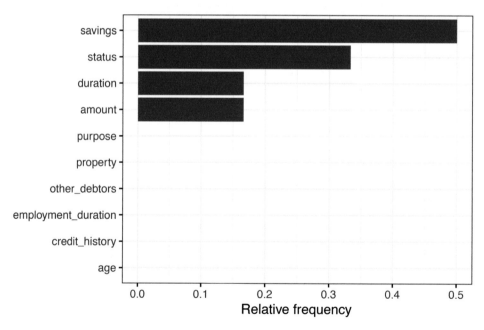

We can see that "status" and "savings" were changed most frequently in the counterfactuals. To see *how* the features were changed, we can visualize the counterfactuals for two features on a two-dimensional ICE plot.

```
cfe_multi$plot_surface(feature_names = c("status", "savings")) +
    theme(axis.text.x = element_text(angle = 15, hjust = .7))
```

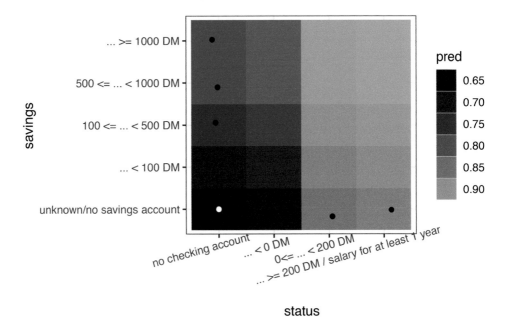

Figure 12.7: Two-dimensional surface plot for the "status" and "savings" variables, higher predictions are lighter. The colors and contour lines indicate the predicted value of the model when "status" and "savings" differ while all other features are set to the true (Charlie's) values. The white point displays the true prediction (Charlie), and the black points are the counterfactuals that only propose changes in the two features.

12.3   The DALEX Package

DALEX (Biecek 2018) implements a similar set of methods as iml, but the architecture of DALEX is oriented towards model comparison. The logic behind working with this package assumes that the process of exploring models is iterative, and in successive iterations, we want to compare different perspectives, including perspectives presented/learned by different models. This logic is commonly referred to as the Rashomon perspective, first described in Breiman (2001b) and more extensively developed and formalized as interactive explanatory model analysis (Baniecki, Parzych, and Biecek 2023).

You can use the DALEX package with any classification and regression model built with mlr3 as well as with other frameworks in R. As we have already explored the methodology behind most of the methods discussed in this section, we will just focus on the implementations of these methods in DALEX using the tsk("german_credit") running example.

Once you become familiar with the philosophy of working with the `DALEX` package, you can use other packages from this family such as `fairmodels` (Wiśniewski and Biecek 2022) for detection and mitigation of biases, `modelStudio` (Baniecki and Biecek 2019) for interactive model exploration, `modelDown` (Romaszko et al. 2019) for the automatic generation of IML model documentation, `survex` (Krzyziński et al. 2023) for the explanation of survival models, or `treeshap` for the analysis of tree-based models.

The analysis of a model is usually an interactive process starting with evaluating a model based on one or more performance metrics, known as a "shallow analysis". In a series of subsequent steps, one can systematically deepen understanding of the model by exploring the importance of single variables or pairs of variables to an in-depth analysis of the relationship between selected variables to the model outcome. See Bücker et al. (2022) for a broader discussion of what the model exploration process looks like.

This explanatory model analysis (EMA) process can focus on a single observation, in which case we speak of local model analysis, or for a set of observations, in which case we refer to global model analysis. Figure 12.8 visualizes an overview of the key functions in these two scenarios that we will discuss in this section. An in-depth description of this methodology can be found in Biecek and Burzykowski (2021).

Explanatory Model Analysis

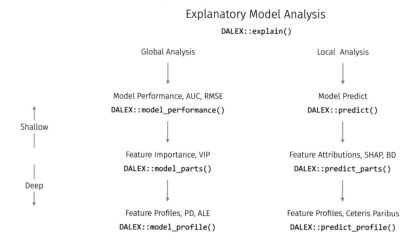

Figure 12.8: Taxonomy of methods for model exploration presented in this section. The left side shows global analysis methods and the right shows local analysis methods. Methods increase in analysis complexity from top to bottom.

As with `iml`, `DALEX` also implements a wrapper that enables a unified interface to its functionality. For models created with the `mlr3` package, we would use `explain_mlr3()`, which creates an S3 `explainer` object, which is a list containing at least: the model object, the dataset that will be used for calculation of explanations, the predict function, the function that calculates residuals, name/label of the model name and other additional information about the model.

```
library(DALEX)
library(DALEXtra)

gbm_exp = DALEXtra::explain_mlr3(lrn_gbm,
  data = credit_x,
  y = as.numeric(credit_y$credit_risk == "bad"),
  label = "GBM Credit",
  colorize = FALSE)

gbm_exp
```

```
Model label:  GBM Credit
Model class:  LearnerClassifGBM,LearnerClassif,Learner,R6
Data head  :
  age amount                          credit_history duration
1  67   1169 all credits at this bank paid back duly        6
2  49   2096 all credits at this bank paid back duly       12
  employment_duration other_debtors                property
1              >= 7 yrs             none unknown / no property
2    4 <= ... < 7 yrs             none unknown / no property
                purpose                        savings
1 furniture/equipment                ... >= 1000 DM
2               repairs unknown/no savings account
                                          status
1                        no checking account
2 ... >= 200 DM / salary for at least 1 year
```

12.3.1 Global EMA

Global EMA aims to understand how a model behaves on average for a set of observations. In DALEX, functions for global level analysis are prefixed with model_.

The model exploration process starts (Figure 12.8) by evaluating the performance of a model. model_performance() detects the task type and selects the most appropriate measure, as we are using binary classification the function automatically suggests recall, precision, F1-score, accuracy, and AUC; similarly the default plotting method is selected based on the task type, below ROC is selected.

```
perf_credit = model_performance(gbm_exp)
perf_credit
```

```
Measures for:  classification
recall    : 0.3535
precision : 0.614
f1        : 0.4487
accuracy  : 0.7394
auc       : 0.7689

Residuals:
       0%       10%       20%       30%       40%       50%       60%       70%
-0.88117  -0.44188  -0.31691  -0.20743  -0.14601  -0.10782  -0.07089   0.03232
      80%       90%      100%
 0.49779   0.65661   0.94458
```

```
old_theme = set_theme_dalex("ema")
plot(perf_credit, geom = "roc")
```

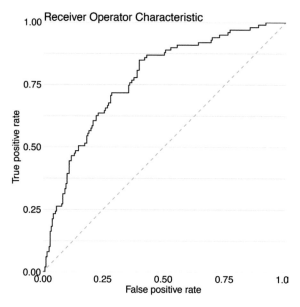

Figure 12.9: Graphical summary of model performance using the Receiver Operator Curve (Section 3.4).

> 💡 Visual Summaries
>
> Various visual summaries may be selected with the `geom` parameter. For the credit risk task, the LIFT curve is a popular graphical summary.

Feature importance methods can be calculated with `model_parts()` and then plotted.

```
gbm_effect = model_parts(gbm_exp)
head(gbm_effect)
```

```
             variable mean_dropout_loss      label
1         _full_model_            0.2311 GBM Credit
2         other_debtors           0.2351 GBM Credit
3               amount            0.2351 GBM Credit
4             property            0.2353 GBM Credit
5                  age            0.2355 GBM Credit
6 employment_duration            0.2403 GBM Credit
```

```
plot(gbm_effect, show_boxplots = FALSE)
```

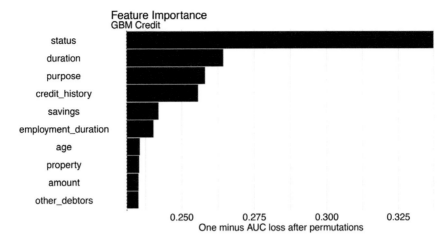

Figure 12.10: Graphical summary of permutation importance of features. The longer the bar, the larger the change in the loss function after permutation of the particular feature and therefore the more important the feature. This plot shows that "status" is the most important feature and "other_debtors" is the least important.

💡 Calculating Importance

The `type` argument in the `model_parts` function allows you to specify how the importance of the features is to be calculated, by the difference of the loss functions (`type = "difference"`), by the quotient (`type = "ratio"`), or without any transformation (`type = "raw"`).

Feature effects can be calculated with `model_profile()` and by default are plotted as PD plots.

```
gbm_profiles = model_profile(gbm_exp)
gbm_profiles
```

```
Top profiles    :
  _vname_    _label_  _x_ _yhat_ _ids_
1 duration GBM Credit   4 0.2052     0
2 duration GBM Credit   6 0.2052     0
3 duration GBM Credit   7 0.2052     0
4 duration GBM Credit   8 0.2052     0
5 duration GBM Credit   9 0.2246     0
6 duration GBM Credit  10 0.2246     0
```

```
plot(gbm_profiles) +
  theme(legend.position = "top") +
  ggtitle("Partial Dependence for GBM Credit model","")
```

From Figure 12.11, we can see that the GBM model has learned a non-monotonic relationship for the feature `amount`.

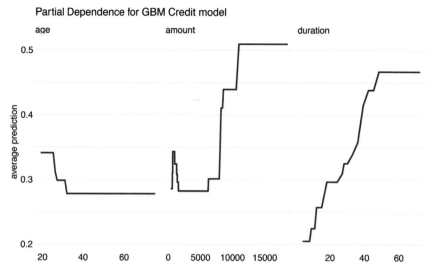

Figure 12.11: Graphical summary of the model's partial dependence profile for three selected variables (age, amount, duration).

💡 Marginal and Accumulated Local Profiles

The `type` argument of the `model_profile()` function also allows *marginal profiles* (with `type = "conditional"`) and *accumulated local profiles* (with `type = "accumulated"`) to be calculated.

12.3.2   Local EMA

Local EMA aims to understand how a model behaves for a single observation. In `DALEX`, functions for local analysis are prefixed with `predict_`. We will carry out the following examples using Charlie again.

Local analysis starts with the calculation of a model prediction (Figure 12.8).

```
predict(gbm_exp, Charlie)
```

```
   bad
0.3654
```

As a next step, we might consider break-down plots, which decompose the model's prediction into contributions that can be attributed to different explanatory variables (see the *Break-down Plots for Additive Attributions* chapter in Biecek and Burzykowski (2021) for more on this method). These are calculated with `predict_parts()`:

```
plot(predict_parts(gbm_exp, new_observation = Charlie))
```

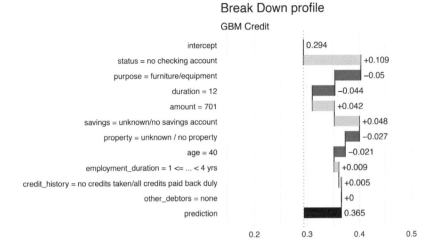

Figure 12.12: Graphical summary of local attributions of features calculated by the break-down method. Positive attributions are shown in green and negative attributions in red. The violet bar corresponds to the model prediction for the explained observation and the dashed line corresponds to the average model prediction.

Looking at Figure 12.12, we can read that the biggest contributors to the final prediction for Charlie were the features `status` and `savings`.

> 💡 Selected Order of Features
>
> The `order` argument allows you to indicate the selected order of the features. This is a useful option when the features have some relative conditional importance (e.g., pregnancy and sex).

The `predict_parts()` function can also be used to plot Shapley values with the SHAP algorithm (Lundberg, Erion, and Lee 2019) by setting `type = "shap"`:

```
plot(predict_parts(gbm_exp, new_observation = Charlie, type = "shap"),
  show_boxplots = FALSE)
```

The results for Break Down and SHAP methods are generally similar. Differences will emerge if there are many complex interactions in the model.

> 💡 Speeding Up Shapley Computation
>
> Shapley values can take a long time to compute. This process can be sped up at the expense of accuracy. The parameters `B` and `N` can be used to tune this trade-off, where `N` is the number of observations on which conditional expectation values are estimated (500 by default) and `B` is the number of random paths used to calculate Shapley values (25 by default).

Finally, we can plot ICE curves using `predict_profile()`:

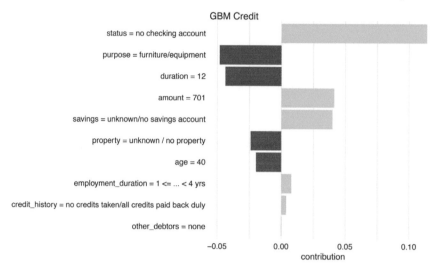

Figure 12.13: Graphical summary of local attributions of features calculated by the SHAP method. Positive attributions are shown in green and negative attributions in red. The most important feature here is the "status" variable and least is "other_debtors".

```
plot(predict_profile(gbm_exp, credit_x[30:40, ]))
```

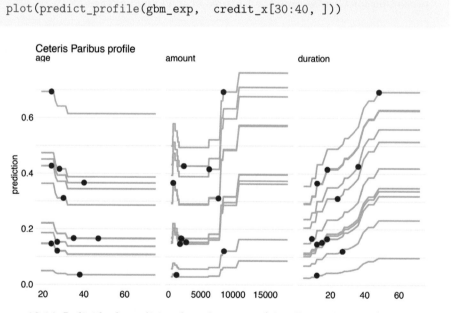

Figure 12.14: Individual conditional explanations (aka Ceteris Paribus) plots for 11 rows in the credit data (including Charlie) for three selected variables (age, amount, duration).

12.4 Conclusions

In this chapter, we learned how to gain post hoc insights into a model trained with `mlr3` by using the most popular approaches from the field of interpretable machine learning. The methods are all model-agnostic and so do not depend on specific model classes. `iml` and `DALEX` offer a wide range of (partly) overlapping methods, while `counterfactuals` focuses solely on counterfactual methods. We demonstrated on `tsk("german_credit")` how these packages offer an in-depth analysis of a GBM model fitted with `mlr3`. As we conclude the chapter we will highlight some limitations in the methods discussed above to help guide your own post hoc analyses.

Correlated Features

If features are correlated, the insights from the interpretation methods should be treated with caution. Changing the feature values of an observation without taking the correlation with other features into account leads to unrealistic combinations of the feature values. Since such feature combinations are also unlikely to be part of the training data, the model will likely extrapolate in these areas (Molnar et al. 2022; Hooker and Mentch 2019). This distorts the interpretation of methods that are based on changing single feature values such as PFI, PD plots, and Shapley values. Alternative methods can help in these cases: conditional feature importance instead of PFI (Strobl et al. 2008; Watson and Wright 2021), accumulated local effect plots instead of PD plots (Apley and Zhu 2020), and the KernelSHAP method instead of Shapley values (Lundberg, Erion, and Lee 2019).

Rashomon Effect

Explanations derived from an interpretation method can be ambiguous. A method can deliver multiple equally plausible but potentially contradicting explanations. This phenomenon is also called the Rashomon effect (Breiman 2001b). This effect can be due to changes in hyperparameters, the underlying dataset, or even the initial seed (Molnar et al. 2022).

High-Dimensional Data

`tsk("german_credit")` is low-dimensional with a limited number of observations. Applying interpretation methods off-the-shelf to higher dimensional datasets is often not feasible due to the enormous computational costs and so recent methods, such as Shapley values that use kernel-based estimators, have been developed to help overcome this. Another challenge is that the high-dimensional IML output generated for high-dimensional datasets can overwhelm users. If the features can be meaningfully grouped, grouped versions of methods, e.g. the grouped feature importance proposed by Au et al. (2022), can be applied.

Table 12.1: Important classes and functions covered in this chapter with underlying class (if applicable), class constructor or function, and important class fields and methods (if applicable).

Class	Constructor/Function	Fields/Methods
`Predictor`	`$new()`	-
`FeatureImp`	`$new(some_predictor)`	`$plot()`
`FeatureEffect`	`$new(some_predictor)`	`$plot()`
`LocalModel`	`$new(some_predictor, some_x)`	`$results()`
`Shapley`	`$new(some_predictor, x.interest)`	`$plot()`
`WhatIfClassif`	`$new(some_predictor)`	`$find_counterfactuals()`
`MOCClassif`	`$new(some_predictor)`	`$find_counterfactuals()`
`explainer`	`explain_mlr3()`	`model_parts();` `model_performance();` `predict_parts()`

12.5   Exercises

The following exercises are based on predictions of the value of soccer players based on their characteristics in the FIFA video game series. They use the 2020 `fifa` data available in DALEX. Solve them with either `iml` or DALEX.

1. Prepare an `mlr3` regression task for the `fifa` data. Select only variables describing the age and skills (i.e., features starting with "skills_") of soccer players. Train a predictive model of your own choice on this task, to predict the value of a soccer player.
2. Use the permutation importance method to calculate feature importance ranking. Which feature is the most important? Do you find the results surprising?
3. Use the partial dependence plot/profile to draw the global behavior of the model for this feature. Is it aligned with your expectations?
4. Choose Robert Lewandowski as a specific example (row number 21 in the original fifa dataset) and calculate and plot the Shapley values. Which feature is locally the most important and has the strongest influence on his valuation as a soccer player?

13

Beyond Regression and Classification

Raphael Sonabend
Imperial College London

Patrick Schratz
Friedrich Schiller University Jena

Damir Pulatov
University of Wyoming

So far, this book has only considered two tasks. In Chapter 2, we introduced deterministic regression as well as deterministic and probabilistic single-label classification (Table 13.1). But our infrastructure also works well for many other tasks, some of which are available in extension packages (Figure 1.1) and some are available by creating pipelines with `mlr3pipelines`. In this chapter, we will take you through just a subset of these new tasks, focusing on the ones that have a stable API. As we work through this chapter, we will refer to the "building blocks" of `mlr3`, this refers to the base classes that must be extended to create new tasks, these are `Prediction`, `Learner`, `Measure`, and `Task`. Table 13.1 summarizes available extension tasks, including the package(s) they are implemented in and a brief description of the task.

Table 13.1: Table of extension tasks that can be used with `mlr3` infrastructure. As we have a growing community of contributors, this list is far from exhaustive and many "experimental" task implementations exist; this list just represents the tasks that have a functioning interface.

Task	Package	Description
Deterministic regression	`mlr3`	Point prediction of a continuous variable.
Deterministic single-label classification	`mlr3`	Prediction of a single class for each observation.
Probabilistic single-label classification	`mlr3`	Prediction of the probability of an observation falling into one or more mutually exclusive categories.
Cost-sensitive classification	`mlr3` and `mlr3pipelines`	Classification predictions with unequal costs associated with misclassifications.
Survival analysis	`mlr3proba`	Time-to-event predictions with possible "censoring".
Density estimation	`mlr3proba`	Unsupervised estimation of probability density functions.

DOI: 10.1201/9781003402848-13

Task	Package	Description
Spatiotemporal analysis	`mlr3spatiotempcv` and `mlr3spatial`	Supervised prediction of data with spatial (e.g., coordinates) and/or temporal outcomes.
Cluster analysis	`mlr3cluster`	Unsupervised estimation of homogeneous clusters of data points.

13.1   Cost-Sensitive Classification

We begin by discussing a task that does not require any additional packages or infrastructure, only the tools we have already learned about from earlier chapters. In "regular" classification, the aim is to optimize a metric (often the misclassification rate) while assuming all misclassification errors are deemed equally severe. A more general approach is cost-sensitive classification, in which costs caused by different kinds of errors may not be equal. The objective of cost-sensitive classification is to minimize the expected costs. We will use `tsk("german_credit")` as a running example.

Cost-sensitive Classification

Imagine you are trying to calculate if giving someone a loan of $5K will result in a profit after one year, assuming they are expected to pay back $6K. To make this calculation, you will need to predict if the person will have good credit. This is a deterministic classification problem where we are predicting whether someone will be in class "Good" or "Bad". Now let us consider some potential costs associated with each prediction and the eventual truth. As cost-sensitive classification is a minimization problem, we assume lower costs correspond to higher profits/positive outcomes, hence we write profits as negative values and losses as positive values:

```
costs = matrix(c(-1, 0, 5, 0), nrow = 2, dimnames =
  list("Predicted Credit" = c("good", "bad"),
    Truth = c("good", "bad")))
costs
```

```
                Truth
Predicted Credit good bad
          good    -1   5
          bad      0   0
```

In this example, if the model predicts that the individual has bad credit (bottom row) then there is no profit or loss, the loan is not provided. If the model predicts that the individual has good credit and indeed the customer repays the loan with interest (top left), then you will make a $1K profit. On the other hand, if they default (top right), you will lose $5K.

13.1.1   Cost-Sensitive Measure

We will now see how to implement a more nuanced approach to classification errors with `msr("classif.costs")`. This measure takes one argument, which is a matrix with row and column names corresponding to the class labels in the task of interest.

Let us put our insurance example into practice, notice that we have already named the cost matrix as required for the measure:

```
library(mlr3verse)

tsk_german = tsk("german_credit")

msr_costs = msr("classif.costs", costs = costs)
msr_costs
```

```
<MeasureClassifCosts:classif.costs>: Cost-sensitive Classification
* Packages: mlr3
* Range: [-Inf, Inf]
* Minimize: TRUE
* Average: macro
* Parameters: normalize=TRUE
* Properties: -
* Predict type: response
```

```
learners = lrns(c("classif.log_reg", "classif.featureless",
    "classif.ranger"))
bmr = benchmark(benchmark_grid(tsk_german, learners,
    rsmp("cv", folds = 3)))
bmr$aggregate(msr_costs)[, c(4, 7)]
```

```
          learner_id classif.costs
1:      classif.log_reg        0.1791
2: classif.featureless        0.8002
3:      classif.ranger        0.2331
```

In this experiment, we find that the logistic regression learner happens to perform best as it minimizes the expected costs (and maximizes expected profits) and the featureless learner performs the worst. All losses result in positive costs, which means each model results in us losing money. To improve our models, we will now turn to thresholding.

13.1.2 Thresholding

As we have discussed in Chapter 2, thresholding is a method to fine-tune the probability at which an observation will be predicted as one class label or another. Currently in our running example, the models above will predict a customer has good credit (in the class "Good") if the probability of good credit is greater than 0.5. Here, this might not be a sensible approach as we would likely act more conservatively and reject more credit applications with a higher threshold due to the non-uniform costs. This is highlighted in the `"threshold"` `autoplot` (Figure 13.1), which plots `msr("classif.costs")` over all possible thresholds.

```
prediction = lrn("classif.log_reg",
    predict_type = "prob")$train(tsk_german)$predict(tsk_german)
autoplot(prediction, type = "threshold", measure = msr_costs)
```

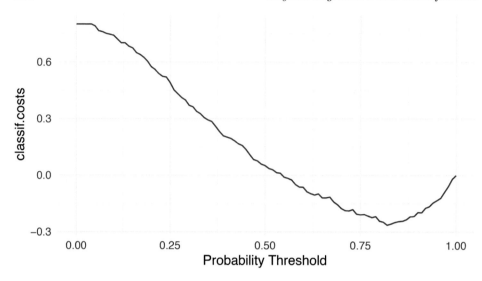

Figure 13.1: Changing values of cost-sensitive measure as the prediction threshold is changed.

As expected, the optimal threshold is greater than 0.5 which means the optimal model should predict "bad" credit more often than not.

The optimal threshold can be automated by making use of mlr3tuning (Chapter 4) and mlr3pipelines (Chapter 7) to tune po("tunethreshold"). Continuing the same example:

```
po_cv = po("learner_cv", lrn("classif.log_reg", predict_type = "prob"))
graph =  po_cv %>>% po("tunethreshold", measure = msr_costs)

learners = list(as_learner(graph), lrn("classif.log_reg"))
bmr = benchmark(benchmark_grid(tsk_german, learners,
   rsmp("cv", folds = 3)))
bmr$aggregate(msr_costs)[, c(4, 7)]
```

```
                    learner_id classif.costs
1: classif.log_reg.tunethreshold      -0.1060
2:               classif.log_reg       0.1481
```

By using po("learner_cv") for internal resampling and po("tunethreshold") to find the optimal threshold we have improved our model performance considerably and can now even expect a profit.

13.2   Survival Analysis

Survival analysis is a field of statistics concerned with trying to predict/estimate the time until an event takes place. This predictive problem is unique as survival models are trained and tested on data that may include "censoring", which occurs when the event of interest does *not* take place. Survival analysis can be hard to explain in the abstract, so as a working example consider a marathon runner in a race. Here the "survival problem" is trying to predict the time when the marathon runner finishes the race. However, if the event of interest does not take place (e.g., the marathon runner gives up and does not finish the race), they are said to be censored. Instead of throwing away information about censored events, survival analysis datasets include a status variable that provides information about the "status" of an observation. So in our example, we might write the runner's outcome as $(4, 1)$ if they finish the race at four hours, otherwise, if they give up at two hours we would write $(2, 0)$.

The key to modeling in survival analysis is that we assume there exists a hypothetical time the marathon runner would have finished if they had not been censored, it is then the job of a survival learner to estimate what the true survival time would have been for a similar runner, assuming they are *not* censored (see Figure 13.2). Mathematically, this is represented by the hypothetical event time, Y, the hypothetical censoring time, C, the observed outcome time, $T = \min(Y, C)$, the event indicator $\Delta = (T = Y)$, and as usual some features, X. Learners are trained on (T, Δ) but, critically, make predictions of Y from previously unseen features. This means that unlike classification and regression, learners are trained on two variables, (T, Δ), which, in R, is often captured in a `Surv` object. Relating to our example above, the runner's outcome would then be $(T = 4, \Delta = 1)$ or $(T = 2, \Delta = 0)$. Another example is in the code below, where we randomly generate six survival times and six event indicators, an outcome with a + indicates the outcome is censored, otherwise, the event of interest occurred.

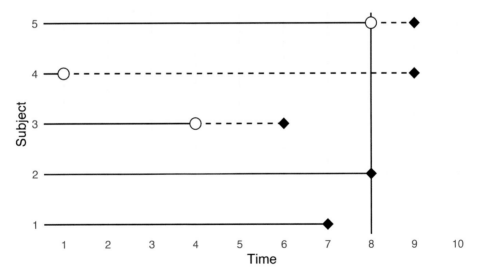

Figure 13.2: Plot illustrating different censoring types. Dead and censored subjects (y-axis) over time (x-axis). Black diamonds indicate true death times and white circles indicate censoring times. Vertical line is the study end time. Subjects 1 and 2 die in the study time. Subject 3 is censored in the study and (unknown) dies within the study time. Subject 4 is censored in the study and (unknown) dies after the study. Subject 5 dies after the end of the study. Figure and caption from R. E. B. Sonabend (2021).

```
library(survival)
Surv(runif(6), rbinom(6, 1, 0.5))
```

[1] 0.5523+ 0.2905 0.4404+ 0.1184 0.9216+ 0.7326

Readers familiar with survival analysis will recognize that the description above applies specifically to "right censoring". Currently, this is the only form of censoring available in the mlr3 universe, hence restricting our discussion to that setting. For a good introduction to survival analysis see Collett (2014) or for machine learning in survival analysis specifically see R. Sonabend and Bender (2023).

For the remainder of this section, we will look at how mlr3proba (R. Sonabend et al. 2021) extends the building blocks of mlr3 for survival analysis. We will begin by looking at objects used to construct machine learning tasks for survival analysis, then we will turn to the learners we have implemented to solve these tasks, before looking at measures for evaluating survival analysis predictions, and then finally we will consider how to transform prediction types.

13.2.1 TaskSurv

As we saw in the introduction to this section, survival algorithms require two targets for training, this means the new TaskSurv object expects two targets. The simplest way to create a survival task is to use as_task_surv(), as in the following code chunk. Note this has more arguments than as_task_regr() to reflect multiple target and censoring types, time and event arguments expect strings representing column names where the "time" and "event" variables are stored, type refers to the censoring type (currently only right censoring supported so this is the default). as_task_surv() coerces the target columns into a Surv object. In this section we will use the rats dataset as a running example, this dataset looks at predicting if a drug treatment was successful in preventing 150 rats from developing tumors. The dataset, by its own admission, is not perfect and should generally be treated as "dummy" data, which is good for examples but not real-world analysis.

```
library(mlr3verse)
library(mlr3proba)
library(survival)

tsk_rats = as_task_surv(survival::rats, time = "time",
  event = "status", type = "right", id = "rats")

tsk_rats$head()
```

```
   time status litter rx sex
1:  101      0      1  1   f
2:   49      1      1  0   f
3:  104      0      1  0   f
4:   91      0      2  1   m
5:  104      0      2  0   m
6:  102      0      2  0   m
```

Plotting the task with `autoplot` results in a Kaplan-Meier plot (Figure 13.3), which is a non-parametric estimator of the probability of survival for the average observation in the training set.

`autoplot(tsk_rats)`

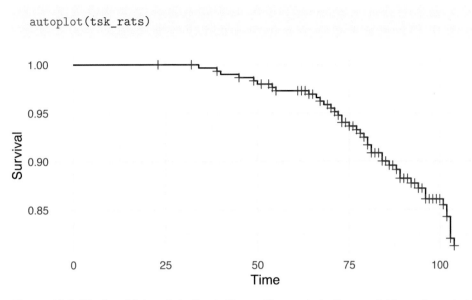

Figure 13.3: Kaplan-Meier plot of `tsk("rats")`. x-axis is time variable and y-axis is survival function, S(T), defined by 1− F(T) where F is the cumulative distribution function. Crosses indicate points where censoring takes place.

As well as creating your own tasks, you can load any of the tasks shipped with `mlr3proba`:

```
as.data.table(mlr_tasks)[task_type == "surv"]
```

```
              key                 label task_type nrow ncol properties
1:           actg              ACTG 320      surv 1151   13
2:           gbcs  German Breast Cancer      surv  686   10
3:          grace            GRACE 1000      surv 1000    8
4:           lung           Lung Cancer      surv  228   10
5:           rats                  Rats      surv  300    5
6:   unemployment  Unemployment Duration     surv 3343    6
7:           whas Worcester Heart Attack     surv  481   11
7 variables not shown: [lgl, int, dbl, chr, fct, ord, pxc]
```

13.2.2   LearnerSurv, PredictionSurv, and Predict Types

The interface for `LearnerSurv` and `PredictionSurv` objects is identical to the regression and classification settings discussed in Chapter 2. Similarly to these settings, survival learners are constructed with `lrn()`.

`mlr3proba` has a different predict interface to `mlr3` as all possible types of prediction ("predict types") are returned when possible for all survival models – i.e., if a model *can* compute a particular predict type then *it will be* returned in `PredictionSurv`. The reason for this design decision is that all these predict types can be transformed to one another and it is therefore computationally simpler to return all at once instead of rerunning models to change predict type. In survival analysis, the following predictions can be made:

- `response` – Predicted survival time.
- `distr` – Predicted survival distribution, either discrete or continuous.
- `lp` – Linear predictor calculated as the fitted coefficients multiplied by the test data.
- `crank` – Continuous risk ranking.

We will go through each of these prediction types in more detail and with examples to make them less abstract. We will use `lrn("surv.coxph")` trained on `tsk("rats")` as a running example, for this model, all predict types except `response` can be computed.

```
tsk_rats = tsk("rats")
split = partition(tsk_rats)
prediction_cph = lrn("surv.coxph")$train(tsk_rats, split$train)$
  predict(tsk_rats, split$test)
prediction_cph
```

```
<PredictionSurv> for 99 observations:
    row_ids time status    crank       lp     distr
          8  102  FALSE  -0.1577  -0.1577  <list[1]>
         16   98  FALSE  -1.9549  -1.9549  <list[1]>
         24   76  FALSE  -2.7150  -2.7150  <list[1]>
---
        241   72   TRUE   0.8827   0.8827  <list[1]>
        247   73   TRUE   0.8897   0.8897  <list[1]>
        249   66   TRUE   0.1226   0.1226  <list[1]>
```

predict_type = "response"

Counterintuitively for many, the `response` prediction of predicted survival times is the least common predict type in survival analysis. The likely reason for this is due to the presence of censoring. We rarely observe the true survival time for many observations and therefore it is unlikely any survival model can confidently make predictions for survival times. This is illustrated in the code below.

In the example below, we train and predict from a survival SVM (`lrn("surv.svm")`), note we use `type = "regression"` to select the algorithm that optimizes survival time predictions and `gamma.mu = 1e-3` is selected arbitrarily as this is a required parameter (this parameter should usually be tuned). We then compare the predictions from the model to the true data.

```
library(mlr3extralearners)
prediction_svm = lrn("surv.svm", type = "regression", gamma.mu = 1e-3)$
  train(tsk_rats, split$train)$predict(tsk_rats, split$test)
data.frame(pred = prediction_svm$response[1:3],
  truth = prediction_svm$truth[1:3])
```

```
     pred truth
1 88.19   102+
2 87.60    98+
3 87.21    76+
```

As can be seen from the output, our predictions are all less than the true observed time, which means we know our model underestimated the truth. However, because each of the true values are censored times, we have absolutely no way of knowing if these predictions are slightly bad or absolutely terrible, (i.e., the true survival times could be $105, 99, 92$, or they could be $300, 1000, 200$). Hence, with no realistic way to evaluate these models, survival time predictions are rarely useful.

predict_type = "distr"

Unlike regression in which deterministic/point predictions are most common, in survival analysis distribution predictions are much more common. You will therefore find that the majority of survival models in `mlr3proba` will make distribution predictions by default. These predictions are implemented using the `distr6` package, which allows visualization and evaluation of survival curves (defined as $1-$ cumulative distribution function). Below we extract the first three `$distr` predictions from our running example and calculate the probability of survival at $t = 77$.

```
  prediction_cph$distr[1:3]$survival(77)
```

```
     [,1]   [,2]   [,3]
77 0.9213 0.9865 0.9937
```

The output indicates that there is a 92.1%, 98.7%, 99.4%, chance of the first three predicted rats being alive at time 77 respectively.

predict_type = "lp"

`lp`, often written as η in academic writing, is computationally the simplest prediction and has a natural analog in regression modeling. Readers familiar with linear regression will know that when fitting a simple linear regression model, $Y = X\beta$, we are estimating the values for β, and the estimated linear predictor (lp) is then $X\hat{\beta}$, where $\hat{\beta}$ are our estimated coefficients. In simple survival models, the linear predictor is the same quantity (but estimated in a slightly more complicated way). The learner implementations in `mlr3proba` are primarily machine-learning focused and few of these models have a simple linear form, which means that `lp` cannot be computed for most of these. In practice, when used for prediction, `lp` is a proxy for a relative risk/continuous ranking prediction, which is discussed next.

predict_type = "crank"

The final prediction type, `crank`, is the most common in survival analysis and perhaps also the most confusing. Academic texts will often refer to "risk" predictions in survival analysis (hence why survival models are often known as "risk prediction models"), without defining what "risk" means. Often, risk is defined as $\exp(\eta)$ as this is a common quantity found in simple linear survival models. However, sometimes risk is defined as $\exp(-\eta)$, and sometimes it can be an arbitrary quantity that does not have a meaningful interpretation. To prevent this confusion in `mlr3proba`, we define the predict type `crank`, which stands for **c**ontinuous **rank**ing. This is best explained by example; continuing from the previous we output the first three `crank` predictions.

```
prediction_cph$crank[1:3]
```

```
     1       2       3
-0.1577 -1.9549 -2.7150
```

The output tells us that the first rat is at the lowest risk of death (smaller values represent lower risk) and the third rat is at the highest risk. The distance between predictions also tells us that the difference in risk between the second and third rats is smaller than the difference between the first and second. The actual values themselves are meaningless and therefore comparing `crank` values between samples (or papers or experiments) is not meaningful.

The `crank` prediction type is informative and common in practice because it allows identifying observations at lower/higher risk to each other, which is useful for resource allocation, e.g., which patient should be given an expensive treatment, and clinical trials, e.g., are people in a treatment arm at lower risk of disease X than people in the control arm.

⚠ Interpreting Survival Risk

The interpretation of "risk" for survival predictions differs across R packages and sometimes even between models in the same package. In `mlr3proba` there is one consistent interpretation of `crank`: lower values represent a lower risk of the event taking place and higher values represent higher risk.

13.2.3 MeasureSurv

Survival models in `mlr3proba` are evaluated with `MeasureSurv` objects, which are constructed in the usual way with `msr()`.

In general survival measures can be grouped into the following:

1. Discrimination measures – Quantify if a model correctly identifies if one observation is at higher risk than another. Evaluate `crank` and/or `lp` predictions.
2. Calibration measures – Quantify if the average prediction is close to the truth (all definitions of calibration are unfortunately vague in a survival context). Evaluate `crank` and/or `distr` predictions.
3. Scoring rules – Quantify if probabilistic predictions are close to true values. Evaluate `distr` predictions.

```
as.data.table(mlr_measures)[
  task_type == "surv", c("key", "predict_type")][1:5]
```

```
              key predict_type
1:        surv.brier         distr
2:   surv.calib_alpha         distr
3:    surv.calib_beta            lp
4: surv.chambless_auc            lp
5:        surv.cindex         crank
```

There is not a consensus in the literature around the "best" survival measures to use to evaluate models. We recommend RCLL (right-censored logloss) (`msr("surv.rcll")`) to evaluate the quality of `distr` predictions, concordance index (`msr("surv.cindex")`) to evaluate a model's discrimination, and D-Calibration (`msr("surv.dcalib")`) to evaluate a model's calibration.

Using these measures, we can now evaluate our predictions from the previous example.

```
prediction_cph$score(msrs(c("surv.rcll", "surv.cindex", "surv.dcalib")))
```

```
 surv.rcll surv.cindex surv.dcalib
    4.0879      0.8593      0.7463
```

The model's performance seems okay as the RCLL and DCalib are relatively low 0 and the C-index is greater than 0.5 however it is very hard to determine the performance of any survival model without comparing it to some baseline (usually the Kaplan-Meier).

13.2.4   Composition

Throughout `mlr3proba` documentation we refer to "native" and "composed" predictions. We define a "native" prediction as the prediction made by a model without any post-processing, whereas a "composed" prediction is returned after post-processing.

13.2.4.1   Internal Composition

`mlr3proba` makes use of composition internally to return a `"crank"` prediction for every learner. This is to ensure that we can meaningfully benchmark all models according to at least one criterion. The package uses the following rules to create `"crank"` predictions:

1. If a model returns a "risk" prediction then `crank = risk` (we may multiply this by −1 to ensure the "low-value low-risk" interpretation).
2. Else if a model returns a `response` prediction then we set `crank = -response`.
3. Else if a model returns a `lp` prediction then we set `crank = lp` (or `crank = -lp` if needed).
4. Else if a model returns a `distr` prediction then we set `crank` as the sum of the cumulative hazard function (see R. Sonabend, Bender, and Vollmer (2022) for full discussion as to why we picked this method).

13.2.4.2   Explicit Composition and Pipelines

At the start of this section, we mentioned that it is possible to transform prediction types between each other. In `mlr3proba` this is possible with "compositor" pipelines

(Chapter 7). There are several pipelines implemented in the package but two in particular focus on predict type transformation:

1. `pipeline_crankcompositor()` – Transforms a `"distr"` prediction to `"crank"`
2. `pipeline_distrcompositor()` – Transforms a `"lp"` prediction to `"distr"`

In practice, the second pipeline is more common as we internally use a version of the first pipeline whenever we return predictions from survival models (so only use the first pipeline to overwrite these ranking predictions), and so we will just look at the second pipeline.

In the example below we load the `rats` dataset, remove factor columns, and then partition the data into training and testing. We construct the `distrcompositor` pipeline around a survival GLMnet learner (`lrn("surv.glmnet")`) which by default can only make predictions for `"lp"` and `"crank"`. In the pipeline, we specify that we will estimate the baseline distribution with a Kaplan-Meier estimator (`estimator = "kaplan"`) and that we want to assume a proportional hazards form for our estimated distribution (`form = "ph"`). We then train and predict in the usual way and in our output we can now see a `distr` prediction.

```
library(mlr3verse)
library(mlr3extralearners)

tsk_rats = tsk("rats")$select(c("litter", "rx"))
split = partition(tsk_rats)

learner = lrn("surv.glmnet")

# no distr output
learner$train(tsk_rats, split$train)$predict(tsk_rats, split$test)
```

```
<PredictionSurv> for 99 observations:
    row_ids time status crank.1    lp.1
          9  104  FALSE  0.0249  0.0249
         10   91  FALSE  0.7997  0.7997
         15  104  FALSE  0.0415  0.0415
---
        236   81   TRUE  0.6558  0.6558
        249   66   TRUE  0.6890  0.6890
        289  103   TRUE  1.5717  1.5717
```

```
graph_learner = as_learner(ppl(
  "distrcompositor",
  learner = learner,
  estimator = "kaplan",
  form = "ph"
))

# now with distr
graph_learner$train(tsk_rats, split$train)$predict(tsk_rats, split$test)
```

```
<PredictionSurv> for 99 observations:
    row_ids time status crank.1   lp.1     distr
          9  104 FALSE  0.0249 0.0249 <list[1]>
         10   91 FALSE  0.7997 0.7997 <list[1]>
         15  104 FALSE  0.0415 0.0415 <list[1]>
---
        236   81  TRUE  0.6558 0.6558 <list[1]>
        249   66  TRUE  0.6890 0.6890 <list[1]>
        289  103  TRUE  1.5717 1.5717 <list[1]>
```

Mathematically, we have done the following:

1. Assume our estimated distribution will have the form $h(t) = h_0(t) \exp(\eta)$ where h is the hazard function and h_0 is the baseline hazard function.
2. Estimate $\hat{\eta}$ prediction using GLMnet.
3. Estimate $\hat{h}_0(t)$ with the Kaplan-Meier estimator.
4. Put this all together as $h(t) = \hat{h}_0(t) \exp(\hat{\eta})$.

For more detail about prediction types and composition we recommend Kalbfleisch and Prentice (2011).

13.2.5 Putting It All Together

Finally, we will put all the above into practice in a small benchmark experiment. We first load `tsk("grace")` (which only has numeric features) and sample 500 rows randomly. We then select the RCLL, D-Calibration, and C-index to evaluate predictions, set up the same pipeline we used in the previous experiment, and load a Cox PH and Kaplan-Meier estimator. We run our experiment with three-fold CV and aggregate the results.

```
library(mlr3extralearners)

tsk_grace = tsk("grace")
tsk_grace$filter(sample(tsk_grace$nrow, 500))
msr_txt = c("surv.rcll", "surv.cindex", "surv.dcalib")
measures = msrs(msr_txt)

graph_learner = as_learner(ppl(
  "distrcompositor",
  learner = lrn("surv.glmnet"),
  estimator = "kaplan",
  form = "ph"
))
graph_learner$id = "Coxnet"
learners = c(lrns(c("surv.coxph", "surv.kaplan")), graph_learner)

bmr = benchmark(benchmark_grid(tsk_grace, learners,
  rsmp("cv", folds = 3)))
bmr$aggregate(measures)[, c("learner_id", ..msr_txt)]
```

```
    learner_id surv.rcll surv.cindex surv.dcalib
1:  surv.coxph     5.220       0.8250        5.050
2: surv.kaplan     5.438       0.5000        4.152
3:      Coxnet     5.212       0.8264       12.164
```

In this small experiment, Coxnet and Cox PH have the best discrimination, the Kaplan-Meier baseline has the best calibration, and Coxnet and Cox PH have similar overall predictive accuracy (with the lowest RCLL).

13.3   Density Estimation

Density estimation is a learning task to estimate the unknown distribution from which a univariate dataset is generated or put more simply to estimate the probability density (or mass) function for a single variable. As with survival analysis, density estimation is implemented in `mlr3proba`, as both can make probability distribution predictions (hence the name "**mlr3proba**bilistic"). Unconditional density estimation (i.e., estimation of a target without any covariates) is viewed as an unsupervised task, which means the "truth" is never known. For a good overview of density estimation see Silverman (1986).

The package `mlr3proba` extends `mlr3` with the following objects for density estimation:

- `TaskDens` to define density tasks.
- `LearnerDens` as the base class for density estimators.
- `PredictionDens` for density predictions.
- `MeasureDens` as a specialized class for density performance measures.

We will consider each in turn.

13.3.1   TaskDens

As density estimation is an unsupervised task, there is no target for prediction. In the code below we construct a density task using `as_task_dens()` which takes one argument, a **data.frame** type object with exactly one column (which we will use to estimate the underlying distribution).

```
tsk_dens = as_task_dens(data.table(x = rnorm(1000)))
tsk_dens
```

```
<TaskDens:data.table(x = rnorm(1000))> (1000 x 1)
* Target: -
* Properties: -
* Features (1):
  - dbl (1): x
```

As with other tasks, we have included a couple of tasks that come shipped with `mlr3proba`:

```
as.data.table(mlr_tasks)[task_type == "dens", c(1:2, 4:5)]
```

```
          key                label nrow ncol
1: faithful Old Faithful Eruptions  272    1
2:   precip    Annual Precipitation   70    1
```

13.3.2   LearnerDens and PredictionDens

Density learners may return the following prediction types:

1. `distr` – probability distribution
2. `pdf` – probability density function
3. `cdf` – cumulative distribution function

All learners will return a `distr` and `pdf` prediction but only some can make `cdf` predictions. Again, the `distr` predict type is implemented using `distr6`. In the code below we train and "predict" with a histogram learner and then plot the estimated probability density function (Figure 13.4), which closely matches the underlying Normally-distributed data.

```
lrn_hist = lrn("dens.hist")
prediction = lrn_hist$train(tsk_dens, 1:900)$predict(tsk_dens, 901:1000)
x = seq.int(-2, 2, 0.01)
df = data.frame(x = x, y = prediction$distr$pdf(x))
ggplot(df, aes(x = x, y = y)) + geom_line() + theme_minimal()
```

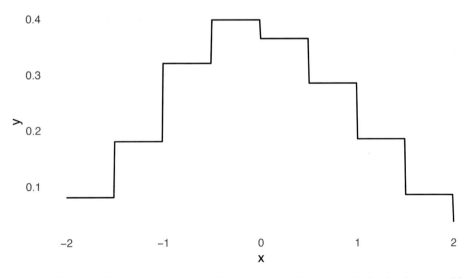

Figure 13.4: Predicted density from the histogram learner, which closely resembles the underlying N(0, 1) data.

The `pdf` and `cdf` predict types are simply wrappers around `distr$pdf` and `distr$cdf` respectively:

```
prediction = lrn_hist$train(tsk_dens, 1:10)$predict(tsk_dens, 11:13)
# pdf and cdf columns in output
prediction
```

```
<PredictionDens> for 3 observations:
 row_ids pdf     cdf               distr
      11 0.4 0.1803 <Distribution[39]>
      12 0.0 0.0000 <Distribution[39]>
      13 0.2 0.9963 <Distribution[39]>
```

```
# comparing cdf from prediction to $cdf method from distr
cbind(prediction$distr$cdf(tsk_dens$data()$x[11:13]),
  prediction$cdf[1:3])
```

```
        [,1]    [,2]
[1,]  0.1803 0.1803
[2,]  0.0000 0.0000
[3,]  0.9963 0.9963
```

13.3.3   MeasureDens and Putting It All Together

At the time of publication, the only measure implemented in `mlr3proba` for density estimation is logloss, which is defined in the same way as in classification, $L(y) = -\log(\hat{f}_Y(y))$, where $\hat{f}_Y$ is our estimated probability density function. Putting this together with the above we are now ready to train a density learner, estimate a distribution, and evaluate our estimation:

```
msr_logloss = msr("dens.logloss")
msr_logloss
```

```
<MeasureDensLogloss:dens.logloss>: Log Loss
* Packages: mlr3, mlr3proba
* Range: [0, Inf]
* Minimize: TRUE
* Average: macro
* Parameters: eps=1e-15
* Properties: -
* Predict type: pdf
```

```
prediction$score(msr_logloss)
```

```
dens.logloss
      12.35
```

This output is most easily interpreted when compared to other learners in a benchmark experiment, so let us put everything together to conduct a small benchmark study on `tsk("faithful")` task using some of the integrated density learners:

```
library(mlr3extralearners)
tsk_faithful = tsk("faithful")
learners = lrns(c("dens.hist", "dens.pen", "dens.kde"))
measure = msr("dens.logloss")
bmr = benchmark(benchmark_grid(tsk_faithful, learners,
  rsmp("cv", folds = 3)))
bmr$aggregate(measure)
```

```
autoplot(bmr, measure = measure)
```

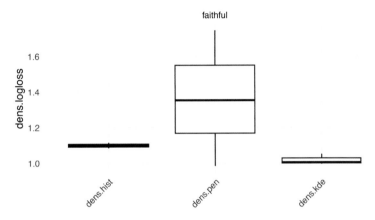

Figure 13.5: Three boxplots comparing performance of dens.hist, dens.pen, and dens.kde on `tsk("faithful")`.

The results (Figure 13.5) of this experiment indicate that the sophisticated Penalized Density Estimator does not outperform the baseline histogram, but the Kernel Density Estimator has at least consistently better (i.e., lower) logloss results.

13.4   Cluster Analysis

Cluster analysis is another unsupervised task implemented in `mlr3`. The objective of cluster analysis is to group data into clusters, where each cluster contains similar observations. The similarity is based on specified metrics that are task and application-dependent. Unlike classification where we try to predict a class for each observation, in cluster analysis there is no "true" label or class to predict.

The package `mlr3cluster` extends `mlr3` with the following objects for cluster analysis:

- `TaskClust` to define clustering tasks
- `LearnerClust` as the base class for clustering learners
- `PredictionClust` as the specialized class for `Prediction` objects
- `MeasureClust` as the specialized class for performance measures

We will consider each in turn.

13.4.1   TaskClust

Similarly to density estimation (Section 13.3), there is no target for prediction and so no `truth` field in TaskClust. By example, we will look at the `ruspini` dataset, which has 75 rows and two columns and was first introduced in Ruspini (1970) to illustrate different clustering techniques. The observations in the dataset form four natural clusters (Figure 13.6). In the code below we construct a cluster task using `as_task_clust()` which only takes one argument, a `data.frame` type object.

```
library(mlr3verse)
library(cluster)
tsk_ruspini = as_task_clust(ruspini)
tsk_ruspini
```

```
<TaskClust:ruspini> (75 x 2)
* Target: -
* Properties: -
* Features (2):
  - int (2): x, y
```

```
tsk_ruspini$data(1:3) # print first 3 rows
```

```
      x  y
1:    4 53
2:    5 63
3:   10 59
```

```
autoplot(tsk_ruspini)
```

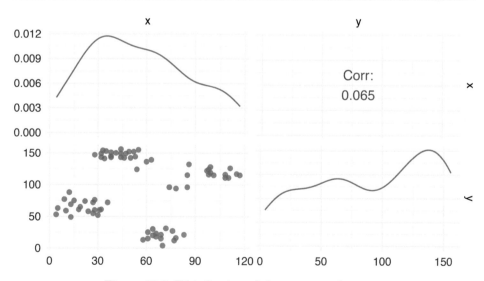

Figure 13.6: Distribution of the `ruspini` dataset.

Technically, we did not need to create a new task for the `ruspini` dataset since it is already included in the package, along with one other task:

```
as.data.table(mlr_tasks)[task_type == "clust", c(1:2, 4:5)]
```

```
        key      label nrow ncol
1:   ruspini    Ruspini   75    2
2: usarrests US Arrests   50    4
```

13.4.2   LearnerClust and PredictionClust

As with density estimation, we refer to `training` and `predicting` for clustering to be consistent with the `mlr3` interface, but strictly speaking, this should be `clustering` and `assigning` (the latter we will return to shortly). Two `predict_types` are available for clustering learners:

1. `"partition"` – Estimate of which cluster an observation falls into
2. `"prob"` – Probability of an observation belonging to each cluster

Similarly to classification, prediction types of clustering learners are either deterministic (`"partition"`) or probabilistic (`"prob"`).

Below we construct a C-Means clustering learner with `"prob"` prediction type and three clusters (`centers = 3`), train it on the `ruspini` dataset and then return the cluster assignments (`$assignments`) for six random observations.

```
lrn_cmeans = lrn("clust.cmeans", predict_type = "prob", centers = 3)
lrn_cmeans
```

```
<LearnerClustCMeans:clust.cmeans>: Fuzzy C-Means Clustering Learner
* Model: -
* Parameters: centers=3
* Packages: mlr3, mlr3cluster, e1071
* Predict Types:  partition, [prob]
* Feature Types: logical, integer, numeric
* Properties: complete, fuzzy, partitional
```

```
lrn_cmeans$train(tsk_ruspini)
lrn_cmeans$assignments[sample(tsk_ruspini$nrow, 6)]
```

```
[1] 1 2 1 3 1 3
```

As clustering is unsupervised, it often does not make sense to use `predict` for new data however this is still possible using the `mlr3` interface.

```
# using different data for estimation (rare use case)
lrn_cmeans$train(tsk_ruspini, 1:30)$predict(tsk_ruspini, 31:32)
```

```
<PredictionClust> for 2 observations:
 row_ids partition prob.1  prob.2  prob.3
      31         1 0.9663 0.01815 0.01552
      32         1 0.9782 0.01178 0.01002
```

```
# using same data as for estimation (common use case)
prediction = lrn_cmeans$train(tsk_ruspini)$predict(tsk_ruspini)
autoplot(prediction, tsk_ruspini)
```

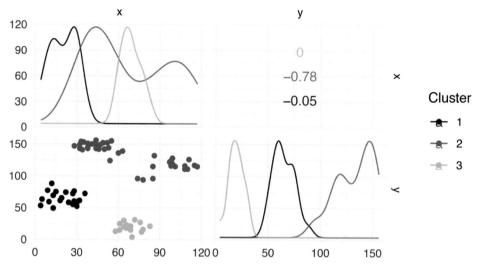

Figure 13.7: Distribution of the estimated clusters.

While two prediction types are possible, there are some learners where "prediction" can never make sense, for example in hierarchical clustering. In hierarchical clustering, the goal is to build a hierarchy of nested clusters by either splitting large clusters into smaller ones or merging smaller clusters into bigger ones. The final result is a tree or dendrogram which can change if a new data point is added. For consistency, mlr3cluster offers a `predict` method for hierarchical clusters but with a warning: Hierarchical Clustering

```
lrn_hclust = lrn("clust.hclust", k = 2)
lrn_hclust$train(tsk_ruspini)$predict(tsk_ruspini)
```

```
Warning: Learner 'clust.hclust' doesn't predict on new data and
predictions may not make sense on new data
```

```
<PredictionClust> for 75 observations:
    row_ids partition
          1         1
          2         1
          3         1
---
         73         1
         74         1
         75         1
```

```
autoplot(lrn_hclust) + theme(axis.text = element_text(size = 5.5))
```

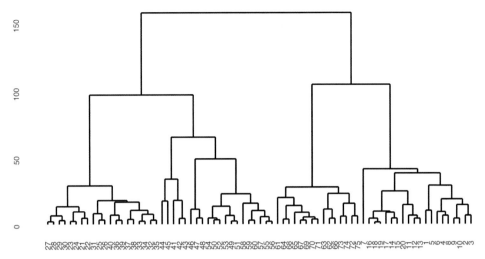

Figure 13.8: Dendrogram representing hierarchical clustering of the `ruspini` dataset. y-axis is similarity of points such that the lower observations (x-axis) are connected, the greater their similarity. The top split represents the separation of the two clusters.

In this case, the `predict` method simply cuts the dendrogram into the number of clusters specified by `k` parameter of the learner.

13.4.3 MeasureClust

As previously discussed, unsupervised tasks do not have ground truth data to compare to in model evaluation. However, we can still measure the quality of cluster assignments by quantifying how closely objects within the same cluster are related (cluster cohesion) as well as how distinct different clusters are from each other (cluster separation). There are a few built-in evaluation metrics available to assess the quality of clustering, which can be found by searching the `mlr_measures` dictionary.

Two common measures are the within sum of squares (WSS) measure (`msr("clust.wss")`) and the silhouette coefficient (`msr("clust.silhouette")`). WSS calculates the sum of squared differences between observations and centroids, which is a quantification of cluster cohesion (smaller values indicate the clusters are more compact). The silhouette coefficient quantifies how well each point belongs to its assigned cluster versus neighboring clusters, where scores closer to 1 indicate well clustered and scores closer to -1 indicate poorly clustered. Note that the silhouette measure in `mlr3cluster` returns the mean silhouette score across all observations and when there is only a single cluster, the measure simply outputs 0.

Putting this together with the above we can now score our cluster estimation (note we must pass the `task` to `$score`):

```
measures = msrs(c("clust.wss", "clust.silhouette"))

prediction$score(measures, task = tsk_ruspini)
```

```
clust.wss clust.silhouette
5.116e+04          6.414e-01
```

The very high WSS and middling mean silhouette coefficient indicate that our clusters could do with a bit more work.

Often reducing an unsupervised task to a quantitative measure may not be useful (given no ground truth) and instead visualization (discussed next) may be a more effective tool for assessing the quality of the clusters.

13.4.4 Visualization

As clustering is an unsupervised task, visualization can be essential not just for "evaluating" models but also for determining if our learners are performing as expected for our task. This section will look at visualizations for supporting clustering choices and following that we will consider plots for evaluating model performance.

13.4.4.1 Visualizing Clusters

It is easy to rely on clustering measures to assess the quality of clustering however this should be done with care as choosing between models may come down to other decisions such as how clusters are formed. By example, consider data generated by `mlbench.spirals`, which results in two individual lines that spiral around each other (Figure 13.9).

```
spirals = mlbench::mlbench.spirals(n = 300, sd = 0.01)
tsk_spirals = as_task_clust(as.data.frame(spirals$x))
autoplot(tsk_spirals)
```

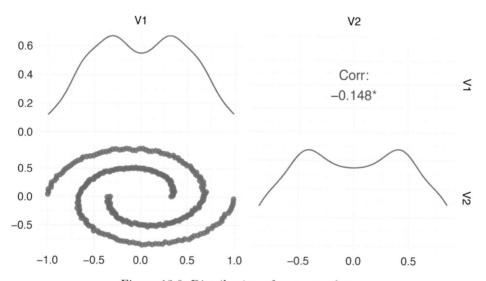

Figure 13.9: Distribution of `spirals` data.

Now let us see what happens when fit two clustering learners on this data:

```
learners = list(
  lrn("clust.kmeans"),
  lrn("clust.dbscan", eps = 0.1)
)

bmr = benchmark(benchmark_grid(tsk_spirals, learners, rsmp("insample")))
bmr$aggregate(msr("clust.silhouette"))[, c(4, 7)]
```

```
      learner_id clust.silhouette
1: clust.kmeans          0.37209
2: clust.dbscan          0.02943
```

We can see that K-means clustering gives us a higher average silhouette score and so we might conclude that a K-means learner with two centroids is a better choice than the DBSCAN method. However, now take a look at the cluster assignment plots in Figure 13.10 (`autoplot.PredictionClust` is available but we do not use it here so we can highlight two particular plots).

```
library(patchwork)
# get K-Means and DBSCAN partitions
pred_kmeans = as.factor(bmr$resample_result(1)$prediction()$partition)
pred_dbscan = as.factor(bmr$resample_result(2)$prediction()$partition)
# plot
df_kmeans = cbind(tsk_spirals$data(), clust = pred_kmeans)
df_dbscan = cbind(tsk_spirals$data(), clust = pred_dbscan)
map = aes(x = V1, y = V2, color = clust)
p_kmeans = ggplot(df_kmeans, map) + ggtitle("K-means")
p_dbscan = ggplot(df_dbscan, map) + ggtitle("DBSCAN")

p_kmeans + p_dbscan + plot_layout(guides = "collect") & geom_point() &
  theme_minimal() & ggplot2::scale_colour_viridis_d(end = 0.8)
```

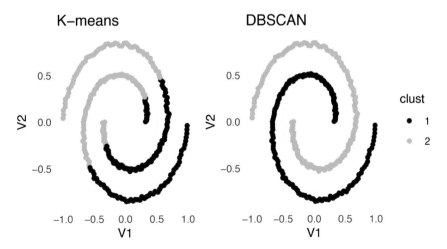

Figure 13.10: Comparing estimated clusters from `lrn("clust.kmeans")` and `lrn("clust.dbscan")`. Both create two distinct clusters that are separated in different ways.

The two learners arrived at two different results to cleanly separate clusters – the K-means algorithm assigned points that are part of the same line into two different clusters whereas DBSCAN assigned each line to its own cluster. Which one of these approaches is correct? The answer is it depends on your specific task and the goal of cluster analysis. If we had only relied on the silhouette score, then the details of how the clustering was performed would have been masked and we would have been unable to decide which method was appropriate for the task.

13.4.4.2   PCA and Silhouette Plots

The two most important plots implemented in `mlr3viz` to support the evaluation of cluster learners are PCA and silhouette plots.

Principal components analysis (PCA) is a commonly used dimension reduction method in ML to reduce the number of variables in a dataset or to visualize the most important "components", which are linear transformations of the dataset features. Components are considered more important if they have higher variance (and therefore more predictive power). In the context of clustering, by plotting observations against the first two components, and then coloring them by cluster, we could visualize our high-dimensional dataset and we would expect to see observations in distinct groups.

Since our running example only has two features, PCA does not make sense to visualize the data. So we will use a task based on the `USArrests` dataset instead. By plotting the result of PCA (Figure 13.11), we see that our model has done a good job of separating observations into two clusters along the first two principal components.

```
tsk_usarrests = tsk("usarrests")
prediction = lrn("clust.kmeans")$train(tsk_usarrests)$
  predict(tsk_usarrests)
autoplot(prediction, tsk_usarrests, type = "pca")
```

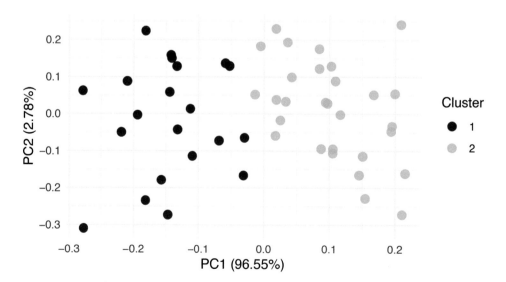

Figure 13.11: First two principal components using PCA on `tsk("usarrests")`.

Silhouette plots visually assess the quality of the estimated clusters by visualizing if observations in a cluster are well-placed both individually and as a group. The plots include a dotted line which visualizes the average silhouette coefficient across all data points and each data point's silhouette value is represented by a bar colored by their assigned cluster. In our particular case, the average silhouette index is 0.59. If the average silhouette value for a given cluster is below the average silhouette coefficient line then this implies that the cluster is not well defined.

Continuing with our new example, we find (Figure 13.12) that a lot of observations are actually below the average line and close to zero, and therefore the quality of our cluster assignments is not very good, meaning that many observations are likely assigned to the wrong cluster.

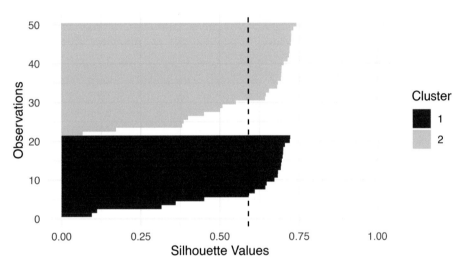

Figure 13.12: Silhouette plot from predictions made by `lrn("clust.kmeans")` on `tsk("usarrests")`.

13.4.5 Putting It All Together

Finally, we conduct a small benchmark study using `tsk("usarrests")` and a few integrated cluster learners:

```
tsk_usarrests = tsk("usarrests")
learners = list(
  lrn("clust.featureless"),
  lrn("clust.kmeans", centers = 4L),
  lrn("clust.cmeans", centers = 3L)
)
measures = list(msr("clust.wss"), msr("clust.silhouette"))
bmr = benchmark(benchmark_grid(tsk_usarrests, learners,
  rsmp("insample")))
bmr$aggregate(measures)[, c(4, 7, 8)]
```

learner_id	clust.wss	clust.silhouette
1: clust.featureless	355808	0.0000
2: clust.kmeans	34729	0.5012
3: clust.cmeans	47964	0.5319

The C-means and K-means algorithms are both considerably better than the featureless baseline but further analysis (and visualizations) would be required to decide which of those two is suitable for our needs.

13.5   Spatial Analysis

The final task we will discuss in this book is spatial analysis. Spatial analysis can be a subset of any other machine learning task (e.g., regression or classification) and is defined by the presence of spatial information in a dataset, usually stored as coordinates that are often named "x" and "y" or "lat" and "lon" (for "latitude" and "longitude" respectively.)

Spatial analysis is its own task as spatial data must be handled carefully due to the complexity of "autocorrelation". Where correlation is defined as a statistical association *between two* variables, autocorrelation is a statistical association *within one* variable. In ML terms, in a dataset with features and observations, correlation occurs when two or more features are statistically associated in some way, whereas autocorrelation occurs when two or more observations are statistically associated across one feature. Autocorrelation, therefore, violates one of the fundamental assumptions of ML that all observations in a dataset are independent, which results in lower confidence about the quality of a trained machine learning model and the resulting performance estimates (Hastie, Friedman, and Tibshirani 2001).

Autocorrelation

Autocorrelation is present in spatial data as there is implicit information encoded in coordinates, such as whether two observations (e.g., cities, countries, continents) are close together or far apart. By example, let us imagine we are predicting the number of cases of a disease two months after an outbreak in Germany (Figure 13.13). Outbreaks radiate outwards from an epicenter and therefore countries closer to Germany will have higher numbers of cases and countries further away will have lower numbers (Figure 13.13, bottom). Thus, looking at the data spatially shows clear signs of autocorrelation across nearby observations. Note in this example the autocorrelation is radial but in practice, this will not always be the case.

Unlike other tasks we have looked at in this chapter, there is no underlying difference between the implemented learners or measures. Instead, we provide additional resampling methods in `mlr3spatiotempcv` to account for the similarity in the train and test sets during resampling that originates from spatiotemporal autocorrelation.

Throughout this section we will use the `ecuador` dataset and task as a working example.

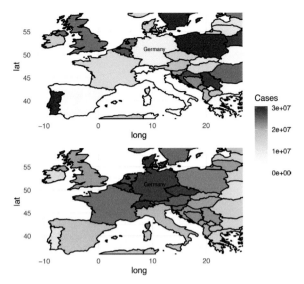

Figure 13.13: Heatmaps where darker countries indicate higher number of cases and lighter countries indicate lower number of cases of imaginary Disease X with epicenter in Germany. The top map imagines a world in which there is no spatial autocorrelation and the number of cases of a disease is randomly distributed. The bottom map shows a more accurate world in which the number of cases radiate outwards from the epicenter (Germany).

13.5.1 TaskClassifST and TaskRegrST

To make use of spatial resampling methods, we have implemented two extensions of `TaskClassif` and `TaskRegr` to accommodate spatial data, `TaskClassifST` and `TaskRegrST` respectively. Below we only show classification examples but regression follows trivially.

```
library(mlr3spatial)
library(mlr3spatiotempcv)

# create task from `data.frame`
tsk_ecuador = as_task_classif_st(ecuador, id = "ecuador_task",
  target = "slides", positive = "TRUE",
  coordinate_names = c("x", "y"), crs = "32717")

# or create task from 'sf' object
data_sf = sf::st_as_sf(ecuador, coords = c("x", "y"), crs = "32717")
tsk_ecuador = as_task_classif_st(data_sf, target = "slides",
  positive = "TRUE")
tsk_ecuador
```

```
<TaskClassifST:data_sf> (751 x 11)
* Target: slides
* Properties: twoclass
* Features (10):
  - dbl (10): carea, cslope, dem, distdeforest, distroad,
    distslidespast, hcurv, log.carea, slope, vcurv
```

```
* Coordinates:
           X        Y
  1: 712882 9560002
  2: 715232 9559582
  3: 715392 9560172
  4: 715042 9559312
  5: 715382 9560142
  ---
747: 714472 9558482
748: 713142 9560992
749: 713322 9560562
750: 715392 9557932
751: 713802 9560862
```

Once a task is created, you can train and predict as normal.

```
lrn("classif.rpart")$train(tsk_ecuador)$predict(tsk_ecuador)
```

```
<PredictionClassif> for 751 observations:
    row_ids truth response
          1  TRUE     TRUE
          2  TRUE     TRUE
          3  TRUE     TRUE
---
        749 FALSE    FALSE
        750 FALSE    FALSE
        751 FALSE     TRUE
```

However, as discussed above, it is best to use the specialized resampling methods to achieve bias-reduced estimates of model performance.

13.5.2   Spatiotemporal Cross-Validation

Before we look at the spatial resampling methods implemented in `mlr3spatiotempcv` we will first show what can go wrong if non-spatial resampling methods are used for spatial data. Below we benchmark a decision tree on `tsk("ecuador")` using two different repeated cross-validation resampling methods, the first ("NSpCV" (non-spatial cross-validation)) is a non-spatial resampling method from `mlr3`, the second ("SpCV" (spatial cross-validation)) is from `mlr3spatiotempcv` and is optimized for spatial data. The example highlights how "NSpCV" makes it appear as if the decision tree is performing better than it is with considerably higher estimated performance, however, this is an overconfident prediction due to the autocorrelation in the data.

```
lrn_rpart = lrn("classif.rpart", predict_type = "prob")
rsmp_nsp = rsmp("repeated_cv", folds = 3, repeats = 2, id = "NSpCV")
rsmp_sp = rsmp("repeated_spcv_coords", folds = 3, repeats = 2,
  id = "SpCV")

design = benchmark_grid(tsk_ecuador, lrn_rpart, c(rsmp_nsp, rsmp_sp))
bmr = benchmark(design)
bmr$aggregate(msr("classif.acc"))[, c(5, 7)]
```

```
    resampling_id classif.acc
1:          NSpCV       0.6718
2:           SpCV       0.5842
```

In the above example, applying non-spatial resampling results in train and test sets that are very similar due to the underlying spatial autocorrelation. Hence there is little difference from testing a model on the same data it was trained on, which should be avoided for an honest performance result (see Chapter 2). In contrast, the spatial method has accommodated autocorrelation and the test data is less correlated (though some association will remain) with the training data. Visually this can be seen using `autoplot()` methods. In Figure 13.14 we visualize how the task is partitioned according to the spatial resampling method (Figure 13.14, left) and non-spatial resampling method (Figure 13.14, right). There is a clear separation in space for the respective partitions when using the spatial resampling whereas the train and test splits overlap a lot (and are therefore more correlated) using the non-spatial method.

```
library(patchwork)

(autoplot(rsmp_sp, tsk_ecuador, fold_id = 1, size = 0.7) +
  ggtitle("Spatial Resampling") +
  autoplot(rsmp_nsp, tsk_ecuador, fold_id = 1, size = 0.7) +
  ggtitle("Non-spatial Resampling")) +
  plot_layout(guides = "collect") &
  theme_minimal() &
  theme(axis.text = element_text(size = 4), legend.position = "bottom")
```

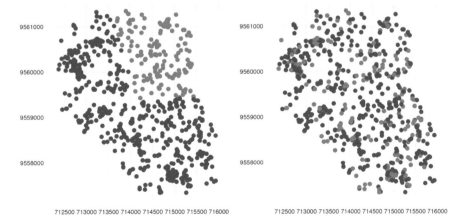

Figure 13.14: Scatterplots show separation of train (blue) and test (orange) data for the first fold of the first repetition of the cross-validation. Left is spatial resampling where train and test data are clearly separated. Right is non-spatial resampling where there is overlap in train and test data.

Now we have seen why spatial resampling matters we can take a look at what methods are available in `mlr3spatiotempcv`. The resampling methods we have added can be categorized into:

- Blocking – Create rectangular blocks in 2D or 3D space.
- Buffering – Create buffering zones to remove observations between train and test sets.
- Spatiotemporal clustering – Clusters based on coordinates (and/or time points).
- Feature space clustering – Clusters based on feature space and not necessarily spatiotemporal.
- Custom (partitioning) – Grouped by factor variables.

The choice of method may depend on specific characteristics of the dataset and there is no easy rule to pick one method over another, full details of different methods can be found in Schratz et al. (2021) – the paper deliberately avoids recommending one method over another as the "optimal" choice is highly dependent on the predictive task, autocorrelation in the data, and the spatial structure of the sampling design. The documentation for each of the implemented methods includes details of each method as well as references to original publications.

> 💡 Spatio*temporal* Resampling
>
> We have focused on spatial analysis but referred to "spatiotemporal" and "spatiotemp". The spatial-only resampling methods discussed in this section can be extended to temporal analysis (or spatial and temporal analysis combined) by setting the `"time"` col_role in the task (Section 2.6) – this is an advanced topic that may be added in future editions of this book. See the `mlr3spatiotempcv` visualization vignette at https://mlr3spatiotempcv.mlr-org.com/articles/spatiotemp-viz.html for specific details about 3D spatiotemporal visualization.

13.5.3   Spatial Prediction

Until now we have looked at resampling to accommodate spatiotemporal *features*, but what if you want to make spatiotemporal *predictions*? In this case, the goal is to make classification or regression predictions at the pixel level, i.e., for an area, defined by the geometric resolution, of a raster image.

To enable these predictions we have created a new function, `predict_spatial()`, to allow spatial predictions using any of the following spatial data classes:

- `stars` (from package `stars`)
- `SpatRaster` (from package `terra`)
- `RasterLayer` (from package `raster`)
- `RasterStack` (from package `raster`)

In the example below, we load the `leipzig_points` dataset for training and coerce this to a spatiotemporal task with `as_task_classif_st`, and we load the `leipzig_raster` raster. Both files are included as example data in `mlr3spatial`.

```
library(mlr3spatial)
library(sf)
library(terra, exclude = "resample")

# load sample points
leipzig_vector = sf::read_sf(system.file("extdata",
  "leipzig_points.gpkg", package = "mlr3spatial"),
  stringsAsFactors = TRUE)
# create training data
tsk_leipzig = as_task_classif_st(leipzig_vector, target = "land_cover")

# load raster image
leipzig_raster = terra::rast(system.file("extdata", "leipzig_raster.tif",
  package = "mlr3spatial"))
```

Now we can continue as normal to train and predict with a classification learner, in this case, a random forest.

```
lrn_ranger = lrn("classif.ranger")$train(tsk_leipzig)
prediction = predict_spatial(leipzig_raster, lrn_ranger,
  format = "terra")
prediction
```

```
class        : SpatRaster
dimensions   : 206, 154, 1  (nrow, ncol, nlyr)
resolution   : 10, 10  (x, y)
extent       : 731810, 733350, 5692030, 5694090  (xmin, xmax, ymin, ymax)
coord. ref.  : WGS 84 / UTM zone 32N (EPSG:32632)
source       : file3aaf41170ae0.tif
categories   : categories
name         : land_cover
min value    :      forest
max value    :       water
```

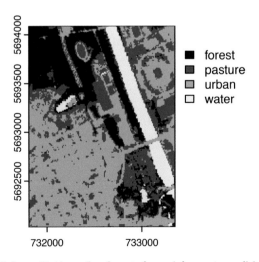

Figure 13.15: Spatial predictions for forest (purple), pasture (blue), urban (green), and water (yellow) categories.

In this example, we specified the creation of a `terra` object, which can be visualized with in-built plotting methods.

```
plot(prediction, col = c("#440154FF", "#443A83FF", "#31688EFF",
  "#21908CFF", "#35B779FF", "#8FD744FF", "#FDE725FF"))
```

13.6　Conclusion

In this chapter, we explored going beyond regression and classification to see how classes in `mlr3` can be used to implement other ML tasks. Cost-sensitive classification extends the "normal" classification setting by assuming that costs associated with false negatives and false positives are unequal. Running cost-sensitive classification experiments is possible using only features in `mlr3`. Survival analysis, available in `mlr3proba`, can be thought of as a regression problem when the outcome may be censored, which means it may never be observed within a given time frame. The final task in `mlr3proba` is density estimation, the unsupervised task concerned with estimating univariate probability distributions. Using `mlr3cluster`, you can perform cluster analysis on observations, which involves grouping observations according to similarities in their variables. Finally, with `mlr3spatial` and `mlr3spatiotempcv`, it is possible to perform spatial analysis to make predictions using coordinates as features and to make spatial predictions. The `mlr3` interface is highly extensible, which means future ML tasks can (and will) be added to our universe and we will add these to this chapter of the book in future editions.

Table 13.2: Important classes and functions covered in this chapter with underlying class (if applicable), class constructor or function, and important class fields and methods (if applicable).

Class	Constructor/Function	Fields/Methods
MeasureClassifCosts	msr("classif.costs")	-
PipeOpTuneThreshold	po("tunethreshold")	-
TaskSurv	as_task_surv()	$data()
PipeOpDistrCompositor	po("distrcompose")	-
TaskDens	as_task_dens()	$data()
TaskClust	as_task_clust()	$data()
TaskClassifST	as_task_classif_st()	$data()
-	predict_spatial()	

13.7　Exercises

1. Run a benchmark experiment on `tsk("german_credit")` with `lrn("classif.featureless")`, `lrn("classif.log_reg")`, and `lrn("classif.ranger")`. Tune the prediction thresholds of all learners by encapsulating them in a `po("learner_cv")` (with two-fold

CV), followed by a po("tunethreshold"). Use msr("classif.costs", costs = costs), where the costs matrix is as follows: true positive is -10, true negative is -1, false positive is 2, and false negative is 3. Use this measure in po("tunethreshold") and when evaluating your benchmark experiment.

2. Train and test a survival forest using lrn("surv.rfsrc") (from mlr3extralearners). Run this experiment using tsk("rats") and partition(). Evaluate your model with the RCLL measure.

3. Estimate the density of the "precip" task from the mlr3proba package using lrn("dens.hist"), evaluate your estimation with the logloss measure. As a stretch goal, look into the documentation of distr6 to learn how to analyse your estimated distribution further.

4. Run a benchmark clustering experiment on the "wine" dataset without a label column. Compare the performance of k-means learner with k equal to 2, 3 and 4 using the silhouette measure and the insample resampling technique. What value of k would you choose based on the silhouette scores?

14

Algorithmic Fairness

Florian Pfisterer
Ludwig-Maximilians-Universität München

In this chapter, we will explore algorithmic fairness in automated decision-making and how we can build fair and unbiased (or at least less biased) predictive models. Methods to help audit and resolve bias in `mlr3` models are implemented in `mlr3fairness`. We will begin by first discussing some of the theory behind algorithmic fairness and then show how this is implemented in `mlr3fairness`.

Automated decision-making systems based on data-driven models are becoming increasingly common but without proper auditing, these models may result in negative consequences for individuals, especially those from underprivileged groups. The proliferation of such systems in everyday life has made it important to address the potential for biases in these models. As a real-world example, historical and sampling biases have led to better quality medical data for patients from White ethnic groups when compared with other ethnic groups. If a model is trained primarily on data from White patients, then the model may appear "good" with respect to a given performance metric (e.g., classification error) when in fact the model could simultaneously be making good predictions for White patients while making bad or even harmful predictions for other patients (J. Huang et al. 2022). As ML-driven systems are used for highly influential decisions, it is vital to develop capabilities to analyze and assess these models not only with respect to their robustness and predictive performance but also with respect to potential biases.

As we work through this chapter we will use the `"adult_train"` and `"adult_test"` tasks from `mlr3fairness`, which contain a subset of the `Adult` dataset (Dua and Graff 2017). This is a binary classification task to predict if an individual earns more than $50,000 per year and is useful for demonstrating biases in data.

```
library(mlr3fairness)
tsk_adult_train = tsk("adult_train")
tsk_adult_train
```

```
<TaskClassif:adult_train> (30718 x 13)
* Target: target
* Properties: twoclass
* Features (12):
  - fct (7): education, marital_status, occupation, race,
    relationship, sex, workclass
  - int (5): age, capital_gain, capital_loss, education_num,
    hours_per_week
* Protected attribute: sex
```

DOI: 10.1201/9781003402848-14

14.1 Bias and Fairness

Bias

Group
Fairness

In the context of fairness, bias refers to disparities in how a model treats individuals or groups. In this chapter, we will concentrate on a subset of bias definitions, those concerning group fairness. For example, in the adult dataset, it can be seen that adults in the group "Male" are significantly more likely to earn a salary greater than $50K per year when compared to the group "Female".

```
sex_salary = table(tsk_adult_train$data(cols = c("sex", "target")))
round(proportions(sex_salary), 2)
```

```
        target
sex      <=50K >50K
  Female  0.29 0.04
  Male    0.46 0.21
```

```
chisq.test(sex_salary)
```

```
    Pearson's Chi-squared test with Yates' continuity correction

data:  sex_salary
X-squared = 1440, df = 1, p-value <2e-16
```

Sensitive
Attribute

In this example, we would refer to the "sex" variable as a sensitive attribute. The goal of group fairness is then to ascertain if decisions are fair across groups defined by a sensitive attribute. The sensitive attribute in a task is set with the `"pta"` (**p**rotected **a**ttribute) column role (Section 2.6).

```
tsk_adult_train$set_col_roles("sex", add_to = "pta")
```

If more than one sensitive attribute is specified, then fairness will be based on observations at the intersections of the specified groups. In this chapter we will only focus on group fairness, however, one could also consider auditing individual fairness, which assesses fairness at an individual level, and causal fairness, which incorporates causal relationships in the data and propose metrics based on a directed acyclic graph (Barocas, Hardt, and Narayanan 2019; Mitchell et al. 2021). While we will only focus on metrics for binary classification here, most metrics discussed naturally extend to more complex scenarios, such as multi-class classification, regression, and survival analysis (Mehrabi et al. 2021; R. Sonabend et al. 2022).

14.2 Group Fairness Notions

It is necessary to choose a notion of group fairness before selecting an appropriate fairness metric to measure algorithmic bias.

Model predictions are said to be bias-transforming (Wachter, Mittelstadt, and
Russell 2021), or to satisfy independence, if the predictions made by the model
are independent of the sensitive attribute. This group includes the concept of
"Demographic Parity", which tests if the proportion of positive predictions (PPV)
is equal across all groups. Bias-transforming methods (i.e., those that test for
independence) do not depend on labels and can help detect biases arising from
different base rates across populations.

A model is said to be bias-preserving, or to satisfy separation, if the predictions
made by the model are independent of the sensitive attribute *given the true label*.
In other words, the model should make roughly the same amount of right/wrong
predictions in each group. Several metrics fall under this category, such as "equalized
odds", which tests if the TPR and FPR is equal across groups. Bias-preserving
metrics (which test for separation) test if errors made by a model are equal across
groups but might not account for bias in the labels (e.g., if outcomes in the real
world may be biased such as different rates of arrest for people from different ethnic
groups).

Choosing a fairness notion will depend on the model's purpose and its societal
context. For example, if a model is being used to predict if a person is guilty of
something then we might want to focus on false positive or false discovery rates
instead of true positives. Whichever metric is chosen, we are essentially condensing
systemic biases and prejudices into a few numbers, and all metrics are limited with
none being able to identify all biases that may exist in the data. For example, if
societal biases lead to disparities in an observed quantity (such as school exam
scores) for individuals with the same underlying ability, these metrics may not
identify existing biases.

To see these notions in practice, let A be a binary sensitive group taking values 0
and 1 and let M be a fairness metric. Then to measure independence we would
simply calculate the difference between these values and test if the result is less
than some threshold, ϵ.

$$|\Delta_M| = |M_{A=0} - M_{A=1}| < \epsilon$$

If we used TPR as our metric M then if $|\Delta_M| > \epsilon$ (e.g., $\epsilon = 0.05$) we would conclude
that predictions from our model violate the equality of opportunity metric and do
not satisfy separation. If we chose accuracy or PPV for M, then we would have
concluded that the model predictions do not satisfy independence.

In `mlr3fairness` we can construct a fairness metric from any `Measure` by constructing `msr("fairness", base_measure, range)` with our metric of choice passed to
`base_measure` as well as the possible range the metric can take (i.e., the range in
differences possible based on the base measure):

```
fair_tpr = msr("fairness", base_measure = msr("classif.tpr"),
  range = c(0, 1))
fair_tpr
```

```
<MeasureFairness:fairness.tpr>
* Packages: mlr3, mlr3fairness
* Range: [0, 1]
```

* Minimize: TRUE
* Average: macro
* Parameters: list()
* Properties: requires_task
* Predict type: response

We have implemented several Measures in mlr3fairness that simplify this step for you, these are named fairness.<base_measure>, for example for TPR: msr("fairness.tpr") would run the same code as above.

14.3 Auditing a Model For Bias

With our sensitive attribute set and the fairness metric selected, we can now train a Learner and test for bias. Below we use a random forest and evaluate the absolute difference in true positive rate across groups "Male" and "Female":

```
tsk_adult_test = tsk("adult_test")
lrn_rpart = lrn("classif.rpart", predict_type = "prob")
prediction = lrn_rpart$train(tsk_adult_train)$predict(tsk_adult_test)
prediction$score(fair_tpr, tsk_adult_test)
```

```
fairness.tpr
   0.06034
```

With an ϵ value of 0.05 we would conclude that there is bias present in our model, however, this value of ϵ is arbitrary and should be decided based on context. As well as using fairness metrics to evaluate a single model, they can also be used in larger benchmark experiments to compare bias across multiple models.

Visualizations can also help better understand discrepancies between groups or differences between models. fairness_prediction_density() plots the sub-group densities across group levels and compare_metrics() scores predictions across multiple metrics:

```
library(patchwork)
library(ggplot2)

p1 = fairness_prediction_density(prediction, task = tsk_adult_test)
p2 = compare_metrics(prediction,
  msrs(c("fairness.fpr", "fairness.tpr", "fairness.eod")),
  task = tsk_adult_test
)

(p1 + p2) *
  theme_minimal() *
  scale_fill_viridis_d(end = 0.8, alpha = 0.8) *
  theme(
    axis.text.x = element_text(angle = 15, hjust = .7),
    legend.position = "bottom"
  )
```

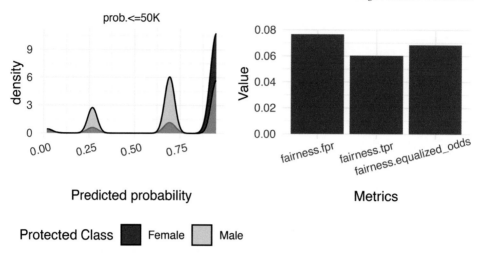

Figure 14.1: Fairness prediction density plot (left) showing the density of predictions for the positive class split by "Male" and "Female" individuals. The metrics comparison barplot (right) displays the model's scores across the specified metrics.

In this example (Figure 14.1), we can see the model is more likely to predict "Female" observations as having a lower salary. This could be due to systemic prejudices seen in the data, i.e., women are more likely to have lower salaries due to societal biases, or could be due to bias introduced by the algorithm. As the right plot indicates that all fairness metrics exceed 0.05, this supports the argument that the algorithm may have introduced further bias (with the same caveat about the 0.05 threshold).

14.4 Fair Machine Learning

If we detect that our model is unfair, then a natural next step is to mitigate such biases. `mlr3fairness` comes with several options to address biases in models, which broadly fall into three categories (Caton and Haas 2020):

1. Preprocessing data – The underlying data is preprocessed in some way to address bias in the data before it is passed to the `Learner`.
2. Employing fair models – Some algorithms can incorporate fairness considerations directly, for example, generalized linear model with fairness constraints (`lrn("classif.fairzlrm")`).
3. Postprocessing model predictions – Heuristics/algorithms are applied to the predictions to mitigate biases present in the predictions.

All methods often slightly decrease predictive performance and it can therefore be useful to try all approaches to empirically see which balance predictive performance and fairness. In general, all biases should be addressed at their root cause (or as close to it) as possible as any other intervention will be suboptimal.

Pre- and postprocessing schemes can be integrated using `mlr3pipelines` (Chapter 7). We provide two examples below, first preprocessing to balance observation weights with `po("reweighing_wts")` and second post-processing predictions using `po("EOd")`. The latter enforces the equalized odds fairness definition by stochastically flipping specific predictions. We also test `lrn("classif.fairzlrm")` against the other methods.

```
# load learners
lrn_rpart = lrn("classif.rpart", predict_type = "prob")
lrn_rpart$id = "rpart"
l1 = as_learner(po("reweighing_wts") %>>% lrn("classif.rpart"))
l1$id = "reweight"

l2 = as_learner(po("learner_cv", lrn("classif.rpart")) %>>%
  po("EOd"))
l2$id = "EOd"

# preprocess by collapsing factors
l3 = as_learner(po("collapsefactors") %>>% lrn("classif.fairzlrm"))
l3$id = "fairzlrm"

# load task and subset by rows and columns
task = tsk("adult_train")
task$set_col_roles("sex", "pta")$
  filter(sample(task$nrow, 500))$
  select(setdiff(task$feature_names, "education_num"))

# run experiment
lrns = list(lrn_rpart, l1, l2, l3)
bmr = benchmark(benchmark_grid(task, lrns, rsmp("cv", folds = 5)))
meas = msrs(c("classif.acc", "fairness.eod"))
bmr$aggregate(meas)[,
  .(learner_id, classif.acc, fairness.equalized_odds)]
```

	learner_id	classif.acc	fairness.equalized_odds
1:	rpart	0.836	0.1981
2:	reweight	0.828	0.1860
3:	EOd	0.826	0.1969
4:	fairzlrm	0.814	0.1987

We can study the result using built-in plotting functions, below we use `fairness_accuracy_tradeoff()`, to compare classification accuracy (default accuracy measure for the function) and equalized odds (`msr("fairness.eod")`) across cross-validation folds.

```
fairness_accuracy_tradeoff(bmr, fairness_measure = msr("fairness.eod"),
  accuracy_measure = msr("classif.ce")) +
  ggplot2::scale_color_viridis_d("Learner") +
  ggplot2::theme_minimal()
```

Looking at the table of results and Figure 14.2, the reweighting method appears to yield marginally better fairness metrics than the other methods though the difference is unlikely to be significant. So in this case, we would likely conclude that introducing bias mitigation steps did not improve algorithmic fairness.

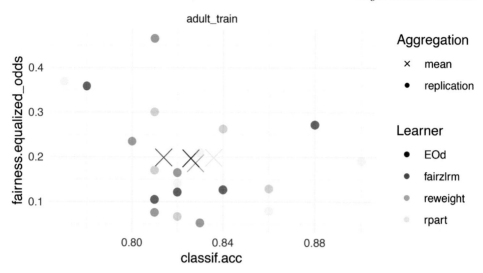

Figure 14.2: Comparison of learners with respect to classification accuracy (x-axis) and equalized odds (y-axis) across (dots) and aggregated over (crosses) folds.

As well as manually computing and analyzing fairness metrics, one could also make use of `mlr3tuning` (Chapter 4) to automate the process with respect to one or more metrics (Section 5.2).

14.5   Conclusion

The functionality introduced above is intended to help users investigate their models for biases and potentially mitigate them. Fairness metrics can not be used to prove or guarantee fairness. Deciding whether a model is fair requires additional investigation, for example, understanding what the measured quantities represent for an individual in the real world and what other biases might exist in the data that could lead to discrepancies in how, for example, covariates or the label are measured.

The simplicity of fairness metrics means they should only be used for exploratory purposes, and practitioners should not solely rely on them to make decisions about employing a machine learning model or assessing whether a system is fair. Instead, practitioners should look beyond the model and consider the data used for training and the process of data and label acquisition. To help in this process, it is important to provide robust documentation for data collection methods, the resulting data, and the models resulting from this data. Informing auditors about those aspects of a deployed model can lead to a better assessment of a model's fairness. Questionnaires for machine learning models and data sets have been previously proposed in the literature and are available in `mlr3fairness` from automated report templates (`report_modelcard()` and `report_datasheet()`) using R markdown for data sets and machine learning models. In addition, `report_fairness()` provides a template for a fairness report inspired by the Aequitas Toolkit (Saleiro et al. 2018). Fairness Report

We hope that pairing the functionality available in `mlr3fairness` with additional exploratory data analysis, a solid understanding of the societal context in which the decision is made and integrating additional tools (e.g. interpretability methods seen in Chapter 12), might help to mitigate or diminish unfairness in systems deployed in the future.

Table 14.1: Important classes and functions covered in this chapter with underlying class (if applicable), class constructor or function, and important class fields and methods (if applicable).

Class	Constructor/Function	Fields/Methods
MeasureFairness	`msr("fairness", ...)`	-
-	`fairness_prediction_density()`	
-	`compare_metrics()`	-
PipeOpReweighingWeights	`po("reweighing_wts")`	-
PipeOpEOd	`po("EOd")`	-
-	`fairness_accuracy_tradeoff()`	
-	`report_fairness()`	-

14.6 Exercises

1. Train a model of your choice on `tsk("adult_train")` and test it on `tsk("adult_test")`, use any measure of your choice to evaluate your predictions. Assume our goal is to achieve parity in false omission rates across the protected "sex" attribute. Construct a fairness metric that encodes this and evaluate your model. To get a deeper understanding, look at the `groupwise_metrics` function to obtain performance in each group.

2. Improve your model by employing pipelines that use pre- or post-processing methods for fairness. Evaluate your model along the two metrics and visualize the resulting metrics. Compare the different models using an appropriate visualization.

3. Add "race" as a second sensitive attribute to your dataset. Add the information to your task and evaluate the initial model again. What changes? Again study the `groupwise_metrics`.

4. In this chapter we were unable to reduce bias in our experiment. Using everything you have learned in this book, see if you can successfully reduce bias in your model. Critically reflect on this exercise, why might this be a bad idea?

References

Apley, Daniel W., and Jingyu Zhu. 2020. "Visualizing the Effects of Predictor Variables in Black Box Supervised Learning Models." *Journal of the Royal Statistical Society Series B: Statistical Methodology* 82 (4): 1059–86. https://doi.org/10.1111/rssb.12377.

Au, Quay, Julia Herbinger, Clemens Stachl, Bernd Bischl, and Giuseppe Casalicchio. 2022. "Grouped Feature Importance and Combined Features Effect Plot." *Data Mining and Knowledge Discovery* 36 (4): 1401–50. https://doi.org/10.1007/s10618-022-00840-5.

Bagnall, Anthony, Jason Lines, Aaron Bostrom, James Large, and Eamonn Keogh. 2017. "The Great Time Series Classification Bake Off: A Review and Experimental Evaluation of Recent Algorithmic Advances." *Data Mining and Knowledge Discovery* 31: 606–60. https://doi.org/10.1007/s10618-016-0483-9.

Baniecki, Hubert, and Przemyslaw Biecek. 2019. "modelStudio: Interactive Studio with Explanations for ML Predictive Models." *Journal of Open Source Software* 4 (43): 1798. https://doi.org/10.21105/joss.01798.

Baniecki, Hubert, Dariusz Parzych, and Przemyslaw Biecek. 2023. "The Grammar of Interactive Explanatory Model Analysis." *Data Mining and Knowledge Discovery*, 1573–756X. https://doi.org/10.1007/s10618-023-00924-w.

Barocas, Solon, Moritz Hardt, and Arvind Narayanan. 2019. *Fairness and Machine Learning: Limitations and Opportunities.* fairmlbook.org.

Bengtsson, Henrik. 2020. "Future 1.19.1 – Making Sure Proper Random Numbers Are Produced in Parallel Processing." https://www.jottr.org/2020/09/22/push-for-statistical-sound-rng/.

———. 2022. "Please Avoid detectCores() in Your R Packages." https://www.jottr.org/2022/12/05/avoid-detectcores/.

Bergstra, James, and Yoshua Bengio. 2012. "Random Search for Hyper-Parameter Optimization." *Journal of Machine Learning Research* 13: 281–305. https://jmlr.org/papers/v13/bergstra12a.html.

Biecek, Przemyslaw. 2018. "DALEX: Explainers for Complex Predictive Models in R." *Journal of Machine Learning Research* 19 (84): 1–5. https://jmlr.org/papers/v19/18-416.html.

Biecek, Przemyslaw, and Tomasz Burzykowski. 2021. *Explanatory Model Analysis.* Chapman; Hall/CRC, New York. https://ema.drwhy.ai/.

Binder, Martin, Florian Pfisterer, and Bernd Bischl. 2020. "Collecting Empirical Data about Hyperparameters for Data Driven AutoML." In *Proceedings of the 7th ICML Workshop on Automated Machine Learning (AutoML 2020).* https://www.automl.org/wp-content/uploads/2020/07/AutoML_2020_paper_63.pdf.

Binder, Martin, Florian Pfisterer, Michel Lang, Lennart Schneider, Lars Kotthoff, and Bernd Bischl. 2021. "mlr3pipelines – Flexible Machine Learning Pipelines in R." *Journal of Machine Learning Research* 22 (184): 1–7. https://jmlr.org/papers/v22/21-0281.html.

Bischl, Bernd, Martin Binder, Michel Lang, Tobias Pielok, Jakob Richter, Stefan Coors, Janek Thomas, et al. 2023. "Hyperparameter Optimization: Foundations, Algorithms, Best Practices, and Open Challenges." *Wiley Interdisciplinary Reviews: Data Mining and Knowledge Discovery*, e1484. Volume 13, Issue 2. https://doi.org/10.1002/widm.1484.

Bischl, Bernd, Giuseppe Casalicchio, Matthias Feurer, Pieter Gijsbers, Frank Hutter, Michel Lang, Rafael Gomes Mantovani, Jan N. van Rijn, and Joaquin Vanschoren. 2021. "OpenML Benchmarking Suites." In *Thirty-Fifth Conference on Neural Information Processing Systems Datasets and Benchmarks Track (Round 2)*. https://openreview.net/forum?id=OCrD8ycKjG.

Bischl, Bernd, Michel Lang, Lars Kotthoff, Julia Schiffner, Jakob Richter, Erich Studerus, Giuseppe Casalicchio, and Zachary M. Jones. 2016. "mlr: Machine Learning in R." *Journal of Machine Learning Research* 17 (170): 1–5. https://jmlr.org/papers/v17/15-066.html.

Bischl, Bernd, Michel Lang, Olaf Mersmann, Jörg Rahnenführer, and Claus Weihs. 2015. "BatchJobs and BatchExperiments: Abstraction Mechanisms for Using r in Batch Environments." *Journal of Statistical Software* 64 (11): 1–25. https://doi.org/10.18637/jss.v064.i11.

Bischl, Bernd, Olaf Mersmann, Heike Trautmann, and Claus Weihs. 2012. "Resampling Methods for Meta-Model Validation with Recommendations for Evolutionary Computation." *Evolutionary Computation* 20 (2): 249–75. https://doi.org/10.1162/EVCO_a_00069 .

Bishop, Christopher M. 2006. *Pattern Recognition and Machine Learning.* Springer.

Bommert, Andrea, Xudong Sun, Bernd Bischl, Jörg Rahnenführer, and Michel Lang. 2020. "Benchmark for Filter Methods for Feature Selection in High-Dimensional Classification Data." *Computational Statistics & Data Analysis* 143: 106839. https://doi.org/10.1016/j.csda.2019.106839.

Breiman, Leo. 1996. "Bagging Predictors." *Machine Learning* 24 (2): 123–40. https://doi.org/10.1007/BF00058655.

———. 2001a. "Random Forests." *Machine Learning* 45: 5–32. https://doi.org/10.1023/A:1010933404324.

———. 2001b. "Statistical Modeling: The Two Cultures (with Comments and a Rejoinder by the Author)." *Statistical Science* 16 (3): 199-231. https://doi.org/10.1214/ss/1009213726.

Bücker, Michael, Gero Szepannek, Alicja Gosiewska, and Przemyslaw Biecek. 2022. "Transparency, Auditability, and Explainability of Machine Learning Models in Credit Scoring." *Journal of the Operational Research Society* 73 (1): 70–90. https://doi.org/10.1080/01605682.2021.1922098.

Byrd, Richard H., Peihuang Lu, Jorge Nocedal, and Ciyou Zhu. 1995. "A Limited Memory Algorithm for Bound Constrained Optimization." *SIAM Journal on Scientific Computing* 16 (5): 1190–1208. https://doi.org/10.1137/0916069.

Caton, S., and C. Haas. 2020. "Fairness in Machine Learning: A Survey." *Arxiv* 2010.04053 [cs.LG]. https://doi.org/10.48550/arXiv.2010.04053.

Chandrashekar, Girish, and Ferat Sahin. 2014. "A Survey on Feature Selection Methods." *Computers and Electrical Engineering* 40 (1): 16–28. https://doi.org/10.1016/j.compeleceng.2013.11.024.

Chen, Tianqi, and Carlos Guestrin. 2016. "XGBoost: A Scalable Tree Boosting System." In *Proceedings of the 22nd ACM SIGKDD International Conference on Knowledge Discovery and Data Mining*, 785–94. https://doi.org/10.1145/2939672.2939785.

Collett, David. 2014. *Modelling Survival Data in Medical Research*. 3rd ed. CRC. https://doi.org/10.1201/b18041.

Couronné, Raphael, Philipp Probst, and Anne-Laure Boulesteix. 2018. "Random Forest Versus Logistic Regression: A Large-Scale Benchmark Experiment." *BMC Bioinformatics* 19: 1–14. https://doi.org/10.1186/s12859-018-2264-5.

Dandl, Susanne, Christoph Molnar, Martin Binder, and Bernd Bischl. 2020. "Multi-Objective Counterfactual Explanations." In *Parallel Problem Solving from Nature PPSN XVI*, 448–69. Springer International Publishing. https://doi.org/10.1007/978-3-030-58112-1_31.

Davis, Jesse, and Mark Goadrich. 2006. "The Relationship Between Precision-Recall and ROC Curves." In *Proceedings of the 23rd International Conference on Machine Learning*, 233–40. https://doi.org/10.1145/1143844.1143874.

De Cock, Dean. 2011. "Ames, Iowa: Alternative to the Boston Housing Data as an End of Semester Regression Project." *Journal of Statistics Education* 19 (3). https://doi.org/10.1080/10691898.2011.11889627.

Demšar, Janez. 2006. "Statistical Comparisons of Classifiers over Multiple Data Sets." *Journal of Machine Learning Research* 7 (1): 1–30. https://jmlr.org/papers/v7/demsar06a.html.

Ding, Yufeng, and Jeffrey S Simonoff. 2010. "An Investigation of Missing Data Methods for Classification Trees Applied to Binary Response Data." *Journal of Machine Learning Research* 11 (6): 131–70. https://www.jmlr.org/papers/v11/ding10a.html.

Dobbin, Kevin K., and Richard M. Simon. 2011. "Optimally Splitting Cases for Training and Testing High Dimensional Classifiers." *BMC Medical Genomics* 4 (1): 31. https://doi.org/10.1186/1755-8794-4-31.

Dua, Dheeru, and Casey Graff. 2017. "UCI Machine Learning Repository." University of California, Irvine, School of Information; Computer Sciences. https://archive.ics.uci.edu/ml.

Eddelbuettel, Dirk. 2020. "Parallel Computing with R: A Brief Review." *WIREs Computational Statistics* 13 (2) e1515. https://doi.org/10.1002/wics.1515.

Feurer, Matthias, and Frank Hutter. 2019. "Hyperparameter Optimization." In *Automated Machine Learning: Methods, Systems, Challenges*, edited by Frank Hutter, Lars Kotthoff, and Joaquin Vanschoren, 3–33. Cham: Springer International Publishing. https://doi.org/10.1007/978-3-030-05318-5_1.

Feurer, Matthias, Jost Springenberg, and Frank Hutter. 2015. "Initializing Bayesian Hyperparameter Optimization via Meta-Learning." In *Proceedings of the AAAI Conference on Artificial Intelligence*. Vol. 29. 1. https://doi.org/10.1609/aaai.v29i1.9354.

Fisher, Aaron, Cynthia Rudin, and Francesca Dominici. 2019. "All Models Are Wrong, but Many Are Useful: Learning a Variable's Importance by Studying an Entire Class of Prediction Models Simultaneously." https://doi.org/10.48550/arxiv.1801.01489.

Friedman, Jerome H. 2001. "Greedy Function Approximation: A Gradient Boosting Machine." *The Annals of Statistics* 29 (5): 1189-1232. https://doi.org/10.1214/aos/1013203451.

Garnett, Roman. 2022. *Bayesian Optimization*. Cambridge University Press. https://bayesoptbook.com/.

Gijsbers, Pieter, Marcos L. P. Bueno, Stefan Coors, Erin LeDell, Sébastien Poirier, Janek Thomas, Bernd Bischl, and Joaquin Vanschoren. 2022. "AMLB: An AutoML Benchmark." arXiv. https://doi.org/10.48550/ARXIV.2207.12560.

Goldstein, Alex, Adam Kapelner, Justin Bleich, and Emil Pitkin. 2015. "Peeking Inside the Black Box: Visualizing Statistical Learning with Plots of Individual Conditional Expectation." *Journal of Computational and Graphical Statistics* 24 (1): 44–65. https://doi.org/10.1080/10618600.2014.907095.

Gower, John C. 1971. "A General Coefficient of Similarity and Some of Its Properties." *Biometrics*, 857–71. https://doi.org/10.2307/2528823.

Grinsztajn, Leo, Edouard Oyallon, and Gael Varoquaux. 2022. "Why Do Tree-Based Models Still Outperform Deep Learning on Typical Tabular Data?" In *Thirty-Sixth Conference on Neural Information Processing Systems Datasets and Benchmarks Track*. https://openreview.net/forum?id=Fp7__phQszn.

Guidotti, Riccardo. 2022. "Counterfactual Explanations and How to Find Them: Literature Review and Benchmarking." *Data Mining and Knowledge Discovery*, 1–55. https://doi.org/10.1007/s10618-022-00831-6.

Guidotti, Riccardo, Anna Monreale, Salvatore Ruggieri, Franco Turini, Fosca Giannotti, and Dino Pedreschi. 2018. "A Survey of Methods for Explaining Black Box Models." *ACM Computing Surveys (CSUR)* 51 (5): 1–42. https://doi.org/10.1145/3236009.

Guyon, Isabelle, and André Elisseeff. 2003. "An Introduction to Variable and Feature Selection." *Journal of Machine Learning Research* 3 (Mar): 1157–82. https://www.jmlr.org/papers/v3/guyon03a.html.

Hand, David J, and Robert J Till. 2001. "A Simple Generalisation of the Area Under the ROC Curve for Multiple Class Classification Problems." *Machine Learning* 45: 171–86. https://doi.org/10.1023/A:1010920819831.

Hansen, Nikolaus, and Anne Auger. 2011. "CMA-ES: Evolution Strategies and Covariance Matrix Adaptation." In *Proceedings of the 13th Annual Conference Companion on Genetic and Evolutionary Computation*, 991–1010. https://doi.org/10.1145/2001858.2002123.

Hastie, Trevor, Jerome Friedman, and Robert Tibshirani. 2001. *The Elements of Statistical Learning*. Springer New York. https://doi.org/10.1007/978-0-387-21606-5.

Hooker, Giles, and Lucas K. Mentch. 2019. "Please Stop Permuting Features: An Explanation and Alternatives." https://doi.org/10.48550/arxiv.1905.03151.

Horn, Daniel, Tobias Wagner, Dirk Biermann, Claus Weihs, and Bernd Bischl. 2015. "Model-Based Multi-Objective Optimization: Taxonomy, Multi-Point Proposal, Toolbox and Benchmark." In *Evolutionary Multi-Criterion Optimization*, edited by António Gaspar-Cunha, Carlos Henggeler Antunes, and Carlos Coello Coello, 64–78. https://doi.org/10.1007/978-3-319-15934-8_5.

Huang, D., T. T. Allen, W. I. Notz, and N. Zheng. 2012. "Erratum to: Global Optimization of Stochastic Black-Box Systems via Sequential Kriging Meta-Models." *Journal of Global Optimization* 54 (2): 431–31. https://doi.org/10.1007/s10898-011-9821-z.

Huang, Jonathan, Galal Galal, Mozziyar Etemadi, and Mahesh Vaidyanathan. 2022. "Evaluation and Mitigation of Racial Bias in Clinical Machine Learning Models: Scoping Review." *JMIR Med Inform* 10 (5) e36388. https://doi.org/10.2196/36388.

Hutter, Frank, Lars Kotthoff, and Joaquin Vanschoren, eds. 2019. *Automated Machine Learning - Methods, Systems, Challenges*. Springer.

"Introduction to Data.table." 2023. https://cran.r-project.org/web/packages/data.table/vignettes/datatable-intro.html.

James, Gareth, Daniela Witten, Trevor Hastie, and Robert Tibshirani. 2014. *An Introduction to Statistical Learning: With Applications in R.* Springer Publishing Company, Incorporated. https://doi.org/10.1007/978-1-4614-7138-7.

Jamieson, Kevin, and Ameet Talwalkar. 2016. "Non-Stochastic Best Arm Identification and Hyperparameter Optimization." In *Proceedings of the 19th International Conference on Artificial Intelligence and Statistics*, edited by Arthur Gretton and Christian C. Robert, 51:240–48. Proceedings of Machine Learning Research. Cadiz, Spain: PMLR. https://proceedings.mlr.press/v51/jamieson16.html.

Japkowicz, Nathalie, and Mohak Shah. 2011. *Evaluating Learning Algorithms: A Classification Perspective.* Cambridge University Press. https://doi.org/10.1017/CBO9780511921803.

Jones, Donald R., Cary D. Perttunen, and Bruce E. Stuckman. 1993. "Lipschitzian Optimization Without the Lipschitz Constant." *Journal of Optimization Theory and Applications* 79 (1): 157–81. https://doi.org/10.1007/BF00941892.

Jones, Donald R., Matthias Schonlau, and William J. Welch. 1998. "Efficient Global Optimization of Expensive Black-Box Functions." *Journal of Global Optimization* 13 (4): 455–92. https://doi.org/10.1023/A:1008306431147.

Kalbfleisch, John D, and Ross L Prentice. 2011. *The Statistical Analysis of Failure Time Data.* Vol. 360. John Wiley & Sons. https://doi.org/10.1002/9781118032985.

Karl, Florian, Tobias Pielok, Julia Moosbauer, Florian Pfisterer, Stefan Coors, Martin Binder, Lennart Schneider, et al. 2022. "Multi-Objective Hyperparameter Optimization–an Overview." *arXiv Preprint arXiv:2206.07438.* https://doi.org/10.48550/arXiv.2206.07438.

Kim, Ji-Hyun. 2009. "Estimating Classification Error Rate: Repeated Cross-Validation, Repeated Hold-Out and Bootstrap." *Computational Statistics & Data Analysis* 53 (11): 3735–45. https://doi.org/10.1016/j.csda.2009.04.009.

Kim, Jungtaek, and Seungjin Choi. 2021. "On Local Optimizers of Acquisition Functions in Bayesian Optimization." In *Machine Learning and Knowledge Discovery in Databases*, edited by Frank Hutter, Kristian Kersting, Jefrey Lijffijt, and Isabel Valera, 675–90. https://doi.org/10.1007/978-3-030-67661-2_40.

Knowles, Joshua. 2006. "ParEGO: A Hybrid Algorithm with on-Line Landscape Approximation for Expensive Multiobjective Optimization Problems." *IEEE Transactions on Evolutionary Computation* 10 (1): 50–66. https://doi.org/10.1109/TEVC.2005.851274.

Kohavi, Ron. 1995. "A Study of Cross-Validation and Bootstrap for Accuracy Estimation and Model Selection." In *Proceedings of the 14th International Joint Conference on Artificial Intelligence – Volume 2*, 1137–43. IJCAI'95. San Francisco, CA, USA: Morgan Kaufmann Publishers Inc.

Kohavi, Ron, and George H. John. 1997. "Wrappers for Feature Subset Selection." *Artificial Intelligence* 97 (1): 273–324. https://doi.org/10.1016/S0004-3702(97)00043-X.

Krzyziński, Mateusz, Mikołaj Spytek, Hubert Baniecki, and Przemysław Biecek. 2023. "SurvSHAP(t): Time-Dependent Explanations of Machine Learning Survival Models." *Knowledge-Based Systems* 262: 110234. https://doi.org/10.1016/j.knosys.2022.110234.

Kuehn, Daniel, Philipp Probst, Janek Thomas, and Bernd Bischl. 2018. "Automatic Exploration of Machine Learning Experiments on OpenML." https://arxiv.org/abs/1806.10961.

Lang, Michel. 2017. "checkmate: Fast Argument Checks for Defensive R Programming." *The R Journal* 9 (1): 437–45. https://doi.org/10.32614/RJ-2017-028.

Lang, Michel, Martin Binder, Jakob Richter, Patrick Schratz, Florian Pfisterer, Stefan Coors, Quay Au, Giuseppe Casalicchio, Lars Kotthoff, and Bernd Bischl. 2019. "mlr3: A Modern Object-Oriented Machine Learning Framework in R." *Journal of Open Source Software*, December. https://doi.org/10.21105/joss.01903.

Lang, Michel, Bernd Bischl, and Dirk Surmann. 2017. "batchtools: Tools for R to Work on Batch Systems." *The Journal of Open Source Software* 2 (10): 135. https://doi.org/10.21105/joss.00135.

LeCun, Yann, Léon Bottou, Yoshua Bengio, and Patrick Haffner. 1998. "Gradient-Based Learning Applied to Document Recognition." *Proceedings of the IEEE* 86 (11): 2278–2324. https://doi.org/10.1109/5.726791.

Li, Lisha, Kevin Jamieson, Giulia DeSalvo, Afshin Rostamizadeh, and Ameet Talwalkar. 2018. "Hyperband: A Novel Bandit-Based Approach to Hyperparameter Optimization." *Journal of Machine Learning Research* 18 (185): 1–52. https://jmlr.org/papers/v18/16-558.html.

Lindauer, Marius, Katharina Eggensperger, Matthias Feurer, André Biedenkapp, Difan Deng, Carolin Benjamins, Tim Ruhkopf, René Sass, and Frank Hutter. 2022. "SMAC3: A Versatile Bayesian Optimization Package for Hyperparameter Optimization." *Journal of Machine Learning Research* 23 (54): 1–9. https://www.jmlr.org/papers/v23/21-0888.html.

Lipton, Zachary C. 2018. "The Mythos of Model Interpretability: In Machine Learning, the Concept of Interpretability Is Both Important and Slippery." *Queue* 16 (3): 31–57. https://doi.org/10.1145/3236386.3241340.

López-Ibáñez, Manuel, Jérémie Dubois-Lacoste, Leslie Pérez Cáceres, Mauro Birattari, and Thomas Stützle. 2016. "The irace Package: Iterated Racing for Automatic Algorithm Configuration." *Operations Research Perspectives* 3: 43–58. https://doi.org/10.1016/j.orp.2016.09.002.

Lundberg, Scott M., Gabriel G. Erion, and Su-In Lee. 2019. "Consistent Individualized Feature Attribution for Tree Ensembles." arXiv. https://doi.org/10.48550/arxiv.1802.03888.

Mehrabi, Ninareh, Fred Morstatter, Nripsuta Saxena, Kristina Lerman, and Aram Galstyan. 2021. "A Survey on Bias and Fairness in Machine Learning." *ACM Comput. Surv.* 54 (6): 1-35. https://doi.org/10.1145/3457607.

Micci-Barreca, Daniele. 2001. "A Preprocessing Scheme for High-Cardinality Categorical Attributes in Classification and Prediction Problems." *ACM SIGKDD Explorations Newsletter* 3 (1): 27–32. https://doi.org/10.1145/507533.507538.

Mitchell, Shira, Eric Potash, Solon Barocas, Alexander D'Amour, and Kristian Lum. 2021. "Algorithmic Fairness: Choices, Assumptions, and Definitions." *Annual Review of Statistics and Its Application* 8: 141–63. https://doi.org/10.1146/annurev-statistics-042720-125902.

Molinaro, Annette M, Richard Simon, and Ruth M Pfeiffer. 2005. "Prediction Error Estimation: A Comparison of Resampling Methods." *Bioinformatics* 21 (15): 3301–7. https://doi.org/10.1093/bioinformatics/bti499.

Molnar, Christoph. 2022. *Interpretable Machine Learning: A Guide for Making Black Box Models Explainable.* 2nd ed. https://christophm.github.io/interpretable-ml-book.

Molnar, Christoph, Bernd Bischl, and Giuseppe Casalicchio. 2018. "iml: An R Package for Interpretable Machine Learning." *JOSS* 3 (26): 786. https://doi.org/10.21105/joss.00786.

Molnar, Christoph, Gunnar König, Julia Herbinger, Timo Freiesleben, Susanne Dandl, Christian A. Scholbeck, Giuseppe Casalicchio, Moritz Grosse-Wentrup, and Bernd Bischl. 2022. "General Pitfalls of Model-Agnostic Interpretation

Methods for Machine Learning Models." In *xxAI – Beyond Explainable AI: International Workshop, Held in Conjunction with ICML 2020, July 18, 2020, Vienna, Austria, Revised and Extended Papers*, edited by Andreas Holzinger, Randy Goebel, Ruth Fong, Taesup Moon, Klaus-Robert Müller, and Wojciech Samek, 39–68. Cham: Springer International Publishing. https://doi.org/10.1007/978-3-031-04083-2_4.

Morales-Hernández, Alejandro, Inneke Van Nieuwenhuyse, and Sebastian Rojas Gonzalez. 2022. "A Survey on Multi-Objective Hyperparameter Optimization Algorithms for Machine Learning." *Artificial Intelligence Review*, 56: 8043–8093. https://doi.org/10.1007/s10462-022-10359-2.

Niederreiter, Harald. 1988. "Low-Discrepancy and Low-Dispersion Sequences." *Journal of Number Theory* 30 (1): 51–70. https://doi.org/10.1016/0022-314X(88)90025-X.

Pargent, Florian, Florian Pfisterer, Janek Thomas, and Bernd Bischl. 2022. "Regularized Target Encoding Outperforms Traditional Methods in Supervised Machine Learning with High Cardinality Features." *Computational Statistics* 37 (5): 2671–92. https://doi.org/10.1007/s00180-022-01207-6.

Poulos, Jason, and Rafael Valle. 2018. "Missing Data Imputation for Supervised Learning." *Applied Artificial Intelligence* 32 (2): 186–96. https://doi.org/10.1080/08839514.2018.1448143.

Provost, Foster, and Tom Fawcett. 2013. *Data Science for Business: What You Need to Know about Data Mining and Data-Analytic Thinking.* O'Reilly Media.

R Core Team. 2019. *R: A Language and Environment for Statistical Computing.* Vienna, Austria: R Foundation for Statistical Computing. https://www.R-project.org/.

Ribeiro, Marco, Sameer Singh, and Carlos Guestrin. 2016. ""Why Should I Trust You?": Explaining the Predictions of Any Classifier." In *Proceedings of the 2016 Conference of the North American Chapter of the Association for Computational Linguistics: Demonstrations*, 97–101. San Diego, California: Association for Computational Linguistics. https://doi.org/10.18653/v1/N16-3020.

Romaszko, Kamil, Magda Tatarynowicz, Mateusz Urbański, and Przemysław Biecek. 2019. "modelDown: Automated Website Generator with Interpretable Documentation for Predictive Machine Learning Models." *Journal of Open Source Software* 4 (38): 1444. https://doi.org/10.21105/joss.01444.

Ruspini, Enrique H. 1970. "Numerical Methods for Fuzzy Clustering." *Information Sciences* 2 (3): 319–50. https://doi.org/10.1016/S0020-0255(70)80056-1.

Saleiro, Pedro, Benedict Kuester, Abby Stevens, Ari Anisfeld, Loren Hinkson, Jesse London, and Rayid Ghani. 2018. "Aequitas: A Bias and Fairness Audit Toolkit." *arXiv Preprint arXiv:1811.05577.* https://doi.org/10.48550/arXiv.1811.05577.

Schmidberger, Markus, Martin Morgan, Dirk Eddelbuettel, Hao Yu, Luke Tierney, and Ulrich Mansmann. 2009. "State of the Art in Parallel Computing with R." *Journal of Statistical Software* 31 (1): 1-27. https://doi.org/10.18637/jss.v031.i01.

Schratz, Patrick, Marc Becker, Michel Lang, and Alexander Brenning. 2021. "mlr3spatiotempcv: Spatiotemporal Resampling Methods for Machine Learning in R," October. https://arxiv.org/abs/2110.12674.

Silverman, Bernard W. 1986. *Density Estimation for Statistics and Data Analysis.* Vol. 26. CRC press.

Simon, Richard. 2007. "Resampling Strategies for Model Assessment and Selection." In *Fundamentals of Data Mining in Genomics and Proteomics*, edited by Werner Dubitzky, Martin Granzow, and Daniel Berrar, 173–86. Boston, MA: Springer US. https://doi.org/10.1007/978-0-387-47509-7_8.

Snoek, Jasper, Hugo Larochelle, and Ryan P Adams. 2012. "Practical Bayesian Optimization of Machine Learning Algorithms." In *Advances in Neural Information Processing Systems*, edited by F. Pereira, C. J. Burges, L. Bottou, and K. Q. Weinberger. Vol. 25. https://proceedings.neurips.cc/paper_files/paper/2012/file/05311655a15b75fab86956663e1819cd-Paper.pdf.

Sonabend, Raphael Edward Benjamin. 2021. "A Theoretical and Methodological Framework for Machine Learning in Survival Analysis: Enabling Transparent and Accessible Predictive Modelling on Right-Censored Time-to-Event Data." PhD, University College London (UCL). https://discovery.ucl.ac.uk/id/eprint/10129352/.

Sonabend, Raphael, and Andreas Bender. 2023. *Machine Learning in Survival Analysis.* https://www.mlsabook.com.

Sonabend, Raphael, Andreas Bender, and Sebastian Vollmer. 2022. "Avoiding C-Hacking When Evaluating Survival Distribution Predictions with Discrimination Measures." Edited by Zhiyong Lu. *Bioinformatics* 38 (17): 4178–84. https://doi.org/10.1093/bioinformatics/btac451.

Sonabend, Raphael, Franz J Király, Andreas Bender, Bernd Bischl, and Michel Lang. 2021. "mlr3proba: An R Package for Machine Learning in Survival Analysis." *Bioinformatics*, February. https://doi.org/10.1093/bioinformatics/btab039.

Sonabend, Raphael, Florian Pfisterer, Alan Mishler, Moritz Schauer, Lukas Burk, Sumantrak Mukherjee, and Sebastian Vollmer. 2022. "Flexible Group Fairness Metrics for Survival Analysis." In *DSHealth 2022 Workshop on Applied Data Science for Healthcare at KDD2022.* https://arxiv.org/abs/2206.03256.

Stein, Michael. 1987. "Large Sample Properties of Simulations Using Latin Hypercube Sampling." *Technometrics* 29 (2): 143–51. https://doi.org/10.2307/1269769.

Strobl, Carolin, Anne-Laure Boulesteix, Thomas Kneib, Thomas Augustin, and Achim Zeileis. 2008. "Conditional Variable Importance for Random Forests." *BMC Bioinformatics* 9 (1): 307. https://doi.org/10.1186/1471-2105-9-307.

Štrumbelj, Erik, and Igor Kononenko. 2013. "Explaining Prediction Models and Individual Predictions with Feature Contributions." *Knowledge and Information Systems* 41 (3): 647–65. https://doi.org/10.1007/s10115-013-0679-x.

Thornton, Chris, Frank Hutter, Holger H. Hoos, and Kevin Leyton-Brown. 2013. "Auto-WEKA." In *Proceedings of the 19th ACM SIGKDD International Conference on Knowledge Discovery and Data Mining.* ACM. https://doi.org/10.1145/2487575.2487629.

Tsallis, Constantino, and Daniel A Stariolo. 1996. "Generalized Simulated Annealing." *Physica A: Statistical Mechanics and Its Applications* 233 (1-2): 395–406. https://doi.org/10.1016/S0378-4371(96)00271-3.

Vanschoren, Joaquin, Jan N. van Rijn, Bernd Bischl, and Luis Torgo. 2013. "OpenML: Networked Science in Machine Learning." *SIGKDD Explorations* 15 (2): 49–60. https://doi.org/10.1145/2641190.2641198.

Wachter, Sandra, Brent Mittelstadt, and Chris Russell. 2017. "Counterfactual Explanations Without Opening the Black Box: Automated Decisions and the GDPR." *SSRN Electronic Journal.* https://doi.org/10.2139/ssrn.3063289.

———. 2021. "Why Fairness Cannot Be Automated: Bridging the Gap Between EU Non-Discrimination Law and AI." *Computer Law & Security Review* 41: 105567. https://doi.org/https://doi.org/10.1016/j.clsr.2021.105567.

Watson, David S, and Marvin N Wright. 2021. "Testing Conditional Independence in Supervised Learning Algorithms." *Machine Learning* 110 (8): 2107–29. https://doi.org/10.1007/s10994-021-06030-6.

Wexler, James, Mahima Pushkarna, Tolga Bolukbasi, Martin Wattenberg, Fernanda Viégas, and Jimbo Wilson. 2019. "The What-If Tool: Interactive Probing of Machine Learning Models." *IEEE Transactions on Visualization and Computer Graphics* 26 (1): 56–65. https://doi.org/10.1109/TVCG.2019.2934619.

Wickham, Hadley, and Garrett Grolemund. 2017. *R for Data Science: Import, Tidy, Transform, Visualize, and Model Data.* 1st ed. O'Reilly Media. https://r4ds.had.co.nz/.

Williams, Christopher KI, and Carl Edward Rasmussen. 2006. *Gaussian Processes for Machine Learning.* Vol. 2. 3. MIT press Cambridge, MA.

Wiśniewski, Jakub, and Przemysław Biecek. 2022. "The R Journal: Fairmodels: A Flexible Tool for Bias Detection, Visualization, and Mitigation in Binary Classification Models." *The R Journal* 14: 227–43. https://doi.org/10.32614/RJ-2022-019.

Wolpert, David H. 1992. "Stacked Generalization." *Neural Networks* 5 (2): 241–59. https://doi.org/10.1016/S0893-6080(05)80023-1.

Xiang, Yang, Sylvain Gubian, Brian Suomela, and Julia Hoeng. 2013. "Generalized Simulated Annealing for Global Optimization: The GenSA Package." *R Journal* 5 (1): 13. https://doi.org/10.32614/RJ-2013-002.

Index

$edges, 167
$pipeops, 167
%»%, 166
AcqOptimizer, 134
AutoFSelector, 158
AutoTuner, 98
DALEX, 273
FSelector, 152, 154, 155
Filter, 148
GraphLearner, 169
Graph, 163, 166
$plot(), 167
LearnerClassif, 37, 41
Learner, 22
$aggregate(), 60
$encapsulate, 226
$fallback, 229
$model, 23
$param_set, 27
$predict(), 22, 24
$predict_newdata(), 25
$predict_type, 26
$set_values(), 29
$train(), 22
MeasureClassif, 37, 42
Measure, 33
$param_set, 35
$predict_type, 42
Objective, 126
ParamDbl, 27
ParamFct, 27
ParamInt, 27
ParamLgl, 27
ParamSet, 27
$deps, 30
ParamUty, 27
PipeOpLearnerCV, 182
PipeOpSelect, 175
PipeOp, 163, 164
PredictionClassif, 43
$confusion, 43
$set_threshold(), 43
PredictionRegr, 24

Prediction, 24, 33
$score(), 34, 43
ResampleResult, 59, 72
Resampling, 58
$instantiate(), 59
Selector, 175
TaskClassif, 37, 38
$positive, 40
TaskGenerator, 103
TaskRegr, 17
Task
$cbind(), 21
$filter(), 20
$rbind(), 21
$select(), 20
$set_col_roles(), 48
Terminator, 88
TunerMbo, 138
Tuner, 90
acqf(), 133
acqo(), 135
as_benchmark_aggr(), 256
as_learner(), 169
as_resamplings(), 246
as_task_classif, 39
as_task_regr(), 17
as_tasks(), 246
auto_fselector, 158
auto_tuner(), 101
batchmark(), 249
batchtools, 248
bbotk, 126, 144
benchmark(), 37, 70
chunk(), 251
counterfactuals, 269
flt(), 148
fs(), 155
fselect(), 153
fsi(), 155
getStatus(), 252
iml, 260
loadRegistry(), 252
loop_function, 135

lrn(), 22
mlr3batchmark, 248, 249, 258
mlr3benchmark, 258
mlr3cluster, 284, 299, 314
mlr3db, 232
mlr3extralearners, 50, 236
mlr3fairness, 316, 322
mlr3filters, 3, 144, 147
mlr3fselect, 3, 144, 152, 231
mlr3hyperband, 90, 116, 123
mlr3learners, 50
mlr3mbo, 90, 116, 126, 144
mlr3misc, 215
mlr3oml, 242, 258
mlr3pipelines, 3, 163, 172, 174, 194, 196, 209, 236, 283, 286, 320
mlr3proba, 283, 288, 314
mlr3spatial, 284, 312, 314
mlr3spatiotempcv, 284, 308, 314
mlr3tuningspaces, 87
mlr3tuning, 3, 86, 90, 116, 144, 172, 174, 185, 231, 236, 286, 322
mlr3verse, 1
mlr3viz, 10, 11, 18, 25, 306
mlr3, 1, 3, 6, 10, 11, 15, 50, 52, 56, 90, 163, 213, 240, 283, 316
mlr_filters, 148
mlr_fselectors, 155
mlr_graphs, 178
mlr_learners, 22, 50
mlr_measures, 32
mlr_pipeops, 164
mlr_tasks, 16
mlr, 11
msr()/msrs(), 32, 34
ocl(), 246
odt(), 242
opt(), 128, 135, 137
otsk(), 244
paradox, 130
partition(), 24
po(), 164
ppl(), 178
ppl(robustify), 203
resample(), 59, 70
rsmp(), 58
srlrn(), 132
testJob(), 250
trm(), 88
tsk(), 17
tune(), 101

accuracy, 42, 75
acquisition function, 129, 132
acquisition function optimizer, 134
algorithm, 15
algorithmic fairness, 259, 316
    causal, 317
    individual, 317
AUC, 76
autocorrelation, 308

bagging, 178, 179
baselines, 31, 118
Bayesian optimization, 91, 92, 125, 186
benchmark experiment, 69
benchmark experiments, 70, 219, 232, 240
benchmark study, *see* benchmark experiment
benchmark suites, 246
bias, 317
bias-preserving, 318
bias-transforming, 318
black box, 259
black box optimization, 85, 90, 125
BLAS, 217
boosting, 50
bootstrapping, 57
Brier score, 42

CASH, *see* combined algorithm selection and hyperparameter optimization
ceteris paribus, *see* individual conditional expectation (ICE) curves
chunk, 251
chunking, 215
classification, 1, 15, 37
    binary, 39, 73
    cost-sensitive, 15, 47, 284
    multiclass, 39, 178
classification error, 43
cluster analysis, 299
cluster cohesion, 303
cluster separation, 303
CMA-ES, 91
column roles, 48
combined algorithm selection and hyperparameter optimization, 185
computational job, 248

confusion matrix, 43, 73
control parameters, 35, 93
correlation, 308
counterfactual, 269
Cox Proportional Hazards, 290
cross-validation, 56
    leave-one-out, 57
    repeated k-fold, 57

DAG, *see* Directed Acyclic Graph
data backend, 232
data cleaning, 196
data imputation, 201
data parallelism, 214
debugging, 116, 170, 213, 225, 253
decision tree, 15, 182, 186, 214
demographic parity, 318
dendrogram, 302
density estimation, 296
descriptor selection, *see* feature selection
dictionaries, 10
DIRECT, 134
directed acyclic graph, 168
DuckDB, 232, 235
dummy encoding, *see* encoding,
    treatment

efficient global optimization, 136
embarrassingly parallel, 214, 247
embedded methods, 149
encapsulation, 117, 226
encoding, 198, 203
    impact, 199
    one-hot, 199
    treatment, 199
ensemble, 163
equalized odds, 318
evolutionary strategies, 91, 92
explanatory model analysis, 274
exploratory data analysis, *see* data
    cleaning

F1, 75
factor encoding, 168, 196
FAIR, 242
fairness, *see* algorithmic fairness
fairness report, 322
fallback learner, 117, 228, 229
false negative, 314
false negatives, 44, 45, 73
false positive, 314

false positive rate, 75, 318
false positives, 44, 45, 73
feature effect, 262
feature engineering, 196
feature extraction, 196, 206
feature importance, 147, 149, 261
feature selection, 146, 192, 206
    embedded, 146
    implicit, 146
features, 15
fidelity, 121
    multi-fidelity, 121
fidelity parameters, 191
Filter
    $calculate(), 148
fitting, *see* model training
friedman, 256

Gaussian process, 50, 131, 141
generalization error, 16, 104
generalization performance, 53
generalized linear model, 50, 182, 199
granularity, 214
grid search, 85, 91
group fairness, 317
grouped resampling, 66

hierarchical clustering, 302
high-performance computing, 240, 247
holdout, 54
HPO, 85
hyperband, 91, 122, 191
hyperparameter optimization, *see* HPO,
    125
hyperparameters, 15, 26, 85, 164, 172

imputation, 168, 169, 196, 203
independence, *see* bias-transforming
individual conditional expectation, 263
initial design, 129, 130
intermediate model, 54
interpretability, 259
interpretable machine learning, 259
inverse weighting, 47
iterated racing, 91

jobs, 248

k-nearest neighbors, 50, 182, 205
Kaplan-Meier, 289, 294
KNN, *see* k-nearest neighbors

L-BFGS-B, 134
LIME, 267
linear predictor, 291
logging, 230
logistic regression, 50, 169, 183
logloss, 42
loop functions, 135

machine learning, 15
macro average, 60
mean absolute error, 33
metadata, 16
micro average, 60
missing data, 200
model, 15, 22
model based optimization, *see* Bayesian
        optimization
model coefficients, *see* parameters
model evaluation, 85
model predicting, 22
model training, 15, 16, 22
Monte Carlo cross-validation, *see*
        subsampling
multi-objective counterfactuals, 270
multi-objective optimization, 146
multi-objective tuning, 89, 116, 119
mutators, 20

negative predictive value, 75
nested resampling, 100
neural network, 15, 109, 119, 122

object-oriented programming, 1, 7
one-versus-rest classification, 178
OpenML, 241
optimism of the training error, 100
optimization instance, 126

parallelization, 213
parallelization backend, 214
parallelization overhead, 214
parameters, 85
Pareto front, 119, 139, 270
Pareto optimality, 119, 157
Parquet, 232, 235
partial dependence, 263
PCA, *see* principal component analysis
performance estimation, 53
performance measure, 53
permutation feature importance, 261
pipelines, 151
positive predictive value, 75, 77, 318

post hoc, 259
postprocessing, 196
precision, *see* measures, positive
        predictive value
precision-recall curve, 77
preprocessing, 146, 163, 168, 196, 320
principal component analysis, 165
principal components analysis, 306

random forest, 50, 122, 131, 214, 261
random search, 85, 91
Rashomon, 273, 281
recall, *see* measures, true positive rate
regression, 1, 15
repeated holdout, *see* subsampling
resampling, 56
result assigner, 141
ROC, 73, 75

scale, 166
scheduling system, 248
search space, 87
sensitive attribute, 317
sensitivity, *see* measures, true positive
        rate
separation, *see* bias-preserving
sequential forward selection, 153
Shapley values, 268
single-objective, 89
Slurm, 250
socket cluster, 215
spatial analysis, 308
specificity, *see* measures, true negative
        rate
SQL, 232
SQLite, 232
stacking, 178, 181
state, 164
stratified sampling, 67
subsampling, 57
sugar functions, 10
supervised learning, 15
support vector machine, 15, 30, 50, 85,
        86, 191
    survival, 290
surrogate model, 129, 131, 263
    global, 264
    local, 265
survival analysis, 286
synchronization overhead, 215

target, 15

tasks, 1, 15, 16
test data, 16, 24, 53
threading, 216
thresholding, 37, 44, 285
training data, 15, 16, 24, 53
true negative rate, 74
true negatives, 44, 73
true positive rate, 74, 77, 318
true positives, 44, 73
tuners, 85
tuning, 85, 185
tuning instance, 89
tuning space, *see* search space

unsupervised learning, 15

variable selection, *see* feature selection

What-If, 269

Yeo-Johnson, 187